function spaces

LECTURE NOTES IN PURE AND APPLIED MATHEMATICS

1. *N. Jacobson,* Exceptional Lie Algebras
2. *L.-Å. Lindahl and F. Poulsen,* Thin Sets in Harmonic Analysis
3. *I. Satake,* Classification Theory of Semi-Simple Algebraic Groups
4. *F. Hirzebruch, W. D. Newmann, and S. S. Koh, Differentiable Manifolds and Quadratic Forms*
5. *I. Chavel,* Riemannian Symmetric Spaces of Rank One
6. *R. B. Burckel,* Characterization of C(X) Among Its Subalgebras
7. *B. R. McDonald, A. R. Magid, and K. C. Smith,* Ring Theory: Proceedings of the Oklahoma Conference
8. *Y.-T. Siu,* Techniques of Extension on Analytic Objects
9. *S. R. Caradus, W. E. Pfaffenberger, and B. Yood,* Calkin Algebras and Algebras of Operators on Banach Spaces
10. *E. O. Roxin, P.-T. Liu, and R. L. Sternberg,* Differential Games and Control Theory
11. *M. Orzech and C. Small,* The Brauer Group of Commutative Rings
12. *S. Thomier,* Topology and Its Applications
13. *J. M. Lopez and K. A. Ross,* Sidon Sets
14. *W. W. Comfort and S. Negrepontis,* Continuous Pseudometrics
15. *K. McKennon and J. M. Robertson,* Locally Convex Spaces
16. *M. Carmeli and S. Malin,* Representations of the Rotation and Lorentz Groups: An Introduction
17. *G. B. Seligman,* Rational Methods in Lie Algebras
18. *D. G. de Figueiredo,* Functional Analysis: Proceedings of the Brazilian Mathematical Society Symposium
19. *L. Cesari, R. Kannan, and J. D. Schuur,* Nonlinear Functional Analysis and Differential Equations: Proceedings of the Michigan State University Conference
20. *J. J. Schäffer,* Geometry of Spheres in Normed Spaces
21. *K. Yano and M. Kon,* Anti-Invariant Submanifolds
22. *W. V. Vasconcelos,* The Rings of Dimension Two
23. *R. E. Chandler,* Hausdorff Compactifications
24. *S. P. Franklin and B. V. S. Thomas,* Topology: Proceedings of the Memphis State University Conference
25. *S. K. Jain,* Ring Theory: Proceedings of the Ohio University Conference
26. *B. R. McDonald and R. A. Morris,* Ring Theory II: Proceedings of the Second Oklahoma Conference
27. *R. B. Mura and A. Rhemtulla,* Orderable Groups
28. *J. R. Graef,* Stability of Dynamical Systems: Theory and Applications
29. *H.-C. Wang,* Homogeneous Branch Algebras
30. *E. O. Roxin, P.-T. Liu, and R. L. Sternberg,* Differential Games and Control Theory II
31. *R. D. Porter,* Introduction to Fibre Bundles
32. *M. Altman,* Contractors and Contractor Directions Theory and Applications
33. *J. S. Golan,* Decomposition and Dimension in Module Categories
34. *G. Fairweather,* Finite Element Galerkin Methods for Differential Equations
35. *J. D. Sally,* Numbers of Generators of Ideals in Local Rings
36. *S. S. Miller,* Complex Analysis: Proceedings of the S.U.N.Y. Brockport Conference
37. *R. Gordon,* Representation Theory of Algebras: Proceedings of the Philadelphia Conference
38. *M. Goto and F. D. Grosshans,* Semisimple Lie Algebras
39. *A. I. Arruda, N. C. A. da Costa, and R. Chuaqui,* Mathematical Logic: Proceedings of the First Brazilian Conference
40. *F. Van Oystaeyen,* Ring Theory: Proceedings of the 1977 Antwerp Conference
41. *F. Van Oystaeyen and A. Verschoren,* Reflectors and Localization: Application to Sheaf Theory
42. *M. Satyanarayana,* Positively Ordered Semigroups
43. *D. L Russell,* Mathematics of Finite-Dimensional Control Systems
44. *P.-T. Liu and E. Roxin,* Differential Games and Control Theory III: Proceedings of the Third Kingston Conference, Part A
45. *A. Geramita and J. Seberry,* Orthogonal Designs: Quadratic Forms and Hadamard Matrices
46. *J. Cigler, V. Losert, and P. Michor,* Banach Modules and Functors on Categories of Banach Spaces

47. *P.-T. Liu and J. G. Sutinen,* Control Theory in Mathematical Economics: Proceedings of the Third Kingston Conference, Part B
48. *C. Byrnes,* Partial Differential Equations and Geometry
49. *G. Klambauer,* Problems and Propositions in Analysis
50. *J. Knopfmacher,* Analytic Arithmetic of Algebraic Function Fields
51. *F. Van Oystaeyen,* Ring Theory: Proceedings of the 1978 Antwerp Conference
52. *B. Kadem,* Binary Time Series
53. *J. Barros-Neto and R. A. Artino,* Hypoelliptic Boundary-Value Problems
54. *R. L. Sternberg, A. J. Kalinowski, and J. S. Papadakis,* Nonlinear Partial Differential Equations in Engineering and Applied Science
55. *B. R. McDonald,* Ring Theory and Algebra III: Proceedngs of the Third Oklahoma Conference
56. *J. S. Golan,* Structure Sheaves Over a Noncommutative Ring
57. *T. V. Narayana, J. G. Williams, and R. M. Mathsen,* Combinatorics, Representation Theory and Statistical Methods in Groups: YOUNG DAY Proceedings
58. *T. A. Burton,* Modeling and Differential Equations in Biology
59. *K. H. Kim and F. W. Roush,* Introduction to Mathematical Consensus Theory
60. *J. Banas and K. Goebel,* Measures of Noncompactness in Banach Spaces
61. *O. A. Nielson,* Direct Integral Theory
62. *J. E. Smith, G. O. Kenny, and R. N. Ball,* Ordered Groups: Proceedings of the Boise State Conference
63. *J. Cronin,* Mathematics of Cell Electrophysiology
64. *J. W. Brewer,* Power Series Over Commutative Rings
65. *P. K. Kamthan and M. Gupta,* Sequence Spaces and Series
66. *T. G. McLaughlin,* Regressive Sets and the Theory of Isols
67. *T. L. Herdman, S. M. Rankin III, and H. W. Stech,* Integral and Functional Differential Equations
68. *R. Draper,* Commutative Algebra: Analytic Methods
69. *W. G. McKay and J. Patera,* Tables of Dimensions, Indices, and Branching Rules for Representations of Simple Lie Algebras
70. *R. L. Devaney and Z. H. Nitecki,* Classical Mechanics and Dynamical Systems
71. *J. Van Geel,* Places and Valuations in Noncommutative Ring Theory
72. *C. Faith,* Injective Modules and Injective Quotient Rings
73. *A. Fiacco,* Mathematical Programming with Data Perturbations I
74. *P. Schultz, C. Praeger, and R. Sullivan,* Algebraic Structures and Applications: Proceedings of the First Western Australian Conference on Algebra
75. *L Bican, T. Kepka, and P. Nemec,* Rings, Modules, and Preradicals
76. *D. C. Kay and M. Breen,* Convexity and Related Combinatorial Geometry: Proceedings of the Second University of Oklahoma Conference
77. *P. Fletcher and W. F. Lindgren,* Quasi-Uniform Spaces
78. *C.-C. Yang,* Factorization Theory of Meromorphic Functions
79. *O. Taussky,* Ternary Quadratic Forms and Norms
80. *S. P. Singh and J. H. Burry,* Nonlinear Analysis and Applications
81. *K. B. Hannsgen, T. L. Herdman, H. W. Stech, and R. L. Wheeler,* Volterra and Functional Differential Equations
82. *N. L. Johnson, M. J. Kallaher, and C. T. Long,* Finite Geometries: Proceedings of a Conference in Honor of T. G. Ostrom
83. *G. I. Zapata,* Functional Analysis, Holomorphy, and Approximation Theory
84. *S. Greco and G. Valla,* Commutative Algebra: Proceedings of the Trento Conference
85. *A. V. Fiacco,* Mathematical Programming with Data Perturbations II
86. *J.-B. Hiriart-Urruty, W. Oettli, and J. Stoer,* Optimization: Theory and Algorithms
87. *A. Figa Talamanca and M. A. Picardello,* Harmonic Analysis on Free Groups
88. *M. Harada,* Factor Categories with Applications to Direct Decomposition of Modules
89. *V. I. Istrăţescu,* Strict Convexity and Complex Strict Convexity
90. *V. Lakshmikantham,* Trends in Theory and Practice of Nonlinear Differential Equations
91. *H. L. Manocha and J. B. Srivastava,* Algebra and Its Applications
92. *D. V. Chudnovsky and G. V. Chudnovsky,* Classical and Quantum Models and Arithmetic Problems
93. *J. W. Longley,* Least Squares Computations Using Orthogonalization Methods
94. *L. P. de Alcantara,* Mathematical Logic and Formal Systems
95. *C. E. Aull,* Rings of Continuous Functions
96. *R. Chuaqui,* Analysis, Geometry, and Probability

97. *L. Fuchs and L. Salce,* Modules Over Valuation Domains
98. *P. Fischer and W. R. Smith,* Chaos, Fractals, and Dynamics
99. *W. B. Powell and C. Tsinakis,* Ordered Algebraic Structures
100. *G. M. Rassias and T. M. Rassias,* Differential Geometry, Calculus of Variations, and Their Applications
101. *R.-E. Hoffmann and K. H. Hofmann,* Continuous Lattices and Their Applications
102. *J. H. Lightbourne III and S. M. Rankin III,* Physical Mathematics and Nonlinear Partial Differential Equations
103. *C. A. Baker and L, M. Batten,* Finite Geometrics
104. *J. W. Brewer, J. W. Bunce, and F. S. Van Vleck,* Linear Systems Over Commutative Rings
105. *C. McCrory and T. Shifrin,* Geometry and Topology: Manifolds, Varieties, and Knots
106. *D. W. Kueker, E. G. K. Lopez-Escobar, and C. H. Smith,* Mathematical Logic and Theoretical Computer Science
107. *B.-L. Lin and S. Simons,* Nonlinear and Convex Analysis: Proceedings in Honor of Ky Fan
108. *S. J. Lee,* Operator Methods for Optimal Control Problems
109. *V. Lakshmikantham,* Nonlinear Analysis and Applications
110. *S. F. McCormick,* Multigrid Methods: Theory, Applications, and Supercomputing
111. *M. C. Tangora,* Computers in Algebra
112. *D. V. Chudnovsky and G. V. Chudnovsky,* Search Theory: Some Recent Developments
113. *D. V. Chudnovsky and R. D. Jenks,* Computer Algebra
114. *M. C. Tangora,* Computers in Geometry and Topology
115. *P. Nelson, V. Faber, T. A. Manteuffel, D. L. Seth, and A. B. White, Jr.,* Transport Theory, Invariant Imbedding, and Integral Equations: Proceedings in Honor of G. M. Wing's 65th Birthday
116. *P. Clément, S. Invernizzi, E. Mitidieri, and I. I. Vrabie,* Semigroup Theory and Applications
117. *J. Vinuesa,* Orthogonal Polynomials and Their Applications: Proceedings of the International Congress
118. *C. M. Dafermos, G. Ladas, and G. Papanicolaou,* Differential Equations: Proceedings of the EQUADIFF Conference
119. *E. O. Roxin,* Modern Optimal Control: A Conference in Honor of Solomon Lefschetz and Joseph P. Lasalle
120. *J. C. Díaz,* Mathematics for Large Scale Computing
121. *P. S. Milojević,* Nonlinear Functional Analysis
122. *C. Sadosky,* Analysis and Partial Differential Equations: A Collection of Papers Dedicated to Mischa Cotlar
123. *R. M. Shortt,* General Topology and Applications: Proceedings of the 1988 Northeast Conference
124. *R. Wong,* Asymptotic and Computational Analysis: Conference in Honor of Frank W. J. Olver's 65th Birthday
125. *D. V. Chudnovsky and R. D. Jenks,* Computers in Mathematics
126. *W. D. Wallis, H. Shen, W. Wei, and L. Zhu,* Combinatorial Designs and Applications
127. *S. Elaydi,* Differential Equations: Stability and Control
128. *G. Chen, E. B. Lee, W. Littman, and L. Markus,* Distributed Parameter Control Systems: New Trends and Applications
129. *W. N. Everitt,* Inequalities: Fifty Years On from Hardy, Littlewood and Pólya
130. *H. G. Kaper and M. Garbey,* Asymptotic Analysis and the Numerical Solution of Partial Differential Equations
131. *O. Arino, D. E. Axelrod, and M. Kimmel,* Mathematical Population Dynamics: Proceedings of the Second International Conference
132. *S. Coen,* Geometry and Complex Variables
133. *J. A. Goldstein, F. Kappel, and W. Schappacher,* Differential Equations with Applications in Biology, Physics, and Engineering
134. *S. J. Andima, R. Kopperman, P. R. Misra, J. Z. Reichman, and A. R. Todd,* General Topology and Applications
135. *P Clément, E. Mitidieri, B. de Pagter,* Semigroup Theory and Evolution Equations: The Second International Conference
136. *K. Jarosz,* Function Spaces
137. *J. M. Bayod, N. De Grande-De Kimpe, and J Martínez-Maurica,* *p*-adic Functional Analysis

Additional Volumes in Preparation

function spaces

edited by
Krzysztof Jarosz
Southern Illinois University at Edwardsville
Edwardsville, Illinois

Marcel Dekker, Inc. **New York • Basel • Hong Kong**

Library of Congress Cataloging-in-Publication Data

Function spaces / edited by Krzysztof Jarosz
p. cm. -- (Lecture notes in pure and applied mathematics ; v. 136)
Includes bibliographical references and index.
1. Function spaces. I. Jarosz, Krzysztof. II. Series.
QA323.F85 1991
515'.73--dc20 91-40967
CIP

This book is printed on acid-free paper.

MARCEL DEKKER, INC.
270 Madison Avenue, New York, New York 10016

Current printing (last digit):
10 9 8 7 6 5 4 3 2 1

PRINTED IN THE UNITED STATES OF AMERICA

Preface

The Conference on Function Spaces was held at Southern Illinois University at Edwardsville, from April 19 to 21, 1990. It was sponsored by grants from Southern Illinois University, the National Science Foundation, and the Central States Universities, Incorporated.

Among the contributors to this volume were participants in the conference as well as others who responded to a call for papers. The papers cover a wide range of topics, including spaces of analytic functions, isometries of function spaces, geometry of Banach spaces, and Banach algebras. Some articles contain expositions of known results, others present fresh discoveries, and still others may contain both ingredients. The editor would like to thank everyone who contributed to the Proceedings: the authors, the referees, and Marcel Dekker, Inc.

Krzysztof Jarosz

Contents

Contributors

JOHN AKEROYD Mathematics Department, University of Arkansas, Fayetteville, Arkansas

JONATHAN ARAZY Department of Mathematics, University of Haifa, Haifa, Israel, and Department of Mathematics, University of Kansas, Lawrence, Kansas

HUGO ARIZMENDI Instituto de Matematicas, UNAM, Circuito exterior, Ciudad Universitaria, Mexico D.F., Mexico

NAKHLÉ ASMAR Department of Mathematics, University of Missouri, Columbia, Missouri

EARL BERKSON Department of Mathematics, University of Illinois, Urbana, Illinois

DAVID P. BLECHER Department of Mathematics, University of Houston, Houston, Texas

EGGERT BRIEM Science Institute, University of Iceland, Reykjavik, Iceland

MICHAEL CAMBERN Department of Mathematics, University of California, Santa Barbara, California

ANGEL CARRILLO Instituto de Matematicas, UNAM, Circuito exterior, Ciudad Universitaria, Mexico D.F., Mexico

C.-H. CHU Goldsmiths' College, University of London, London, England

A. J. ELLIS University of Hong Kong, Hong Kong

PER ENFLO Department of Mathematics and Computer Science, Kent State University, Kent, Ohio

HANS G. FEICHTINGER University of Vienna, Department of Mathematics, Vienna, Austria

R. J. FLEMING Department of Mathematics, Central Michigan University, Mt. Pleasant, Michigan

T. A. GILLESPIE Department of Mathematics, University of Edinburgh, Edinburgh, Scotland

PAMELA GORKIN Bucknell University, Lewisburg, Pennsylvania

OSAMU HATORI Department of Mathematics, Tokyo Medical College, Shinjuku-ku, Tokyo, Japan

JOHN A. HOLBROOK Department of Mathematics and Statistics, University of Guelph, Guelph, Ontario, Canada

ZHIBAO HU Department of Mathematics, The University of Iowa, Iowa City, Iowa

SUNWOOK HWANG Department of Mathematics, Soongsil University, Seoul, Korea

B. IOCHUM Université de Provence and Centre de Physique Théorique, C.N.R.S., Marseille, France

J. E. JAMISON Department of Mathematical Sciences, Memphis State University, Memphis, Tennessee

KRZYSZTOF JAROSZ Department of Mathematics and Statistics, Southern Illinois University at Edwardsville, Edwardsville, Illinois

A. KAMIŃSKA Department of Mathematical Sciences, Memphis State University, Memphis, Tennessee

KEITH LEWIS Department of Mathematics, Brown University, Providence, Rhode Island

BOR-LUH LIN Department of Mathematics, The University of Iowa, Iowa City, Iowa

B. A. LOTTO Department of Mathematics, University of California, Davis, Davis, California

JOHN E. McCARTHY Department of Mathematics, Indiana University, Bloomington, Indiana

MICHAEL M. NEUMANN Department of Mathematics and Statistics, Mississippi State University, Mississippi State, Mississippi

VIJAY D. PATHAK Department of Applied Mathematics, M.S. University of Baroda, Baroda, India

LIZHONG PENG Peking University, Beijing, China

R. R. PHELPS Department of Mathematics, University of Washington, Seattle, Washington

N. V. RAO Department of Mathematics, University of Toledo, Toledo, Ohio

RICHARD ROCHBERG Department of Mathematics, Washington University, St. Louis, Missouri

RUDOLF RUPP Universität Karlsruhe, Math. Institut I, Karlsruhe, Germany

ELIAS SAAB University of Missouri, Columbia, Missouri

PAULETTE SAAB University of Missouri, Columbia, Missouri

S. J. SIDNEY Department of Mathematics, The University of Connecticut, Storrs, Connecticut

TOMA V. TONEV Mathematics Department, The University of Toledo, Toledo, Ohio

S. WATANABE Department of Mathematics, Niigata University, Niigata, Japan

JOHN WERMER Department of Mathematics, Brown University, Providence, Rhode Island

KEITH YALE Department of Mathematical Sciences, University of Montana, Missoula, Montana

GENKAI ZHANG University of Stockholm, Stockholm, Sweden

WENYAO ZHANG Department of Mathematics, The University of Iowa, Iowa City, Iowa

A Note Concerning Cyclic Vectors in Hardy and Bergman Spaces

John Akeroyd Mathematics Department, University of Arkansas, Fayetteville, Arkansas

1. INTRODUCTION

Recall that a bounded operator T on a Hilbert space H is said to be cyclic if there exists a vector h in H such that $\{p(T)h : p \text{ is a polynomial}\}$ is dense in H; under these circumstances h is called a cyclic vector for T.

As defined in [AKS], for $0 \le \theta < \pi$, let $V(\theta) = co(\{z : |z| \le \sin(\theta/2)\} \cup \{1\})$, that is, the closed convex hull of $\{z : |z| \le \sin(\theta/2)\} \cup \{1\}$, and let $G(\theta) = \{z : |z| < 1\} \backslash V(\theta)$. In [AKS] the authors showed that the shift M_z on the Hardy space $H^2(G(\theta))$ or the Bergman space $L^2_a(G(\theta))$ is cyclic if $\pi/2 \le \theta < \pi$ and fails to be cyclic if $0 \le \theta < \pi/3$. In this brief paper we "close the gap" by sharpening arguments found in [AKS] to show that M_z on $H^2(G(\theta))$ or $L_a^2(G(\theta))$ is cyclic if $\pi/3 < \theta < \pi$ and is not cyclic if $0 \le \theta \le \pi/3$.

2. THE SHIFT ON CERTAIN HARDY AND BERGMAN SPACES

A version of our first theorem appears in [Ni] without proof (also cf. [SV]; arguments in the more general setting of [SV] are quite complicated).

2.1 Theorem. **The shift M_z on $H^2(G(\theta))$ is cyclic if $\pi/3 < \theta < \pi$ and fails to be cyclic if $0 \le \theta \le \pi/3$.**

Proof (sketch). The reader can refer to [AKS] for the proof that M_z on $H^2(G(\theta))$ is not cyclic if $0 \le \theta \le \pi/3$.

We now suppose that $\pi/3 < \theta < \pi$. Let $\omega := \omega(\cdot, G(\theta), z_o)$ be harmonic measure on $\partial G(\theta)$ evaluated at some z_o in $G(\theta)$. By the "edge-of-the-wedge" estimate [AKS, Lemma 2.8], $d\omega$ is boundedly equivalent to $|z - 1|^{(\frac{2\pi}{\pi - \theta}) - 1} ds$, where ds is arclength measure on $\partial G(\theta)$. Therefore $\frac{1}{|z - 1|^3} \in L^1(d\omega)$. Let $h(z) = \exp\left(\frac{-1}{|z - 1|^3}\right)$ and φ be a conformal map from $\mathbb{D} := \{z : |z| < 1\}$ onto $G(\theta)$. So $h\circ\varphi \in L^\infty(dm)$ and $\log(h\circ\varphi) \in L^1(dm)$, where dm is normalized arclength measure on $\partial\mathbb{D}$. Therefore

$$F(w) = \exp\left(\frac{1}{2\pi}\int_0^{2\pi} \frac{e^{i\theta} + w}{e^{i\theta} - w} \log(h\circ\varphi(e^{i\theta}))\,d\theta\right)$$

is an outer function in $H^\infty(\mathbb{D})$. Let $f(z) = (F\circ\varphi^{-1})(z)$ and $\tilde{f}(z) = (\tilde{F}\circ\varphi^{-1})(z)$, where $\tilde{F}$ are the nontangential boundary values of F. Now $f \in H^\infty(G(\theta))$ and F is outer. Thus, by Beurling's Theorem, $f\cdot H^2(G(\theta))$ is dense in $H^2(G(\theta))$. So, in order to show that M_z on $H^2(G(\theta))$ is cyclic, we need only show that $f\cdot P :=$ $\{fp : p$ is a polynomial$\}$ is dense in $f\cdot H^2(G(\theta))$. In fact, since $\{p(z) + q(1/z) : p$ and q are polynomials$\}$ is dense in $H^2(G(\theta))$, it is sufficient to show that $f\cdot 1/z$ is in the closure of $f\cdot P$ in $H^2(G(\theta))$; that is, there exists a sequence of polynomials $\{p_n\}$ such that $\int |1/z - p_n|^2\cdot|\tilde{f}|^2 d\omega = \int |1/z\,\tilde{f} - p_n\tilde{f}|^2 d\omega \to 0$, as $n \to \infty$. This means that $1/z \in P^2(|\tilde{f}|^2 d\omega)$ (the closure of the polynomials in $L^2(|\tilde{f}|^2 d\omega)$).

Our strategy is to choose g in $L^2(|\tilde{f}|^2 d\omega)$ such that $\int pg|\tilde{f}|^2 d\omega = 0$ for all polynomials p and show that we must have $\int \frac{g(z)}{z}|\tilde{f}(z)|^2 d\omega(z) = 0$. For such a g, define the Cauchy transform $\hat{g}$ on $\mathbb{C}\backslash\partial G(\theta)$ by

$$\hat{g}(\xi) := \int_{\partial G(\theta)} \frac{g(z)\,|\tilde{f}(z)|^2}{z - \xi}\,d\omega(z)$$

($\hat{g}(\xi) = 0$ for all ξ in the unbounded component of $\mathbb{C}\backslash\overline{G(\theta)}$ since $g \perp P^2(|\tilde{f}|^2 d\omega)$).

Now $|\tilde{f}(z)|^2 = \exp(\frac{-2}{|z - 1|^3})$ a.e. on $\partial G(\theta)$. Consequently, $\frac{1}{(z - 1)^n} \in L^2(|\tilde{f}|^2 d\omega)$ for all positive integers n; in fact, $\frac{1}{(z - 1)^n} \in P^2(|\tilde{f}|^2 d\omega)$, since $\frac{1}{(z - (1 + 1/k))^n} \to \frac{1}{(z - 1)^n}$ in $L^2(|\tilde{f}|^2 d\omega)$ as $k \to \infty$ and, by Runge's Theorem, $\frac{1}{(z - (1 + 1/k))^n}$ is uniformly approximable on $\overline{G(\theta)}$ by polynomials. Moreover, since harmonic measure has total mass 1 and the maximum of $(x-1)^{-2n} \cdot e^{\frac{-2}{(x-1)^3}}$ on $(1,\infty)$ is $(\frac{n}{3e})^{\frac{2n}{3}}$,

$$\|(z - 1)^{-n}\|_2 := \{\int |(z - 1)^{-n}|^2 |\tilde{f}|^2 \, d\omega\}^{1/2} \le n^{n/3}, \tag{2.1.1}$$

for all positive integers n.

Notice that given distinct z and ξ in $\mathbb{C}\backslash\{1\}$ and any positive integer n,

$$\frac{1}{z - \xi} = \frac{1}{z - 1} + \dots + \frac{(\xi - 1)^{n-1}}{(z - 1)^n} + \frac{(\xi - 1)^n}{(z - 1)^n (z - \xi)}.$$

Therefore, since $z \to \frac{(\xi - 1)^{k-1}}{(z - 1)^k} \in P^2(|\tilde{f}|^2 d\omega)$ for any positive integer k and $g \perp P^2(|\tilde{f}|^2 d\omega)$, it follows that

$$\hat{g}(\xi) = (\xi - 1)^n \cdot \int \frac{g(z)|\tilde{f}(z)|^2}{(z-1)^n(z-\xi)}\, d\omega(z).$$

Moreover, for z in $\overline{G(\theta)}$ and ξ in $V(\pi/3)$, $|z - \xi| \geq \text{const}|z - 1|$. So, for ξ in $\text{int}(V(\pi/3))$,

$$\text{(2.1.2)} \qquad \begin{aligned} |\hat{g}(\xi)| &\leq \text{const}|\xi - 1|^n \|g\|_2 \|z^{-n-1}\|_2 \\ &\leq \text{const}|\xi - 1|^n M_n, \end{aligned}$$

where $\|\cdot\|_2$ denotes the $L^2(|\tilde{f}|^2 d\omega)$ norm and, by (2.1.1), $\sum_{n=0}^{\infty} M_{3n}^{-1/n} = \infty$. Applying the Cauchy-Schwarz inequality, and rescaling if needed, we may assume that $\{M_{3n}\}$ is log-convex with respect to n (cf. [CM]).

We now make preparations for the application of a result of T. Carleman. Let $\Psi(\xi) = (1 - \xi)^3$ and $E = \Psi(\text{int}(V(\pi/3)))$. Notice that $\Psi(0) = 0$ and ∂E coincides with the imaginary axis near 0. Now $\Psi^{-1}(z) = 1 - z^{1/3}$ and so by (2.1.2),

$$(\hat{g}\circ\Psi^{-1})(z) \leq \text{const}|z|^{n/3} M_n$$

for all positive integers n. In particular,

$$(\hat{g}\circ\Psi^{-1})(z) \leq \text{const}|z|^n M_{3n}$$

for all positive integers n. Since $\sum_{n=0}^{\infty} M_{3n}^{-1/n} = \infty$, $\{M_{3n}\}$ is log-convex and ∂E is "nice" near 0, a classical result of T. Carleman [Ca] gives us that $\hat{g}\circ\Psi^{-1}|_E \equiv 0$. So $\hat{g}|_{\mathrm{int}(V(\pi/3))} \equiv 0$, and therefore $\int \frac{g(z)}{z} |\tilde{f}|^2 d\omega(z) = 0$. It follows that $1/z \in P^2(|\tilde{f}|^2 d\omega)$, and we are done. □

Our next and final result shows that, concerning the existence of cyclic vectors, there is no difference between the Hardy and Bergman space cases for crescents like $G(\theta)$.

2.2 Theorem. The shift M_z on $L_a^2(G(\theta))$ is cyclic if $\pi/3 < \theta < \pi$ and fails to be cyclic if $0 \le \theta \le \pi/3$.

Proof (sketch). In [AKS] the authors showed that M_z on $L_a^2(G(\theta))$ is not cyclic if $0 \le \theta < \pi/3$. Theorem 1 along with [AKS, Theorem 4.7] give us that M_z on $L_a^2(G(\theta))$ is cyclic if $\pi/3 < \theta < \pi$. What remains is the case $\theta = \pi/3$

Choose $\varepsilon > 0$ small enough so that $W := \{z \in G(\pi/3) : \mathrm{dist}(z, \partial G(\pi/3)) > \min(\varepsilon, |z - 1|^{5/4})\}$ is a crescent. Let Ω be the bounded component of $\mathbb{C}\setminus\bar{W}$, $\omega := \omega(\cdot, W, z_\circ)$ be harmonic measure on ∂W evaluated at some $z_\circ$ in W and $\nu := \nu(\cdot, \Omega, \xi_\circ)$ be harmonic measure on $\partial\Omega$ evaluated at some $\xi_\circ$ in Ω. Using the conformal map $\Psi(z) = (z - 1)^3$ along with [Ts, Theorem IX, 9 (ii)] we get that $d\omega|_{\partial\Omega}$ and $d\nu$ are both boundedly equivalent to $|z - 1|^2 ds$ near 1, where ds is arclength measure on $\partial\Omega$. Therefore, by [AKS, Theorem 3.2], the shift M_z on $H^2(W)$ is not cyclic.

Again, by a conformal mapping argument we see that $d\omega$ is boundedly equivalent to $|z - 1|^2 ds$ near 1, where ds here is arclength measure on ∂W. So, from the definition of W we have that $\int_{\partial W} \frac{1}{r_z^2} d\omega(z) < \infty$, where $r_z := \text{dist}(z, \partial G(\pi/3))$. If M_z on $L_a^2(G(\pi/3))$ is cyclic, then there is an f in $L_a^2(G(\pi/3))$ and a sequence of polynomials $\{p_n\}$ such that $\iint_{G(\pi/3)} |1/z - p_n f|^2 dxdy \to 0$, as $n \to \infty$. Applying [AKS, Lemma 2.7] we get that $f \in H^2(W)$ and $\int |1/z - p_n f|^2 d\omega \to 0$, as $n \to \infty$. This means that M_z on $H^2(W)$ is cyclic, which contradicts our earlier discovery. Therefore, M_z on $L_a^2(G(\pi/3))$ is not cyclic. □

ACKNOWLEDGMENT

The author is grateful to the referee for helpful suggestions.

References

[AKS] J. Akeroyd, D. Khavinson and H.S. Shapiro, Remarks Concerning Cyclic Vectors in Hardy and Bergman Spaces, Michigan Math. J., to appear.

[Ca] T. Carleman, Fonctions Quasi Analytiques, Gauthier-Villars, Paris, 1926

[CM] J.A. Cima and A. Matheson, Approximation in the Mean by Polynomials, Rocky Mountain J. Math., 15(1985), 729-738.

[Ni] N.K. Nikol'skii, Outlines for the Computation of the Multiplicities of the Spectra of Orthogonal Sums, J. Soviet Math., 27(1), 1984, 2521-2526.

[SV] B.M. Solomyak and A.L. Volberg, Multiplicity of Analytic Toeplitz Operators, Operator Theory: Advances and Applications, vol. 42, 1989, 87-192.

[Ts] M. Tsuji, Potential Theory in Modern Function Theory, second edition, Chelsea, New York.

Integral Formulas for the Invariant Inner Products in Spaces of Analytic Functions on the Unit Ball

JONATHAN ARAZY Department of Mathematics, University of Haifa, Haifa 31999, ISRAEL

and Department of Mathematics, University of Kansas, Lawrence, KS 66065

Let B denote the open unit ball in $\mathbf{C}^n$. For $\lambda > n$ consider the probability measure on B

(1) $$d\mu_\lambda(z) = \binom{\lambda - 1}{n} (1 - |z|^2)^{\lambda - n - 1} dV(z)$$

where $dV(z)$ is Lebesgue measure, normalised so that $V(B) = 1$, i.e., $dV(z) = \frac{n!}{\pi^n} dm(z)$. Let $G = Aut(B)_0$ denote the connected component of the identity in the group $Aut(B)$ of all biholomorphic automorphism of B. It is known that G acts transitively on B, see for instance [R, Chapter 2]. The unitary group $U(n)$ is clearly a subgroup of G. Let

Supported in part by a grant from NSF No DMS 4572-0703

$S = \partial B$ be the unit sphere and let $d\sigma(\xi)$ denote Lebesgue measure on S, normalized so that $\sigma(S) = 1$. It is the unique $U(n)$-invariant probability measure on S.

For every $\lambda \in \mathbf{C}$ define an action $U^{(\lambda)}$ of G on functions on B via

$$(2) \qquad U^{(\lambda)}(\varphi)f = (f \circ \varphi) \cdot (J\varphi)^{\lambda/(n+1)}, \varphi \in G.$$

Here $J\varphi(z) = det(\varphi'(z))$ is the complex Jacobian of φ at z and we use the principal branch of the power functions. For $\lambda > n$ let $L^2_a(\mu_\lambda)$ denote the closed subspace of $L^2(\mu_\lambda)$ consisting of analytic functions. It is well known and easy to se that $U^{(\lambda)}$ is a unitary action of G on $L^2(\mu_\lambda)$ which preserves $L^2_a(\mu_\lambda)$. $L^2_a(\mu_\lambda)$ is the *weighted Bergman space* (for $\lambda = n + 1$ one gets the ordinary, unweighted, Bergman space). Point evaluations are continuous linear functionals on $L^2_a(\mu_\lambda)$ and the corresponding reproducing kernel is

$$(3) \qquad K_\lambda(z, w) = (1 - (z, w))^{-\lambda}, z, w \in B.$$

$K(z, w) = K_{n+1}(z, w) = (1 - (z, w))^{-(n+1)}$ is the *Bergman kernel*, and $K_\lambda(z, w) = K(z, w)^{\lambda/(n+1)}$. The *Hardy space* $H^2(S)$ is the closure of the analytic polynomials in $L^2(S) = L^2(S, d\sigma)$. Both $L^2(S)$ and $H^2(S)$ are invariant under the action (2) of G with $\lambda = n$. Since $\lim_{\lambda \to n+} \mu_\lambda = \sigma$ in the w^*-topology of the measures, the Hardy space is the limiting case of the weighted Bergman spaces as $\lambda \to n+$. Evaluations at points of B are continuous linear functionals on $H^2(S)$ and the corresponding reproducing kernel is the *Szego kernel*

$$(4) \qquad S(\xi, z) = (1 - (\xi, z))^{-n} = K(\xi, z)^{n/(n+1)}; z \in B, \xi \in S.$$

There are other natural Hilbert spaces of analytic functions on B which are invariant under the $U^{(\lambda)}$-action of G for appropriate choices of λ. Our description of these spaces goes along the lines of [FK] where the general case of a Cartan domain is considered. For all $\lambda \in (0,\infty), K_\lambda(z,w)$ is a positive definite function of z and w (i.e., if $z_j \in B$ and $a_j \in \mathbf{C}$ then $\sum_{k,j=1}^m a_k\bar{a}_j\ K_\lambda(z_k,z_j) \geq 0$). On $span\{K_\lambda(\cdot,w); w \in B\}$ one defines an inner product $(\cdot,\cdot)_\lambda$ by requiring that $(K_\lambda(\cdot,w),K_\lambda(\cdot,z))_\lambda = K_\lambda(z,w)$. Let $\mathcal{H}_\lambda$ denote the completion of this space. Clearly, K_λ is the reproducing kernel for $\mathcal{H}_\lambda$. It is well known that for every $\varphi \in G$ and $z,w \in B$,

$$J\varphi(z)K(\varphi(z),\varphi(w))\overline{J\varphi(w)} = K(z,w). \tag{5}$$

Thus $U^{(\lambda)}$ is an isometric action of G on $\mathcal{H}_\lambda$. Clearly, $\mathcal{H}_\lambda = L^2_a(\mu_\lambda)$ for $\lambda > n$ and $\mathcal{H}_n = H^2(S)$.

The inner product in $\mathcal{H}_\lambda$ can be written in terms of the homogeneous expansions $f = \sum_{m=0}^\infty f_m, g = \sum_{m=0}^\infty g_m$ as

$$(f,g)_\lambda = \sum_{m=0}^\infty \frac{(f_m,g_m)_F}{(\lambda)_m} = \sum_{m=0}^\infty \frac{1}{(\lambda)_m} \sum_{|\alpha|=m} \hat{f}(\alpha)\overline{\hat{g}(\alpha)}\alpha!. \tag{6}$$

Here $(f,g)_F = \frac{1}{\pi^n}\int_{\mathbf{C}^n} f(z)\overline{g(z)}e^{-|z|^2}dm(z)$ is the Fock-Fischer inner product, in which $(z^\alpha,z^\beta)_F = \delta_{\alpha,\beta}\alpha!$ (where $\alpha! = \alpha_1!\alpha_2!\cdots\alpha_n!$), and

$$(\lambda)_m = \lambda(\lambda+1)(\lambda+2)\cdots(\lambda+m-1). \tag{7}$$

The forms (6) are meromorphic in $\lambda \in \mathbf{C}$ and have simple poles precisely at the points of

$\mathbf{N}$ (but they are positive-definite only for $\lambda \in (0,\infty)$). For every $\ell \in \mathbf{N}$ one defines

$$(8) \qquad (f,g)_{-\ell} = (-1)^\ell \lim_{\lambda \to -\ell} (\lambda+\ell)(f,g)_\lambda = \frac{1}{\ell!} \sum_{m=\ell+1}^{\infty} \frac{(f_m, g_m)_F}{(m-\ell-1)!}.$$

Let $\mathcal{H}_{-\ell}$ denote the completion of the polynomials modulo polynomials of degree $\geq \ell$ with respect to $(\cdot,\cdot)_{-\ell}$. It follows that $U^{(-\ell)}$ is an isometric action of G on $\mathcal{H}_{-\ell}$.

For $\lambda \in (-\mathbf{N}) \cup (0,\infty)$ the $U^{(\lambda)}$-invariant Hilbert space $\mathcal{H}_\lambda$ is unique; in particular $\mathcal{H}_\lambda$ is irreducible. This is discussed in [P1] and [Z] for $\lambda = 0$, in [P2] for the unit disk and $\lambda \in -\mathbf{N}$ (but the proof generalizes to the unit ball) and in [AF] for $0 \leq \lambda$ and general Cartan domains.

The case $\ell = 0$ is of special importance, as the group action is in its simplest possible form: $U^{(0)}_{(\varphi)} f = f \circ \varphi$. In this case (8) takes the form

$$(9) \qquad (f,g)_0 = \sum_\alpha \alpha| \frac{\alpha!}{|\alpha|!} \hat{f}(\alpha)\overline{\hat{g}(\alpha)},$$

where $f(z) = \sum_\alpha \hat{f}(\alpha) z^\alpha, g(z) = \sum_\alpha \hat{g}(\alpha) z^\alpha$, see also [Z]. $\mathcal{H}_0$ is the generalization of the Dirichlet space of analytic functions on the unit disk $\mathbf{D}$ in $\mathbf{C}$. In this case the inner product $(\cdot,\cdot)_0$ admits several integral formulas. If $f(z) = \sum_{n=0}^\infty \hat{f}(n) z^n, g(z) = \sum_{n=0}^\infty \hat{g}(n) z^n$, then

$$(10) \qquad (f,g)_0 = \sum_{n=1}^{\infty} n \hat{f}(n)\overline{\hat{g}(n)}$$

$$= \int_{\mathbf{D}} f'(z)\overline{g'(z)}\, dA(z)$$

$$= \int_{\pi} e^{i\theta} f'(e^{i\theta})\overline{g(e^{i\theta})} \frac{d\theta}{2\pi}$$

$$= \int\int_{\mathbf{D}\times\mathbf{D}} \frac{(f(z)-f(w))\overline{(g(z)-g(w))}}{|1-z\bar{w}|^4} dA(z)dA(w).$$

A natural question is to obtain integral formulas, similar to those in $L_a^2(\mu_\lambda)$ $(\lambda > n)$ and $H^2(S)$, for the inner products $(f,g)_\lambda$ for the other points in $(0,\infty)\cup(-\mathbf{N})$. In what follows we use a method of J. Peetre to obtain integral formulas for these inner product. The essence of this method is an analytic continuation (in the parameter λ) of the integrals $\int_B f\bar{g}d\mu_\lambda$ via integration by parts in the radial direction. Our results are a developed version of the original short manuscript [P3], and they are presented here with the permission of J. Peetre.

To describe the results, recall that the *radial derivative* of a differentiable function f is

(11) $$Rf(z) = \frac{\partial}{\partial t} f(tz)_{|t=1}.$$

If f is analytic, then

(12) $$Rf(z) = \sum_{j=1}^{n} z_j \frac{\partial f}{\partial z_j}(z) = \sum_{m=0}^{\infty} m f_m(z).$$

For $k = 1, 2, \cdots$ we denote

(13) $$\mathcal{R}_k = \prod_{j=1}^{k}(R+n-j) = (R+n-k)_k.$$

THEOREM 1. *Let f, g be analytic functions in a neighborhood of $\bar{B}$. Then*

(14) $$(f,g)_\lambda = \frac{(\lambda-1)}{n!}\int_B (\mathcal{R}_{n-1}f)(z)\overline{g(z)}\frac{(1-|z|^2)^{\lambda-2}}{|z|^{2(n-1)}}dV(z); \lambda > 1$$

and

(15) $$(f,g)_\lambda = f(0)g(\bar{0}) + \frac{1}{n!}\int_B (\mathcal{R}_n f)(z)\overline{g(z)}\frac{(1-|z|^2)^{\lambda-1}}{|z|^{2n}}dV(z); \lambda > 0.$$

Moreover, for $\lambda = 1,2,\cdots,n$,

(16) $$(f,g)_\lambda = \frac{(\lambda-1)!}{(n-1)!}\int_S \mathcal{R}_{n-\lambda} f(\xi)\cdot g(\bar{\xi})d\sigma(\xi).$$

THEOREM 2. *Let $\ell = 0,1,2,\cdots$ and let f,g be analytic functions in a neighborhood of $\bar{B}$. Then*

(17) $$(f,g)_{-\ell} = \frac{1}{\ell!}\left[(f_{\ell+1},g_{\ell+1})_F + \frac{1}{n!}\int_B \frac{(\mathcal{R}_{n+\ell+1}f)(z)\overline{g(z)}dV(z)}{|z|^{2(n+\ell+1)}}\right]$$

and

(18) $$(f,g)_{-\ell} = \frac{1}{\ell!(n-1)!}\int_S (\mathcal{R}_{n+\ell}f)(\xi)g(\bar{\xi})d\sigma(\xi).$$

In the special case $\ell = 0$ we get

COROLLARY 1. *Let f,g be analytic functions in a neighborhood of $\bar{B}$. Then*

(19) $$(f,g)_0 = (\nabla f(0),\nabla g(0)) + \frac{1}{n!}\int_B \frac{(\mathcal{R}_{n+1}f)(z)\overline{g(z)}dV(z)}{|z|^{2(n+1)}}$$

and

(20) $$(f,g)_0 = \frac{1}{(n-1)!}\int_S (\mathcal{R}_n f)(\xi)\overline{g(\xi)}d\sigma(\xi).$$

Notice that (19) and (20) are the natural generatizations of (10). A formula similar to (19) was obtained independently by M. Peloso [Pe], using different techniques.

A basic compulational tool is the formula of *integration in polar coordinates* [R,1.4.3]

(21) $$\int_B f(z)dV(z) = 2n\int_0^1 r^{2n-1}dr\int_S f(r\xi)d\sigma(\xi), f \in L^1(B).$$

The *radialization* of the function f is the function

$$f^{\#}(z) = \int_{U(n)} f(uz)du = \int_S f(r\xi)d\sigma(\xi), |z| = r. \tag{22}$$

f is *radial* if $f^{\#} = f$, equivalently: $f(z) = f(uz)$ for all $u \in U(n)$. Let us define

$$\tilde{f}(r) = f^{\#}(z), \text{ where } |z| = r^{1/2}. \tag{23}$$

Thus, $f^{\#}(re_1) = \tilde{f}(r^2)$ and if $f \in L^1(B)$ then

$$\int_B f(z)dV(z) = 2n\int_0^1 r^{2n-1}f^{\#}(re_1)dr = n\int_0^1 r^{n-1}\tilde{f}(r)dr. \tag{24}$$

It is well known that the radial derivative commutes with the action of $U(n)$, i.e., $R(f \circ u) = (Rf) \circ u$ for every differentiable function f and $u \in U(n)$. From this it is obvious that

$$(Rf)^{\#} = R(f^{\#}) = r\frac{\partial f^{\#}}{\partial r} = 2r^2\tilde{f}'(r^2) \tag{25}$$

where $r^{1/2} = |z|$.

Using (7.18) and the well-known fact (see [R,1.4.9], or use (4.12)) that

$$(\xi^{\alpha}, \xi^{\beta})_{L^2(S)} = \delta_{\alpha,\beta}\frac{\alpha!}{(n)_{|\alpha|}}, \tag{26}$$

one can verify all the formulas in Theorems 1 and 2. Let us illustrate this for (14). By bilinearity and orthogonality of mononials, it is enough to take $f(z) = g(z) = z^{\alpha}$. Let $m = |\alpha|$. Using (12) and (13), we get

$$\frac{(\lambda - 1)}{n!}\int_B (\mathcal{R}_{n-1}z^{\alpha})\bar{z^{\alpha}}\frac{(1-|z|^2)^{\lambda-2}}{|z|^{2(n-1)}}dV(z)$$

$$= \frac{(\lambda-1)(m+n-1)!}{n!\,m!} \|\xi^\alpha\|^2_{L^2(S)} 2n \int_0^1 r^{2m+1}(1-r^2)^{\lambda-2} dr$$

$$= \frac{(\lambda-1)(m+n-1)!\alpha!\,n}{n!\,m!\,(n)_m} \cdot \frac{m!}{(\lambda-1)\lambda(\lambda+1)\cdots(m+\lambda-1)}$$

$$= \frac{\alpha!}{(\lambda)_m},$$

in accordance with (6). The other formulas follow similarly, where in Theorem 2 one has to notice that $ker(\mathcal{R}_{n+\ell})$ is the space of polynomials of degree $\leq \ell$. We would like to show something more, namely how one actually arrives at formulas (14)-(18) using integration by parts in the radial direction. This technique may lead to analogous formulas in the context of other Cartan domain.

Let f, g be analytic functions in a neighborhood of $\bar{B}$. For $m = 0, 1, 2, \cdots$ let

$$c(m) = (f_m, g_m)_{L^2(S)} = \sum_{|\alpha|=m} \frac{\hat{f}(\alpha)\overline{\hat{g}(\alpha)}\,\alpha!}{(n)_m} \tag{27}$$

and set $c(m) = 0$ for $m < 0$. Then

$$(f\bar{g})^{\#}(z) = \sum_{m=0}^{\infty} c(m) r^{2m}, \widetilde{(f\bar{g})}(r) = \sum_{m=0}^{\infty} c(m) r^m. \tag{28}$$

For $s \in \mathbf{C}$ with $Re\, s > -1$ and an integer ℓ so that $c(m) = 0$ whenever $m < -\ell$, consider the integral

$$J(s, \ell, f, g) = \int_B f(z)\overline{g(z)} \frac{(1-|z|^2)^s}{|z|^{2(n-\ell-1)}} dV(z). \tag{29}$$

LEMMA. *Let f, g, s and ℓ be as above, then*

$$J(s,\ell,f,g) = \frac{1}{s+1}J(s+1,\ell-1,(R+\ell)f,g) + \frac{n\,c(-\ell)}{s+1}. \tag{30}$$

Proof: Observe first that by (24) and (28),

$$J(s,\ell,f,g) = n\int_0^1 (1-r)^s r^\ell \cdot (\widetilde{f\bar{g}})(r)dr. \tag{31}$$

Next, let $\delta = r\frac{\partial}{\partial r}$ and observe that

$$\delta(\tilde{u}) = \frac{1}{2}\widetilde{Ru} \tag{32}$$

for any differentiable function u. It follows that

$$\delta(f\bar{g}) = \frac{1}{2}(R(f\bar{g}))^\sim = \frac{1}{2}(Rf\cdot\bar{g} + f\cdot\overline{Rg})^\sim = (Rf\cdot\bar{g})^\sim \tag{33}$$

Notice also that

$$(\delta-\ell+1)[(1-r)^{s+1}r^{\ell-1}] = -(s+1)(1-r)^s r^\ell. \tag{34}$$

Integrating by parts (31), we get

$$\begin{aligned} J(s,\ell,f,g) &= -\frac{n}{s+1}\int_0^1 (\delta-\ell+1)[(1-r)^{s+1}r^{\ell-1}](\widetilde{f\bar{g}})(r)dr \\ &= \frac{(\ell-1)}{s+1}J(s+1,\ell-1,f,g) - \frac{n}{s+1}[(1-r)^{s+1}r^\ell\widetilde{f\bar{g}}(r)]_0^1 \\ &\quad +\frac{n}{s+1}\int_0^1 (1-r)^{s+1}r^{\ell-1}(\widetilde{f\bar{g}})(r)dr \\ &\quad +\frac{n}{s+1}\int_0^1 (1-r)^{s+1}r^{\ell-1}\delta(\widetilde{f\bar{g}})(r)\,dr \end{aligned}$$

$$= \frac{\ell}{s+1} J(s+1, \ell-1, f, g) + \frac{n}{s+1} c(-\ell) + \frac{n}{s+1} \int_0^1 (1-r)^{s+1} r^{\ell-1} (Rf \cdot \bar{g})^{\sim}(r)\, dr$$

$$= \frac{1}{s+1} J(s+1, \ell-1, (R+\ell)f, g) + \frac{nc(-\ell)}{s+1}.$$

□

Proof of Theorem 1 and 2: Let f, g be analytic in a neighborhood of $\bar{B}$. We claim first for every $k = 0, 1, \cdots, n-1$ and every $\lambda \in \mathbf{C}$ with $Re\,\lambda > n$,

$$(35) \qquad (f,g)_\lambda = \frac{\prod_{j=1}^{n-k}(\lambda - j)}{n!} \int_B (\mathcal{R}_k f)(z) g(\bar{z}) \frac{(1-|z|^2)^{\lambda+k-n-1}}{|z|^{2k}} dV(z)$$

$$= \frac{\prod_{j=1}^{n-k}(\lambda - j)}{n!} J(\lambda + k - n - 1, n - k - 1, \mathcal{R}_k f, g).$$

Here $\mathcal{R}_k = \prod_{j=1}^{k}(R + n - j)$ for $k \geq 1$ and $\mathcal{R}_0 = I$. The case $k = 0$ is trivial since then (35) reduces to $(f,g)_\lambda = \int_B f\bar{g}d\mu_\lambda$. If (35) holds for $k \leq n-2$, then by the Lemma it holds also for $k+1$.

By analytic continuation it follows that (35) holds also for $\lambda \in \mathbf{C}$ with $Re\,\lambda > n - k$ and thus (35) provides an integral formula for $(f,g)_\lambda$ defined by (6). By taking $k = n-1$ in (35), we obtain (14).

To get (16), write (35) in the form

$$(36) \qquad (f,g)_\lambda = \frac{\prod_{j=1}^{n-k-1}(\lambda - j)}{(n-1)!} (\lambda + k - n) \int (\mathcal{R}_k f \cdot \bar{g})^{\sim}(r) \frac{(1-r)^{\lambda+k-n-1}}{r^k} dr.$$

As $\lambda \to (n-k)_+$, the probability measures $(\lambda+k-n)(1-r)^{\lambda+k-n-1}dr$ converge (in the weak-* topology of measures) to the Dirac measure δ_1 at 1. Thus

$$\begin{aligned}(f,g)_{n-k} &= \lim_{\lambda\to(n-k)_+}(f,g)_\lambda \\ &= \frac{(n-k-1)!}{(n-1)!}(\mathcal{R}_k f\cdot\bar{g})^\sim(1) \\ &= \frac{(n-k-1)!}{(n-1)!}\int_S (\mathcal{R}_k f)(\xi)\overline{g(\xi)}d\sigma(\xi),\end{aligned}$$

and (16) is established.

We claim next that for $\ell = -1, 0, 1, 2, \cdots$ and every $\lambda \in \mathbf{C}$ with $Re\,\lambda > -\ell$,

$$(f,g)_\lambda = \frac{J(\lambda_{\lambda+\ell}-1, -l-1, \mathcal{R}_{n+\ell}f, g)}{n!(\lambda)_\ell} + \sum_{j=0}^{\ell}\frac{(n)_j}{(\lambda)_j}c(j), \tag{37}$$

where the $c(j)'s$ are defined by (27). For $\ell=-1$, (37) reduces to (35). Suppose that (37) holds for some integer $\ell \geq -1$. Notice that

$$(\mathcal{R}_{n+\ell}f\cdot\bar{g})^\sim(r) = \sum_{m=\max(\ell+1,0)}^{\infty}(m-\ell)_{n+\ell}c(m)r^m. \tag{38}$$

Using Lemma 4, (37), and (38), we obtain

$$\begin{aligned}(f,g)_\lambda &= \frac{J(\lambda+\ell, -\ell-2, (R-\ell-1)\mathcal{R}_{n+\ell}f, g)}{n!(\lambda+\ell)(\lambda)_\ell} \\ &\quad + \frac{n\cdot(1)_{n+\ell}\cdot c(\ell+1)}{n!(\lambda)_\ell(\lambda+\ell)} + \sum_{j=0}^{\ell}\frac{(n)_j}{(\lambda)_j}c(j)\end{aligned}$$

$$= \frac{J(\lambda+\ell,-\ell-2,\mathcal{R}_{n+\ell+1}f,g)}{n!(\lambda)_{\ell+1}} + \frac{(n)_{\ell+1}}{(\lambda)_{\ell+1}}c(\ell+1) + \sum_{j=0}^{\ell}\frac{(n)_j}{(\lambda)_j}c(j).$$

Thus (37) holds for $\ell+1$, too, and the inductive proof of (37) is complete.

Formula (15) is just the case $\ell = 0$ in (37). If $\ell = 0,1,2,\cdots$, then (37) leads by the previous arguments to

$$\begin{aligned}(f,g)_{-\ell,1} &= (-1)^{\ell}\lim_{\lambda\to-\ell+0}(\lambda+\ell)(f,g)_{\lambda}\\ &= (-1)^{\ell}\lim_{\lambda\to-\ell+0}\frac{(\lambda+\ell)\int_0^1(\mathcal{R}_{n+\ell}f\cdot\bar{g})^{\sim}(r)r^{-\ell-1}(1-r)^{\lambda+\ell-1}dr}{(n-1)!(\lambda)_{\ell}}\\ &= \frac{1}{(n-1)!\ell!}(\mathcal{R}_{n+\ell}f\bar{g})^{\sim}(1)\\ &= \frac{1}{(n-1)!\ell!}\int_S(\mathcal{R}_{n+\ell}f)(\xi)g(\bar{\xi})d\sigma(\xi).\end{aligned}$$

This proves (18). To prove (17) for $\ell = 0,1,2,\cdots$, use (37) with $\ell+1$ and $Re\,\lambda > -\ell-1$. Thus

$$\begin{aligned}(f,g)_{-\ell,1} &= (-1)^{\ell}\lim_{\lambda\to-\ell+0}(\lambda+\ell)(f,g)_{\lambda}\\ &= (-1)^{\ell}\lim_{\lambda\to-\ell+0}\frac{(\lambda+\ell)J(\lambda+\ell,-\ell-2,\mathcal{R}_{n+\ell+1}f,g)}{n!(\lambda)_{\ell+1}}\end{aligned}$$

$$+(-1)^{\ell} \lim_{\lambda \to -\ell+0} \frac{(\lambda+\ell)(n)_{\ell+1} c(\ell+1)}{(\lambda)_{\ell+1}}$$

$$= \frac{1}{n!\ell!} \int_B (\mathcal{R}_{n+\ell+1} f)(z) \overline{g(z)} \frac{dV(z)}{|z|^{2(n+\ell+1)}}$$

$$+ \frac{(n)_{\ell+1}}{\ell!} c(\ell+1).$$

This proves (17), since $c(\ell+1) = (f_{\ell+1}, g_{\ell+1})_{L^2(S)}$. The proof of Theorems 1 and 2 is complete. □

A variant of the method described above leads to the following results.

THEOREM 3. *Let f, g be analytic functions in a neighborhood of $\bar{B}$.*

(a) *For $\ell = 0, 1, 2, \cdots, n-1$ and $\lambda \in \mathbf{C}$ with $Re\,\lambda > \ell$*

$$(39) \qquad (f, g)_\lambda = \frac{\prod_{j=1}^{\ell}(\lambda - j)}{n!} \int_B (\mathcal{R}_{n-\ell-1}(R+\lambda) f)(z) \cdot \overline{g(z)} \frac{(1-|z|^2)^{\lambda-\ell-1}}{|z|^{2(n-\ell-1)}} dV(z)$$

(b) *For $\ell = 1, 2, 3, \cdots$ and $\lambda \in \mathbf{C}$ with $Re\,\lambda > -\ell$*

$$(40)\ (f, g)_\lambda = \sum_{m=0}^{\ell-1} \frac{(f_m, g_m)_F}{(\lambda)_m} + \frac{1}{n!(\lambda)_\ell} \int_B (\mathcal{R}_{n+\ell-1}(R+\lambda) f)(z) \overline{g(z)} \frac{(1-|z|^2)^{\lambda+\ell-1}}{|z|^{2(n+\ell-1)}} dV(z).$$

Formulas (16) and (18) follow from (39) and (40), respectively, by limiting procedures.

Theorem 5 leads also to the following.

COROLLARY 2. *Let f, g be analytic functions in a neighborhood of $\bar{B}$.*

(a) *For* $\ell = 1, 2, \cdots, n$

(41) $$(f,g)_\ell = \frac{(\ell-1)!}{n!} \int_B (\mathcal{R}_{n-\ell}(R+\ell)f)(z) \cdot \overline{g(z)} \frac{dV(z)}{|z|^{2(n-\ell)}}$$

(b) *For* $\ell = 0, -1, -2, \cdots$

(42) $$(f,g)_{-\ell,1} = \frac{1}{n!\ell!} \int_B (\mathcal{R}_{n+\ell}(R-\ell))f(z) \cdot \overline{g(z)} \frac{dV(z)}{|z|^{2(n+\ell)}}$$

References

[AF] J. Arazy and S.D. Fisher, Invariant Hilbert spaces of analytic functions on bounded symmetric domains, to appear in Integral Equations and Operator Theory.

[FK] J. Faraut and A. Koranyi, Function spaces and reproducing kernels on bunded symmetric domains, J. Funct. Anal. 88 (1990), 64-89.

[P1] J. Peetre, Möbius invariant function spaces in several variables, manuscript (1982).

[P2] J. Peetre, Invariant function spaces connected with the holomorphic discrete series, in P.L. Butzer Anniversary Volume on Approximation and Functional Analysis, International Series of Numerical Mathematics Vol. 65, 1984 Birkhauser Verlag Basel, 119-134.

[P3] J. Peetre, Analytic continuation of norms, manuscript (1986).

[Pe] M. Peloso, Möbins invariant spaces on the unit ball, preprint (1990).

[R] W. Rudin, Function Theory in the Unit Ball of $\mathbf{C}^n$, Springer Verlag, New-York, Heidelberg, Berlin (1980).

[Z] K. Zhu, Möbius invariant Hilbert spaces of holomorphic functions in the unit ball of $\mathbf{C}^n$, to appear in the Transactions of the AMS.

On the Extended Spectral Radius in Some Classes of Locally Convex Algebras

HUGO ARIZMENDI Instituto de Matematicas, UNAM, Circuito exterior, Ciudad Universitaria, Mexico D. F. 04510, Mexico.

ANGEL CARRILLO Instituto de Matematicas, UNAM, Circuito exterior, Ciudad Universitaria, Mexico D. F. 04510, Mexico.

This paper is a report on joint work concerning the extended spectral radius in locally convex algebras.

Let A be a complete complex locally convex algebra with unit and let x belong to A.

In [5] W. Żelazko defines the extended spectral radius $R(x)$ and he poses an open question whether $r_6(x) = R(x)$, where $r_6(x)$ is defined as:

$$r_6(x) = \inf\{\, 0 < r \leq \infty \mid \text{there exists } (a_n),\ n = 0, 1, \ldots,\ a_n \in \mathbb{C} \text{ such that the radius of convergence of } \sum_{n=0}^{\infty} a_n \lambda^n \text{ is } r \text{ and } \sum_{n=0}^{\infty} a_n x^n \text{ converges in } A \,\}.$$

We prove in [2] that $r_6(x) = R(x)$ is false in general.

We first recall some definitions.

A *locally convex algebra* A is a topological algebra which is a locally convex space. The topology of such an algebra can be introduced by means of a family $\{\|x\|_\alpha\}$, $\alpha \in U$, of seminorms such that for each index α there is an index β with

$$\|xy\|_\alpha \leq \|x\|_\beta \|y\|_\beta \tag{1}$$

for all $x, y \in A$.

A locally convex metrizable and complete algebra A is called a B_0*-algebra.* In such cases, the topology of A can be introduced by means of a sequence $(\|x\|_i)$, $i = 1, 2, \dots,$ of seminorms satisfying

$$\|x\|_i \leq \|x\|_{i+1} \tag{2}$$

and

$$\|xy\|_i \leq \|x\|_{i+1} \|y\|_{i+1} \tag{3}$$

for $i = 1, 2, \dots$ and for all $x, y \in A$.

If for a locally convex algebra relations (1) or (3) can be replaced by

$$\|xy\|_\alpha \leq \|x\|_\alpha \|y\|_\alpha \tag{4}$$

for all $x, y \in A$ and $\alpha \in U$, then we say that A is a *locally multiplicative convex* (shortly *m - convex*) *algebra.*

Definition [5]. Let A be a complete complex locally convex algebra with unit e. For $x \in A$, the extended spectrum of x is defined as

$$\Sigma(x) = \sigma(x) \cup \sigma_d(x) \cup \sigma_\infty(x),$$

where

$$\sigma(x) = \Big\{ t \in \mathbb{C} \,\Big|\, x - te \quad \text{is not invertible in} \quad A \Big\},$$

$$\sigma_d(x) = \Big\{ t_0 \in \mathbb{C} \setminus \sigma(x) \,\Big|\, t \mapsto R(t,x) = (te - x)^{-1} \ \text{ is discontinuous at } \ t = t_0 \Big\},$$

and

$$\sigma_\infty(x) = \begin{cases} \emptyset & \text{if} \quad x \mapsto R(1,tx) \quad \text{is continuous at} \quad t = 0\,, \\ \infty & \text{otherwise.} \end{cases}$$

$$R(x) = \sup \{ |t| \ \ | \ t \in \Sigma(x) \}.$$

$\Sigma(x)$ and $R(x)$ are generalizations of the spectrum $\sigma(x)$ of x and the spectral radius $\rho(x)$ of x, respectively. Moreover, if A is a commutative complete complex m-convex algebra, then $\Sigma(x) = \sigma(x)$ and $R(x) = \rho(x)$.

In [5] W. Żelazko gives in this case seven formulas $r_i(x)$, $1 \leq i \leq 7$, on $\rho(x)$. So, it is natural to ask whether $r_i(x) = R(x)$, $1 \leq i \leq 7$, for x belonging to a locally convex algebra.

In the general case there are given only three formulas on $R(x)$, which are denoted by $R_1(x)$, $R_2(x)$ and $R_3(x)$. In particular, $R_1(x)$ is defined as

$$R_1(x) = \sup_{\beta \in \Sigma} \limsup_n \sqrt[n]{\| x^n \|_\beta}$$

where $\{ \| \cdot \|_\beta \}$, $\beta \in \Sigma$, is the set of all continuous seminorms on A.

We are also interested in $R_3(x)$, which is defined as

$R_3(x) = r_7(x) = \inf \ \{ 0 < r \leq \infty \mid$ for all (a_n), $n = 0, 1, \ldots$, $a_n \in \mathbb{C}$ such that radius of convergence of $\sum_{n=0}^{\infty} a_n \lambda^n$ is r, we have $\sum_{n=0}^{\infty} a_n x^n$ converges in $A\}$.

Finally, we define

$$R_*(x) = \sup_{\beta \in \Sigma} \liminf_n \sqrt[2]{\|x^n\|_\beta} \ ,$$

where $\{\| \cdot \|_\beta\}$, $\beta \in \sum$, has the meaning already stated.

In [2] we establish the following

THEOREM. If $\Big(A, (\| \cdot \|)\Big)$ is a complex B_0- algebra with unit, then

$$r_6(x) = R_*(x) \quad \text{for all } x \text{ in } A.$$

Proof. Here, we only prove that $R_*(x) \geq r_6(x)$. It is clear, if $R_*(x) = \infty$. So assume $R_*(x) < \infty$.

We can find two strictly increasing sequences (i_p) and (n_p) of positive integers

such that the sequence $\left(\sqrt[n_p]{\|x^{n_p}\|_{i_p}}\right)$ converges to $R_*(x)$.

Therefore the radius of convergence of $\sum_{p=1}^{\infty} \|x^{n_p}\|_{i_p} \lambda^{n_p}$ is $\frac{1}{R_*(x)}$.

Let $0 < b < \frac{1}{R_*(x)}$. Then $\sum_{p=1}^{\infty} \|x^{n_p}\|_{i_p} b^{n_p} < \infty$.

By relation (2) we have that $\sum_{p=1}^{\infty} \|x^{n_p}\|_i b^{n_p} < \infty$ for $i = 1, 2, \ldots,$

and therefore, the series $\sum_{p=1}^{\infty} b^{n_p} x^{n_p}$ converges in A.

It is obvious that the radius of convergence of the series $\sum_{p=1}^{\infty} b^{n_p} \lambda^{n_p}$ is $\frac{1}{b}$.

So $\frac{1}{b} \geq r_6(x)$.

Letting b tend, from below, to $\frac{1}{R_*(x)}$ we conclude that

$$R_*(x) \geq r_6(x). \square$$

From this theorem it follows that

$$r_6(x) = R_*(x) \leq R_1(x) = R(x),$$

whenever x is in a B_0-algebra with unit.

Finally in [1] a complex B_0-algebra with a generator t is constructed, and it is proved that $R_*(t) = 1$ and $R(t) = \infty$. So in that B_0-algebra it happens that $r_6(t) = R(t)$ is false.

In [2] we also prove that $r_6(x) \geq R_*(x)$ is true for x belonging to a complete complex locally convex algebra with unit. Thus, we have in the general case that

$$R_*(x) \leq \mathrm{r}_6(x) \leq \mathrm{r}_7(x) = R(x). \tag{5}$$

Recently, in [3], we have considered a class of topological algebras with orthogonal basis, namely the class of all Φ-algebras defined by S. El-Helaly and T. Husain in [4], and we have proved that if A is a complex Φ-algebra with unit, then we have

$$r_6(x) = \mathrm{r}_7(x) = R_*(x) = R(x) \tag{6}$$

for all x in A.

In the Remark 3.4 of [4], S. El-Helaly and T. Husain mention that they do not know an example of a complete complex locally convex algebra with an orthogonal basis and a unit which is not a Φ-algebra. Thus, it might happen that (6) is true for all x in a complete complex locally convex algebra with an orthogonal basis and a unit.

It is an open problem to determine what is the relation between $r_6(x)$, $\mathrm{r}_7(x)$, $R_*(x)$, and $R(x)$ in the class of all complete complex locally convex algebras with an orthogonal basis and a unit, and in other classes of complete complex locally convex algebras with unit.

References

[1] H. Arizmendi. *On the spectral radius of a matrix algebra.* Comment. Math. Funct. Approx. Poznań, Poland (in print).

[2] H. Arizmendi and A. Carrillo. *On the extended spectral radius in B_0-algebras.* Ibidem (in print).

[3] H. Arizmendi and A. Carrillo. *On the extended spectral radius in Φ-algebras.* Manuscript.

[4] S. El Helaly and T. Husain. *Orthogonal bases are Schauder bases and a characterization of Φ-algebras.* Pacific J. Math. 132, No. 2, 265–275.

[5] W. Żelazko. *Selected topics in topological algebras.* Aarhus University Lecture Notes Series 31, 1971.

Almost Everywhere Convergence for Transferred Convolution Operators

Nakhlé Asmar, Department of Mathematics, University of Missouri-Columbia, Columbia, Missouri 65211

Earl Berkson, Department of Mathematics, University of Illinois, Urbana, Illinois 61801

T.A. Gillespie, Department of Mathematics, University of Edinburgh, James Clerk Maxwell Building, Edinburgh EH9 3JZ, Scotland

§1. Introduction

In this introductory section we fix some notation and blanket hypotheses which will be in force throughout the subsequent sections. Let G denote a locally compact abelian group with dual group $\hat{G}$, and let R be a representation of G in $L^p(\Omega,\mu)$, where p is a fixed number in $[1,\infty)$ and (Ω,μ) is a σ-finite measure space. We also fix a Haar measure λ on G, and denote the Fourier transform of $k \in L^1(\lambda)$ by $\hat{k}$. We suppose throughout that R is strongly continuous, and that

(1.1) $$c \equiv \sup\left\{ \| R_u \| : u \in G \right\} < \infty .$$

We further assume that for each $f \in L^p(\mu)$, $\left[R_u f\right](x)$ has a version jointly measurable in $(u,x) \in G \times \Omega$. For $k \in L^1(\lambda)$, let T_k denote the convolution operator on $L^p(\lambda)$ corresponding to k, and let $T_k^{\#}: L^p(\mu) \longrightarrow L^p(\mu)$ be the transferred convolution operator defined for each $f \in L^p(\mu)$ by Bochner integration as follows:

$$T_k^{\#} f = \int_G k(u) R_{-u} f \, d\lambda(u) .$$

It is easy to see from Fubini's Theorem that

(1.2) $$\left[T_k^{\#} f\right](x) = \int_G k(u) \left[R_{-u} f\right](x)\, d\lambda(u) \, , \quad \text{for } \mu\text{-a.a. } x \in \Omega \, .$$

Throughout what follows, $\{ k_n \}_{n=1}^{\infty}$ will be a fixed sequence in $L^1(G)$, and we shall write $T_n^{\#}$ in place of $T_{k_n}^{\#}$.

The purpose of this note is to describe certain conditions on R and the sequence $\{ k_n \}_{n=1}^{\infty}$ which are sufficient to insure that for each $f \in L^p(\mu)$, the sequence $\left\{ T_n^{\#} f \right\}$ converges μ-a.e. on Ω. The conditions we use are suggested by those of Calderón [4], but apply to a more general framework. Our viewpoint is inspired by the spirit of [5]. As a particular application of our results, we develop two theorems ((3.9) and (3.12) below) generalizing the discrete ergodic Hilbert transform.

We will suppose henceforth that $\{ k_n \}_{n=1}^{\infty} \subseteq L^1(G)$ satisfies the following assumptions.

(1.3)

(i) $\lim_n \int_G k_n(u)\, d\lambda(u)$ exists as a complex number.

(ii) There is a subset $\mathcal{S}$ of $L^{\infty}(G) \cap L^1(G)$ such that the closure of $\mathcal{S}$ in the $L^1(G)$-norm contains $\left\{ g \in L^1(G) : \text{support } \hat{g} \subseteq \hat{G} \setminus \{0\} \right\}$; and

(iii) for each $g \in \mathcal{S}$, the sequence of convolutions $\{ k_n * g \}$ converges in $L^1(G)$ to a function $\mathcal{T}g \in L^1(G)$.

The further condition that will henceforth be imposed on the representation R is expressed in the following property.

(1.4) There is a subset D of $L^p(\mu) \cap L^{\infty}(\mu)$ such that D is norm-dense in $L^p(\mu)$, and such that for any $f \in D$ and $\epsilon > 0$, there are a μ-measurable set A and a real constant $M \geq 0$ satisfying: (i) $\mu(\Omega \setminus A) < \epsilon$, and, (ii) for each $u \in G$, $\left| \left[R_u f\right](x) \right| \leq M$ for μ-a.a. $x \in A$. (The set A and the constant M depend on f and ϵ.)

(1.5) <u>Some Examples</u>. (i) For a μ-distributionally bounded representation S, the associated representation $R = S^{(p)}$ on $L^p(\mu)$ satisfies (1.4) with $D = L^p(\mu) \cap L^\infty(\mu)$. (See §2 for the definition of distributionally bounded representations, and [3] for their properties used herein.) (ii) An example of a representation which satisfies all our blanket assumptions on R and is not associated with a distributionally bounded representation is provided by the one-parameter group $\{ R_t \}$ of isometries of $L^p(\mathbb{R})$ defined as follows. For $t \in \mathbb{R}$, $f \in L^p(\mathbb{R})$, let

$$\left[R_t f\right](x) \equiv e^{t/p} f(e^t x) .$$

In this instance only (1.4) needs detailed verification. Let $C_{00}(\mathbb{R})$ be the linear space of all continuous complex-valued functions on $\mathbb{R}$ having compact support. Let $f \in C_{00}(\mathbb{R})$, and denote by K the support of f. If $\epsilon > 0$, choose ρ so that $\epsilon < \rho < \infty$, and $K \subseteq [-\rho, \rho]$. For $| x | \geq \epsilon$, and $t \in \mathbb{R}$, we have: $\left[R_t f\right](x) = 0$ if $t > \log(\rho/\epsilon)$; $\left| \left[R_t f\right](x) \right| \leq (\rho/\epsilon)^{1/p} \| f \|_\infty$ if $t \leq \log(\rho/\epsilon)$. Thus $\left| \left[R_t f\right](x) \right| \leq (\rho/\epsilon)^{1/p} \| f \|_\infty$ for all $t \in \mathbb{R}$ and all x with $| x | \geq \epsilon$. So in the present particular set-up (1.4) holds with $D = C_{00}(\mathbb{R})$, but $\{ R_t \}$ is not associated with a distributionally bounded representation because it does not preserve the L^∞-norm for functions belonging to $L^p(\mathbb{R}) \cap L^\infty(\mathbb{R})$.

§2. <u>Almost Everywhere Convergence on Ω</u>

Our first aim is to describe, in the case $1 < p < \infty$, a dense subset of $L^p(\mu)$ such that for each f in the dense subset, the sequence $\left\{ T_n^{\#} f \right\}$ converges μ-a.e. on Ω (Theorem (2.2) below). We set the stage by taking $X = L^p(\mu)$ and $H = G$ in [1, Theorem (2.1)], and applying the latter to our setting. This gives us the following lemma.

(2.1) *Lemma*. *Suppose that* $1 < p < \infty$. *Let*

$$Y = \left\{ f \in L^p(\mu) : \ R_u f = f \ \text{ for all } \ u \in G \right\} ;$$

$$Z = \mathrm{clm} \left\{ T_g^{\#}(L^p(\mu)) : \ g \in L^1(G) \ \text{ and } \ \mathrm{support}(\hat{g}) \subseteq \hat{G} \setminus \{0\} \right\} ,$$

where "clm" *denotes* "closed linear span in $L^p(\mu)$ ". *Then*

$$L^p(\mu) = Y \oplus Z .$$

(2.2) *Theorem*. *Suppose that* $1 < p < \infty$. *Let* $\mathcal{A}$ *denote the span in* $L^p(\mu)$ *of* Y *and* $\bigcup \left\{ T_g^{\#}(D) : \ g \in \mathcal{S} \right\}$. *Then* $\mathcal{A}$ *is norm-dense in* $L^p(\mu)$, *and for each* $f \in \mathcal{A}$, *the sequence* $\left\{ T_n^{\#} f \right\}$ *converges* μ-a.e. *on* Ω .

Proof. Since D in (1.4) is dense in $L^p(\mu)$, the first conclusion is easy to see from Lemma (2.1) together with (1.3)-(ii). If $f \in Y$, then for each $n \in \mathbb{N}$, $T_n^{\#} f = \left[\int_G k_n(u)\, d\lambda(u) \right] f$. Hence by (1.3)-(i), $\left\{ T_n^{\#} f \right\}$ converges pointwise on Ω for $f \in Y$. Hence to complete the proof of the theorem, it suffices by linearity to show that $\left\{ T_n^{\#} T_g^{\#} h \right\}$ converges μ-a.e. on Ω for $g \in \mathcal{S}$, $h \in D$. Since $T_n^{\#} T_g^{\#} = T_{k_n * g}^{\#}$, it follows from (1.2) that for μ-a.a. $x \in \Omega$ we have:

$$\text{(2.3)} \qquad \left[T_n^{\#} T_g^{\#} h \right](x) = \int_G \left[k_n * g \right](u) \left[R_{-u} h \right](x)\, d\lambda(u) , \quad \text{for all } \ n \in \mathbb{N} ,$$

and, in the notation of (1.3)-(iii),

$$\text{(2.4)} \qquad \left[T_{\mathcal{T} g}^{\#} h \right](x) = \int_G \left[\mathcal{T} g \right](u) \left[R_{-u} h \right](x)\, d\lambda(u) .$$

Let B be a σ-compact subset of G such that for each $n \in \mathbb{N}$, $k_n * g$ vanishes on $G \setminus B$, and $\mathcal{T} g$ vanishes on $G \setminus B$. Given $\epsilon > 0$, pick A and M for h and ϵ in accordance with (1.4) . Let $\mathcal{W} = \left\{ (u,x) \in (-B) \times A : \ \left| \left[R_u h \right](x) \right| > M \right\}$. Thus $\mathcal{W}$ is a measurable subset of $(-B) \times A$ whose characteristic function we shall denote by $\chi_{\mathcal{W}}$. By (1.4) and Fubini's Theorem we have

$$0 = \int_{-B} \int_A \chi_{\mathscr{W}}(u,x)\ d\mu(x)\ d\lambda(u) = \int_A \int_{-B} \chi_{\mathscr{W}}(u,x)\ d\lambda(u)\ d\mu(x)\ .$$

Hence for μ-a.a. $x \in A$, we have $\left| \left[R_{-u}h\right](x) \right| \leq M$ for λ-a.a. $u \in B$. Combining this with (2.3) and (2.4), we see that for μ-a.a. $x \in A$, we have for all $n \in \mathbb{N}$:

$$\left| \left[T_n^{\#}T_g^{\#}h\right](x) - \left[T_{\mathscr{T}g}^{\#}h\right](x) \right| \leq M \left\| k_n * g - \mathscr{T}g \right\|_{L^1(G)} .$$

In view of (1.3)-(iii) it now follows that $\left\{ T_n^{\#}T_g^{\#}h \right\}$ converges μ-almost uniformly (and hence μ-a.e. on Ω) to $T_{\mathscr{T}g}^{\#}h$. This completes the proof of Theorem (2.2) .

In order to obtain the μ-a.e. convergence of $\left\{ T_n^{\#}f \right\}$ for all $f \in L^p(\mu)$, we next combine Theorem (2.2) with results from [2] and [3] concerning the transference of maximal estimates. For $r \in [1,\infty)$, let M_r denote the maximal convolution operator on $L^r(G)$ defined by our sequence $\{ k_n \} \subseteq L^1(G)$:

$$M_r g = \sup_n | k_n * g | , \quad \text{for all } g \in L^r(G) .$$

(2.5) <u>Corollary</u>. *Suppose that* $1 < p < \infty$, *and that the representation* R *described in* §1 *is also separation-preserving* [*that is, whenever* $f \in L^p(\mu)$, $g \in L^p(\mu)$, *and* $fg = 0$ μ-a.e. , *then* $(R_u f)(R_u g) = 0$ μ-a.e. *for all* $u \in G$] . *Then if* M_p *is of strong type* (p,p) *on* $L^p(G)$, *it follows that* $\left\{ T_n^{\#}f \right\}$ *converges* μ-a.e. *for all* $f \in L^p(\mu)$.

<u>Proof</u>. By virtue of [2, Théorème 1], the maximal operator defined by the sequence $\left\{ T_n^{\#} \right\}$ of transferred convolution operators is of strong type (p,p) on $L^p(\mu)$. In view of the Banach Principle, the desired conclusion is now immediate from Theorem (2.2).

Since, in general, separation-preserving representations need not transfer weak type maximal estimates [2], we need to consider a special subclass of such representations, in order to obtain an analogue of Corollary (2.5) in the case of weak type estimates for the maximal convolution operator. This subclass consists of the μ-distributionally bounded representations of G introduced in [3], to which the reader is referred for more details than those sketched here.

For a μ-measurable function f, let $\phi(f,\cdot)$ be the distribution function of f defined by:

$$\phi(f;y) = \mu\Big[\{ \omega \in \Omega: \ |f(\omega)| > y \} \Big], \quad \text{for all } y > 0 .$$

Definition. Let $\Gamma(\mu)$ denote the group (under composition) of all one-to-one linear maps of $L^1(\mu) \cap L^\infty(\mu)$ onto itself. A *μ-distributionally bounded representation of* G is a homomorphism S of G into $\Gamma(\mu)$ such that for some positive real constant α we have:

(2.6) $$\phi(S_u f;y) \le \alpha \, \phi(f;y) ,$$

for all $u \in G$, $f \in L^1(\mu) \cap L^\infty(\mu)$, and all $y > 0$.

Henceforth let S denote a μ-distributionally bounded representation of G. For $r \in [1,\infty)$, we obtain, as a consequence of (2.6), a unique representation $S^{(r)}$ of G in $L^r(\mu)$ such that $S_u^{(r)} f = S_u f$ for all $u \in G$ and all $f \in L^1(\mu) \cap L^\infty(\mu)$. The representation $S^{(r)}$ satisfies:

$$\sup \Big\{ \| S_u^{(r)} \| : \ u \in G \Big\} \le \alpha^{1/r} .$$

Moreover, (2.6) remains valid for $S^{(r)}$, and all $f \in L^r(\mu)$, $y > 0$. If $S^{(r)}$ is strongly continuous for some r in the range $1 \le r < \infty$, then $S^{(r)}$ is strongly continuous for all $r \in [1,\infty)$. In this case we say that S is strongly continuous. If S is strongly continuous, then for each $r \in [1,\infty)$ and $\phi \in L^1(G)$, we use Bochner integration to define the bounded linear operator $H_\phi^{(r)}: L^r(\mu) \longrightarrow L^r(\mu)$ by putting

$$H_\phi^{(r)}f = \int_G \phi(u)\, S_{-u}^{(r)}f\, d\lambda(u)\ , \quad \text{for all} \quad f \in L^r(\mu)\ .$$

It is easy to see that if $r \in [1,\infty)$, $t \in [1,\infty)$, and $F \in L^1(\mu) \cap L^\infty(\mu)$, then $H_\phi^{(r)}F = H_\phi^{(t)}F$.

(2.7) Corollary. *Under the blanket assumptions of §1, suppose that* $R = S^{(p)}$ *for some* μ*-distributionally bounded representation* S *of* G *(in particular,* S *is strongly continuous). Then the following propositions are valid.*

(i) *If* $1 < p < \infty$, *and the maximal convolution operator* M_p *is of weak type* (p,p) *on* $L^p(G)$, *then* $\left\{ T_n^\# f \right\}$ *converges* μ-a.e. *for each* $f \in L^p(\mu)$.

(ii) *Suppose that* $p = 1$, M_1 *is of weak type* $(1,1)$ *on* $L^1(G)$, *and* M_r *is of weak type* (r,r) *on* $L^r(G)$ *for some* r *such that* $1 < r < \infty$. *If* $\left[S_u^{(r)}F\right](x)$ *is jointly measurable in* $(u,x) \in G \times \Omega$ *for each* $F \in L^r(\mu)$, *then* $\left\{ T_n^\# f \right\}$ *converges* μ-a.e. *for each* $f \in L^p(\mu)$.

Proof. (i) It follows from [3, Théorème 2] that the maximal operator on $L^p(\mu)$ defined by the sequence $\left\{ T_n^\# \right\}$ is of weak type (p,p) . The desired conclusion in (i) now follows from Theorem (2.2). (ii) It likewise follows from [3, Théorème 2] that the maximal operator on $L^1(\mu)$ $\left[\text{respectively, } L^r(\mu)\right]$ defined by $\left\{ T_n^\# \right\}$ $\left[\text{respectively, } \left\{ H_{k_n}^{(r)} \right\}\right]$ is of weak type $(1,1)$ $\left[\text{respectively, weak type } (r,r)\right]$. The representation $S^{(r)}$ satisfies the blanket assumptions for representations described in §1 . Hence we can apply Theorem (2.2) to $S^{(r)}$ and $\{ k_n \}$ to infer that the sequence $\left\{ H_{k_n}^{(r)} f \right\}$ converges μ-a.e. for each f in a certain dense subset of $L^r(\mu)$. By the

Banach Principle, this sequence converges μ-a.e. for all $f \in L^r(\mu)$ --in particular, for all $f \in L^1(\mu) \cap L^\infty(\mu)$. But for $f \in L^1(\mu) \cap L^\infty(\mu)$, and each positive integer n,

$$H_{k_n}^{(r)} f = H_{k_n}^{(1)} f = T_n^{\#} f .$$

So we have shown that $\left\{ T_n^{\#} f \right\}$ converges μ-a.e. for each $f \in L^1(\mu) \cap L^\infty(\mu)$. Since $L^1(\mu) \cap L^\infty(\mu)$ is dense in $L^1(\mu)$, another application of the Banach Principle completes the proof.

§3. Generalization of the discrete ergodic Hilbert transform

In this section we specialize the previous considerations by taking G to be the additive group $\mathbb{Z}$ of all integers, and the sequence $\{ k_n \}$ to be the sequence $\{ h_N \}_{N=1}^{\infty}$ of truncated discrete Hilbert kernels defined on $\mathbb{Z}$ by putting:

$$h_N(j) = j^{-1} \quad \text{if} \quad 0 < |j| \le N \ ; \quad h_N(j) = 0 \quad \text{otherwise}.$$

We also denote by h the discrete Hilbert kernel defined on $\mathbb{Z}$ by: $h(j) = j^{-1}$ if $j \in \mathbb{Z} \setminus \{0\}$; $h(0) = 0$. It is well-known that (in the notation for the maximal convolution operators adopted in §2) M_1 is of weak type $(1,1)$ on $\ell^1(\mathbb{Z})$, and M_r is of strong type (r,r) on $\ell^r(\mathbb{Z})$ for $1 < r < \infty$ (see, e.g., [6], [7]). However, before we can begin to make use of Corollaries (2.5) and (2.7), it must be established that the sequence $\{ h_N \}$ satisfies the requirements in (1.3). Since (1.3)-(i) is trivially true here, we confine our attention to obtaining (1.3)-(ii),(iii) in the present circumstances.

(3.1) <u>Scholium</u> *(Summation by Parts). Let* $\{ a_j \}_{j=-\infty}^{\infty}$ *and* $\{ b_j \}_{j=-\infty}^{\infty}$ *be two sequences of complex numbers such that for each* $n \in \mathbb{Z}$, $A_n \equiv \sum_{j=-\infty}^{n} a_j$ *is convergent. Then for any* $K \in \mathbb{Z}$, $L \in \mathbb{Z}$, *with* $K < L$, *we have*

$$(3.2) \qquad \sum_{n=K}^{L} a_n b_n = \sum_{n=K}^{L} A_n(b_n - b_{n+1}) + A_L b_{L+1} - A_{K-1} b_K .$$

Proof. Notice that for $n \in \mathbb{Z}$,

$$a_n b_n = A_n(b_n - b_{n+1}) + A_n b_{n+1} - A_{n-1} b_n .$$

Let $\mathcal{S}_0$ consist of all finitely supported sequences of complex numbers $\{ a_j \}_{j=-\infty}^{\infty}$ such that $\sum_{j=-\infty}^{\infty} a_j = 0$. We denote the discrete Hilbert transform of a by Ha. Thus, $Ha = h * a$. We also put $H_N a = h_N * a$ for each positive integer N.

(3.3) Theorem. *Let* $a \equiv \{ a_j \}_{j=-\infty}^{\infty} \in \mathcal{S}_0$. *Then* $Ha \in \ell^1(\mathbb{Z})$, *and*

$$(3.4) \qquad \left\| H_N a - Ha \right\|_{\ell^1(\mathbb{Z})} \longrightarrow 0 , \quad as \quad N \longrightarrow +\infty .$$

Proof. Fix a positive integer υ such that $a_j = 0$ for $|j| > \upsilon$. For $j \in \mathbb{Z}$, we have:

$$\left[Ha\right](j) = \sum_{n=j-\upsilon}^{j+\upsilon} a_{j-n} h(n) .$$

Apply the summation by parts formula (3.2) to this sum to get:

$$(3.5) \qquad \left[Ha\right](j) = \sum_{n=j-\upsilon}^{j+\upsilon} A_n^{(j)}(h(n) - h(n+1)) + A_{j+\upsilon}^{(j)} h(j+\upsilon+1) - A_{j-\upsilon-1}^{(j)} h(j-\upsilon) ,$$

where $A_n^{(j)} = \sum_{\rho=-\infty}^{n} a_{j-\rho}$. In particular, $A_{j+\upsilon}^{(j)} = \sum_{\rho=-j-\upsilon}^{\infty} a_{j+\rho} = 0$, since $a \in \mathcal{S}_0$. Moreover, $A_{j-\upsilon-1}^{(j)} = \sum_{\rho=\upsilon+1}^{\infty} a_\rho = 0$. So (3.5) takes the form:

$$(3.6) \qquad \left[Ha\right](j) = \sum_{n=j-\upsilon}^{j+\upsilon} A_n^{(j)}(h(n) - h(n + 1)) .$$

If $|j| > \upsilon + 1$, then $|n - j| \le \upsilon$ implies $n \ne 0$, $n \ne -1$, and $|n| \ge |j| - \upsilon$. Hence we see from (3.6) that:

$$(3.7) \qquad \left| \left[Ha\right](j) \right| \le (2\upsilon + 1) \left\| a \right\|_{\ell^1(\mathbb{Z})} \left[|j| - \upsilon\right]^{-1} \left[|j| - \upsilon - 1\right]^{-1} ,$$

$$\text{for all } j \in \mathbb{Z} \text{ such that } |j| > \upsilon + 1 .$$

This shows that $Ha \in \ell^1(\mathbb{Z})$.

We now proceed to establish (3.4) by estimating the size of $\left| \left[H_N a\right](j) - \left[Ha\right](j) \right|$ for $j \in \mathbb{Z}$. We can assume without loss of generality that $N > 2\upsilon$. We distinguish three cases for j .

Case I: $|j| \le N - \upsilon$. In this case, $|n - j| \le \upsilon$ implies $|n| \le N$. It follows easily that

$$\left| \left[H_N a\right](j) - \left[Ha\right](j) \right| = 0 .$$

Case II: $|j| > N + \upsilon$. In this case, $|n| \le N$ implies $|j - n| > \upsilon$, and so $\left[H_N a\right](j) = 0$. Hence in Case II we see with the aid of (3.7) that:

$$\left| \left[H_N a\right](j) - \left[Ha\right](j) \right| \le (2\upsilon + 1) \left\| a \right\|_{\ell^1(\mathbb{Z})} \left[|j| - \upsilon\right]^{-1} \left[|j| - \upsilon - 1\right]^{-1} .$$

Case III: $N - \upsilon < |j| \le N + \upsilon$. Since, in this case, $|n| \le \upsilon$ implies $|j - n| > N - 2\upsilon > 0$, we see from

$$\left| \left[H_N a\right](j) \right| \le \left\| a \right\|_{\ell^1(\mathbb{Z})} \sum_{n=-\upsilon}^{\upsilon} | h(j - n) |$$

that

$$\left| \left[H_N a\right](j) \right| \le (2\upsilon + 1) \left\| a \right\|_{\ell^1(\mathbb{Z})} (N - 2\upsilon)^{-1} .$$

Similarly,

$$\left| \left[Ha\right](j) \right| \le (2\upsilon + 1) \left\| a \right\|_{\ell^1(\mathbb{Z})} (N - 2\upsilon)^{-1} .$$

Hence in Case III we have:

$$\left|\left[\mathcal{H}_N a\right](j) - \left[\mathcal{H}a\right](j)\right| \le 2\,(2\nu + 1)\,\|a\|_{\ell^1(\mathbb{Z})}(N - 2\nu)^{-1} .$$

Assembling the facts from Cases I, II, and III, we find that

$$\left\|\mathcal{H}_N a - \left[\mathcal{H}a\right]\right\|_{\ell^1(\mathbb{Z})}$$
$$\le (2\nu + 1)\,\|a\|_{\ell^1(\mathbb{Z})} \sum_{|j| > N+\nu} \left[|j| - \nu\right]^{-1}\left[|j| - \nu - 1\right]^{-1}$$
$$+ 8\nu\,(2\nu + 1)\,\|a\|_{\ell^1(\mathbb{Z})}(N - 2\nu)^{-1} .$$

This readily completes the proof of Theorem (3.3).

(3.8) <u>Proposition</u>. *Let* $\mathcal{B} = \left\{ b = \{ b_j \} \in \ell^1(\mathbb{Z}) : \sum_{j=-\infty}^{\infty} b_j = 0 \right\}$.

Then $\mathcal{B}$ *is the closure in* $\ell^1(\mathbb{Z})$ *of* $\mathcal{S}_0$.

<u>Proof</u>. Clearly $\mathcal{S}_0 \subseteq \mathcal{B}$. Let $b \in \mathcal{B}$. Given $\epsilon > 0$, pick $N \in \mathbb{N}$ such that

$$\left| \sum_{|j| \le N} b_j \right| < \epsilon/2 \; ; \qquad \sum_{|j| > N} |b_j| < \epsilon/2 .$$

Define $a \in \mathcal{S}_0$ by setting $a_j = b_j$ for $|j| \le N$, $a_{N+1} = -\sum_{j=-N}^{N} b_j$,

and $a_j = 0$ for all other values of $j \in \mathbb{Z}$. Then

$$\|b - a\|_{\ell^1(\mathbb{Z})} = \sum_{|j| > N} |b_j - a_j|$$
$$\le \sum_{|j| > N} |b_j| + |a_{N+1}| < \epsilon .$$

This establishes Proposition (3.8).

It is clear from Theorem (3.3) and Proposition (3.8) that in the case where $G = \mathbb{Z}$ and $\{ k_n \}$ is taken to be the sequence $\{ h_N \}_{N=1}^{\infty}$ of truncated discrete Hilbert kernels, the requirements (1.3)-(ii),(iii) are satisfied by taking $\mathcal{S}$ to be the class $\mathcal{S}_0$ defined above. In this case, $\hat{G}$ is the multiplicative group of the unit circle $\mathbb{T}$, and we notice in passing that $\mathcal{S}_0$ is not a subset of $\left\{ g \in \ell^1(\mathbb{Z}): \text{support}(\hat{g}) \subseteq \mathbb{T} \setminus \{1\} \right\}$ (the latter set does not contain the sequence $\{ a_j \} \in \mathcal{S}_0$ defined by $a_j = j$ if $j = \pm 1$, and $a_j = 0$ otherwise).

The stage is now set for applying Corollaries (2.5) and (2.7) to the present set-up. Let p and (Ω,μ) be as described in §1. Let U be a bounded injective linear mapping of $L^p(\mu)$ onto itself such that U is power-bounded $\left[\text{that is, } c \equiv \sup \{ \| U^n \| : n \in \mathbb{Z} \} < \infty \right]$. Notice that the representation of $\mathbb{Z}$ defined by $n \longrightarrow U^n$ is automatically strongly continuous, and automatically enjoys the property that for each $f \in L^p(\mu)$, $\left[U^n f\right](x)$ is jointly measurable in $(n,x) \in \mathbb{Z} \times \Omega$. Moreover, as is well-known, if U is separation-preserving, so is U^n for all $n \in \mathbb{Z}$.

(3.9) <u>Theorem</u>. *Suppose that* $1 < p < \infty$ *and* (Ω,μ) *is a* σ*-finite measure space. Let* $U: L^p(\mu) \longrightarrow L^p(\mu)$ *be an invertible, power-bounded linear operator such that* U *is separation-preserving. Suppose further that the representation of* $\mathbb{Z}$ *given by* $n \longrightarrow U^n$ *satisfies (1.4). Then for each* $f \in L^p(\mu)$

$$\tilde{f}(x) = \lim_{n\to+\infty} \sum_{0 < |k| \le n} k^{-1} \left[U^k f\right](x) \tag{3.10}$$

exists for μ*-a.a.* $x \in \Omega$.

<u>Proof</u>. This now follows directly from Corollary (2.5).

(3.11) <u>Example</u>. Let β be a positive real number, let $p \in (1,\infty)$, and let U be the surjective linear isometry of $L^p(\mathbb{R})$ defined by

$$\left[Uf\right](x) \equiv \beta^{1/p} f(\beta x), \quad \text{for } f \in L^p(\mathbb{R}).$$

In view of the discussion in (1.5)-(ii), the hypotheses of Theorem (3.9) are satisfied, and so the conclusion expressed in (3.10) holds for the present particular operator U.

(3.12) <u>Theorem</u>. *Suppose that* $1 \le p < \infty$ *and* (Ω,μ) *is a* σ*-finite measure space. Let* $U: L^p(\mu) \longrightarrow L^p(\mu)$ *be an invertible, power-bounded linear operator. Suppose further that there is a positive real* *constant* α *such that for each* $f \in L^p(\mu)$, $n \in \mathbb{Z}$, *and* $y > 0$,

$$\mu\left[\left\{ x \in \Omega: \ \left|\left[U^n f\right](x)\right| > y \right\}\right] \leq \alpha\, \mu\left[\left\{ x \in \Omega: \ \left|f(x)\right| > y \right\}\right] .$$

Then for each $f \in L^p(\mu)$

$$\tilde{f}(x) = \lim_{n \to +\infty} \sum_{0 < |k| \leq n} k^{-1} \left[U^k f\right](x)$$

exists for μ-a.a. $x \in \Omega$.

Proof. This theorem follows directly from Corollary (2.7) and the discussion in (1.5)-(i).

(3.13) Example. A simple example of an operator U satisfying the hypotheses of Theorem (3.12) is readily obtained by taking an invertible measure-preserving transformation Φ of the points of Ω , together with a measurable function θ of constant modulus 1 on Ω , and setting

$$\left[Uf\right](x) \equiv \theta(x)\, f(\Phi(x)) , \quad \text{for} \quad f \in L^p(\mu) .$$

(3.14) Remark. A recent result of R. Sato obtained by direct calculations with the discrete averages [8, Theorem 2] shows that Theorem (3.9) remains valid if the requirement (1.4) is deleted from the hypotheses.

Acknowledgements. The authors would like to express their appreciation to the Organizers for the opportunity to present this article in the Conference Proceedings. The first author is grateful to the University of Missouri for its Summer Research Fellowship. The work of the second author was supported by a National Science Foundation grant.

REFERENCES

1. N. Asmar, E. Berkson, and T.A. Gillespie, *Representations of groups with ordered duals and generalized analyticity*, J. Functional Analysis, 90(1990), 206-235.

2. N. Asmar, E. Berkson, and T.A. Gillespie, *Transfert des opérateurs maximaux par des représentations conservant la séparation*, C.R. Acad. Sci. (Paris), t. 309, Série I (1989), 163-166.

3. N. Asmar, E. Berkson, and T.A. Gillespie, *Transfert des inégalités maximales de type faible*, C.R. Acad. Sci. (Paris), t. 310, Série I (1990), 167-170.

4. A.P. Calderón, *Ergodic theory and translation-invariant operators*, Proc. Nat. Acad. of Sciences U.S.A., 59(1968), 349-353.

5. R.R. Coifman and G. Weiss, *Transference methods in analysis*, C.B.M.S. Regional Conference Series in Math., No. 31, Amer. Math. Soc., Providence, R.I., 1977.

6. R. Hunt, B. Muckenhoupt, and R. Wheeden, *Weighted norm inequalities for the conjugate function and Hilbert transform*, Trans. Amer. Math. Soc., 176(1973), 227-251.

7. K. Petersen, *Another proof of the existence of the ergodic Hilbert transform*, Proc. Amer. Math. Soc., 88(1983), 39-43.

8. R. Sato, *On the ergodic Hilbert transform for Lamperti operators*, Proc. Amer. Math. Soc., 99 (1987), 484-488.

Generalizing Grothendieck's Program

David P. Blecher Department of Mathematics, University of Houston, Houston, TX 77204-3476

Introduction.

One may view operator spaces as generalized function spaces. Together with Vern Paulsen we were able to show that the elementary theory of tensor norms of Banach spaces carries over to operator spaces. We suggested that the Grothendieck tensor norm program, which was of course enormously important in the development of Banach space theory, be carried out for operator spaces. Some of this has been done by the authors mentioned above, and by Effros and Ruan. We describe these results and make some conjectures.

1. Background. In its loosest sense, a function space is a subspace X of the space $C(\Omega)$ of continuous (complex-valued) functions on a topological space Ω. In this sense, by the Bourbaki-Alaoglu theorem, the function spaces are exactly the normed linear spaces. An *operator space* is a subspace X of the space $B(\mathcal{H})$ of continuous operators on a (complex) Hilbert space $\mathcal{H}$. We begin by describing the "quantized functional analysis" program of E.G. Effros, for more details see [11]. By the Gelfand-Naimark-Segal theorem [30], since $C(\Omega)$ is a C*-algebra there is a Hilbert space $\mathcal{H}$ such that $C(\Omega) \subset B(\mathcal{H})$, and so it follows that every function space may be regarded as an operator space. However an operator space possesses a certain additional structure, recognition of which has proven rewarding. The appropriate objects and morphisms are then the operator spaces and completely bounded maps respectively, and their representation theory [1,21,10,23,27,8,7] lead to a phenomenal theory much of which is absent if we ignore the additional structure.

If $\mathcal{H}$ is a Hilbert spaces then $B(\mathcal{H})$ is more than a mere Banach space, there is a natural way to assign norms $\|\cdot\|_n$ to the spaces $M_n(B(\mathcal{H}))$ of nxn matrices with entries in $B(\mathcal{H})$. If $X \subset B(\mathcal{H})$, then the spaces $M_n(X)$ of nxn matrices with entries in X inherit the norm $\|\cdot\|_n$. The pairing $(X , \{ \|\cdot\|_n \})$ is technically what we mean by an operator space.

The morphisms between operator spaces are the completely bounded maps. A map $T : X \to Y$ between operator spaces is completely bounded if there is a constant K such that $\|[T(x_{ij})]\|_n \leq K \|[x_{ij}]\|_n$ for all n, and $[x_{ij}] \in M_n(X)$. We define $\|T\|_{cb}$ to be the least K which will suffice. If $\|[T(x_{ij})]\|_n = \|[x_{ij}]\|_n$ always then T is said to be a *complete isometry*. We write $CB(X,Y)$ for the space of completely bounded maps from X to Y, this possesses a norm $\|\cdot\|_{cb}$.

The realization of the first paragraph that every function space is an operator space really describes a functor which we call MIN, it is not hard to see that if X is a function space then MIN(X) is the smallest operator space structure on X, that is the sequence of norms $\|\cdot\|_n^{min}$ obtained in this manner is dominated by every other sequence of norms making X an operator space. If X is a normed space we define MAX(X) to be the biggest operator space structure on X: that is we define

$$\|[x_{ij}]\|_n^{max} = \sup \{ \|[T(x_{ij})]\|_n : T : X \to B(\mathcal{H}) \text{ contractive } \} .$$

Thus MIN and MAX define two natural functors from the category of function spaces and bounded linear maps to the category of operator spaces and completely bounded linear maps. In the intersection of the images of these functors there are probably only three spaces: $\mathbb{C}$, l_2^∞ and l_2^1.

Conjecture: if MIN(X) = MAX(X) then X is isometrically isomorphic to $\mathbb{C}$, l_2^∞ or l_2^1.

Indeed Vern Paulsen has shown in [22] that if MIN(X) = MAX(X) then $\dim(X) \leq 4$. The fact that l_2^∞ and l_2^1 have this property follows from results in [19,4] (see [22] for a simpler proof).

We remark that MIN and MAX are full embeddings, this means that as categories there is no difference between "function spaces", MIN("function spaces") and MAX("function spaces"). The first statement here is really a theorem [21].

In the above sense then operator spaces form a much larger and more complicated category. Effros's program asks what results and notions from functional analysis can we generalize to the larger category? There are, for instance, natural analogues for operator spaces of quotients [27], direct sums ($X \oplus_\infty Y \subset B(\mathcal{H} \oplus \mathcal{K})$), spatial tensor products ($X \otimes_\infty Y \subset B(\mathcal{H} \otimes \mathcal{K})$), etcetera. Until recently the drawback to this generalization was duality: how can the dual of an operator space be an operator space? The solution was independently proposed in [6] and [14]: identify $M_n(X^*) = CB(X,M_n)$. Then X^* is an operator space. It is important to realise that this is the ordinary dual space X^*, with new norms assigned to $M_2(X^*)$, $M_3(X^*)$, ... This is precisely the correct operator space dual, and possesses many desirable properties, for instance

Theorem [4]. If X is a normed space then $MIN(X)^* = MAX(X^*)$ and $MAX(X)^* = MIN(X^*)$.

Armed with these notions we can now attempt to generalize some aspects of functional analysis. In this talk we shall show how some of the elementary aspects of Grothendieck's tensor norm program [18], which was so influential in the development of Banach space theory, may be generalized to operator spaces. This will lead someday to an understanding of the geometry of operator spaces.

We remark that there is no reason why we should only study subspaces of $B(\mathcal{H})$. More generally, everything here makes sense for subspaces of the bounded operators on l^p spaces. Gilles Pisier has done some fascinating work on complete boundedness in this and more general situations, which undoubtably will play a role in a full generalization of the tensor norm program.

For a study of operator space analogues of Grothendieck's approximation property we refer the reader to [15]. Effros and Ruan show there that *not every operator space possesses local reflexivity*. This may be a serious setback to the entire program, perhaps one has to restrict attention to "nuclear" operator spaces (in some appropriate sense).

2. Three tensor norms, and their descendants. We now proceed to define the three major operator space tensor norms, min , max and h ,we shall be brief since these are treated at length in [6]. Suffice it to say that these generalize respectively the injective, projective, and Hilbertian Banach space tensor products of [28,18], denoted by λ, γ, and H (or γ_2 [25]), and behave precisely analogously. The theorem in section 3 below shows that these are generalizations in a very concrete sense.

The operator space injective tensor product min , also known as the spatial tensor product [21], is defined as follows. If X and Y are operator spaces contained in $B(\mathcal{H})$ and $B(\mathcal{K})$ respectively then $X \otimes Y$ may be identified with a subspace of $B(\mathcal{H} \otimes \mathcal{K})$; this assigns an operator space structure to $X \otimes Y$, which is independent of the particular Hilbert spaces on which X and Y are represented. We write this operator space as $X \otimes_{min} Y$; there is another characterization of the matrix norms on $X \otimes_{min} Y$ given in [6].

The operator space projective tensor product $X \otimes_{max} Y$, which was independently and contemporaneously discovered in [6] and [14], may be defined by specifying $CB(X \otimes_{max} Y, B(\mathcal{H}))$ for an arbitary Hilbert space $\mathcal{H}$. A map $\phi : X \otimes_{max} Y \to B(\mathcal{H})$ is completely contractive if and only if $|||[\phi(x_{ij} \otimes y_{kl})]|||_{nm} \le |||[x_{ij}]|||_n \, |||[y_{kl}]|||_m$ whenever $[x_{ij}] \in M_n(X)$, $[y_{kl}] \in M_m(Y)$. Another useful description of the norms on $X \otimes_{max} Y$ is given in [14]. One attractive property max possesses is the following:

Theorem [15]. If $\mathcal{M}$ and $\mathcal{N}$ are von Neumann algebras then $\mathcal{M} \otimes \mathcal{N} = (\mathcal{M}_* \otimes_{max} \mathcal{N}_*)^*$.

The Haagerup tensor product $X \otimes_h Y$ of operator spaces X and Y [11,23] may also be defined by specifying $CB(X \otimes_h Y, B(\mathcal{H}))$ for an arbitary Hilbert space $\mathcal{H}$. A map $\phi : X \otimes_h Y \to B(\mathcal{H})$ is completely contractive if and only if $|||[\Sigma_k \phi(x_{ik} \otimes y_{kj})]|||_n \le |||[x_{ij}]|||_n \, |||[y_{ij}]|||_n$ whenever $[x_{ij}] \in M_n(X)$, $[y_{ij}] \in M_n(Y)$.

The basic canonical correspondences of [28,18] carry over to the larger category, for instance $CB(X,Y^*) = (X \otimes_{max} Y)^*$.

As in [28,18] we study reasonable operator space tensor norms α , by which we mean [6] that α is defined on the tensor product of any pair of operator spaces, and that it behaves well with respect to morphisms between spaces. More precisely, if $[S_{ij}] \in M_n(CB(X_1,X_2))$ and if $[T_{kl}] \in M_m(CB(X_1,X_2))$ then $|||[S_{ij} \otimes_\alpha T_{kl}]|||_{mn} \le |||[S_{ij}]|||_n \, |||[T_{kl}]|||_m$. If α is such a norm we define the dual norm α^* by $X \otimes_{\alpha*} Y \subset (X^* \otimes_\alpha Y^*)^*$

Theorem [6]. The largest reasonable operator space tensor norm is max, and min is the smallest reasonable operator space tensor norm α such that α^* is also reasonable. Also $\max^* = \min$.

The theorem above is the precise analogue of the Banach space case.

We now introduce some new norms, following [18,17]. In what follows we shall only consider reasonable norms, for such a norm α we sometimes write $\alpha_n(U;X \otimes Y)$ for the norm of an element $U \in M_n(X \otimes_\alpha Y)$. We shall say that a uniform norm is *completely injective* if subspaces X_1 of X_2 and Y_1 of Y_2 determine a complete isometry of $X_1 \otimes_\alpha Y_1$ into $X_2 \otimes_\alpha Y_2$. A uniform norm is *completely projective* if complete quotients X_1 of X_2 and Y_1 of Y_2 determine a complete quotient map of (the completions of) $X_2 \otimes_\alpha Y_2$ onto $X_1 \otimes_\alpha Y_1$. The spatial norm min is completely injective, the projective operator space norm max is completely projective, and the Haagerup norm h is both [23,6,16]. It may be seen in several ways that min is not projective and that max is not injective.

A uniform norm α is *associative* if $(X \otimes_\alpha Y) \otimes_\alpha Z = X \otimes_\alpha (Y \otimes_\alpha Z)$ completely isometrically for all operator spaces X , Y and Z . One may show that min , max and h are all associative. Curiously enough, H is not associative (this was communicated to the author by Pisier).

Now there is a natural linear map $t : X \otimes Y \to Y \otimes X$; if α is a uniform norm then we define the *transposed and symmetrized* norms α^t and α^s, of α by $(\alpha^t)_n(U;X \otimes Y) = \alpha_n(t_n(U);Y \otimes X)$ and

$$(\alpha^s)_n(U;X \otimes Y) = \max\{ \alpha_n(U;X \otimes Y), (\alpha^t)_n(U;X \otimes Y) \}.$$

Both these norms are uniform.

We define the associate norm α' of α by

$$(\alpha')_n(U;X \otimes Y) = \inf\{ (\alpha^*)_n (U; E \otimes F) \},$$

where the infimum is taken over finite dimensional subspaces E and F of X and Y respectively, with $U \in M_n(E \otimes F)$.

If α is uniform then α^*, and consequently α', is uniform. Of course if $\alpha \leq \beta$ then $\beta' \leq \alpha'$. If α is a completely injective uniform tensor norm then α' is completely projective, and if α is a completely projective uniform tensor norm then α^* is completely injective and $\alpha^* = \alpha'$. Thus $\max^* = \max' = \min$ [6], and $h^* = h' = h$ [16,5,9].

We say that α is *tensorial* if $\alpha_n(U;X \otimes Y) = \inf\{ \alpha_n(U ; E \otimes F) \}$, where the infimum is taken over finite dimensional subspaces E and F of X and Y respectively, with $U \in M_n(E \otimes F)$. It is easy to see that min, max and the Haagerup norm are tensorial. Of course if α and β are both tensorial, and if $\alpha = \beta$ on finite dimensional operator spaces, then $\alpha = \beta$. From this it follows immediately that if α is tensorial then $\alpha'' = \alpha$. Thus $\min' = \max$.

If α is a uniform norm then we *define* $/\alpha\backslash$ to be the greatest completely injective uniform norm dominated by α. This is again a uniform operator space tensor norm. We note that this also has a concrete representation $(/\alpha\backslash)_n(U;X \otimes Y) = \sup \{ \alpha_{npq}((S \otimes T)_n(U) ; B(\mathcal{H}) \otimes B(\mathcal{K})) \}$, where the infimum ranges over all Hilbert spaces $\mathcal{H}$ and $\mathcal{K}$, all positive integers p and q, and all complete contractions S and T from X and Y to $M_p(B(\mathcal{H}))$ and $M_q(B(\mathcal{K}))$ respectively. One may replace the $B(\mathcal{H})$ spaces here with any class of injective operator spaces with the property that every operator space is contained in some element of the class. In the Banach space theory one may use finite dimensional l^∞ spaces instead of the $B(\mathcal{H})$ spaces, however it does not seem possible here to use finite dimensional spaces. We define $\backslash\alpha/$ to be $(/\alpha'\backslash)'$; it is not difficult to see that this is the least completely projective norm dominating α. There are obvious notions of left and right injectivity and projectivity, and corresponding operations $\backslash$ and $/$ as in the classical theory.

One can generate many new tensor norms if one applies the operations described above repeatedly to the norms min, max and the Haagerup norm, and as in [18] we would like to study their relationships up to equivalence. Grothendieck showed that for the Banach space category there are precisely 14 inequivalent norms obtained in this manner. However for operator spaces there is no reason to suppose that there will not be an infinite number of inequivalent norms, and our "table of norms" may be exceedingly ugly. Indeed it is clear that we have $\min \le \backslash\min/ \le h \le /\max\backslash \le \max$, and since the Haagerup norm is not symmetric, and since min is not projective and max is not injective, these five norms are inequivalent. This is one major difference between the classical Grothendieck theory and our situation. By the Grothendieck inequality we know that $/\gamma\backslash$ is dominated by a constant multiple of $\backslash\lambda/$; here γ and λ are the Banach space projective and injective norms respectively. Again the moral of the story is that the operator space situation is more complicated than the classical case.

Indeed that $/\gamma\backslash$ is equivalent to H is a restatement of the inequality. One may ask what the corresponding inequality should be for operator spaces. The Grothendieck-Pisier-Haagerup inequality [20] does not have an apparent matricial generalization which fits into this context.

Conjecture: $/\max\backslash$ is equivalent to h^s.

3. Connections with Banach space geometry. We recall that λ and γ are the Banach space injective and projective norms respectively. If X and Y are normed spaces then we shall write π^* for the norm on $X \otimes Y$ induced by identifying $(X \otimes_{\pi^*} Y)^* = \Pi_2(Y,X^*)$, where $\Pi_2(\cdot,\cdot)$ is the space of 2-summing operators [25]. Then π^{*t} is the norm induced by identifying $(X \otimes_{\pi^{*t}} Y)^* = \Pi_2(X,Y^*)$. We write H' for the associate norm of Grothendieck's H norm.

Theorem [6,7,16]. Let X and Y be normed spaces. We have

(i) $MIN(X) \otimes_{min} MIN(Y) = MIN(X \otimes_\lambda Y)$

(ii) $MAX(X) \otimes_{max} MAX(Y) = MAX(X \otimes_\gamma Y)$

completely isometrically, and also

(iii) $MIN(X) \otimes_h MIN(Y) = X \otimes_H Y$,

(iv) $MAX(X) \otimes_h MIN(Y) = X \otimes_{\pi^*} Y$,

(v) $MIN(X) \otimes_h MAX(Y) = X \otimes_{\pi^{*t}} Y$,

(vi) $MAX(X) \otimes_h MAX(Y) = X \otimes_{H'} Y$

isometrically.

Thus the Haagerup norm generalizes H , H' , π^* and π^{*t}.

In the Resume [18] Grothendieck stated that $H \le \rho H'$, for some (least) universal constant ρ whose value he was unable to ascertain. In the list of problems at the end he conjectured that $\rho = 1$. This is in fact true and was probably first observed by Kwapien. We are indebted to Professors W. B. Johnson and N. Tomczak-Jaegermann for pointing this out. That $H \le H'$ is evident here: since the operator space norms on MAX(X) dominate the norms on MIN(X) we have a complete contraction $MAX(X) \otimes_h MAX(Y) \to MIN(X) \otimes_h MIN(Y)$. This gives a contraction $X \otimes_{H'} Y \to X \otimes_H Y$, and so $H \le H'$. A more direct approach is to use a Hahn-Banach type seperation argument as in [24] (see also [12]).

Corollary . If X is a normed space such that MIN(X) = MAX(X) then $\Gamma_2(X,Y) = \Pi_2(X,Y)$ and $\Gamma_2(X^*,Y) = \Pi_2(X^*,Y)$ isometrically for any Banach space Y . Also if Y is either X or X^* then H = H' on $X \otimes Y$. If X is finite dimensional then X is at maximal Banach-Mazur distance from the l^2 space of the same dimension.

For $X = l_2^\infty$ or l_2^1 we have $\Gamma_2(X,Y) = \Pi_2(X,Y)$ isometrically for any Banach space Y . Indeed if in addition $Y = l_2^\infty$ or l_2^1 then on $X \otimes Y$ we have $H = \pi^* = \pi^{*t} = H'$.

These results relate to results of G. Pisier in [26] and V. I. Paulsen in [22]. As noted in Section 1 above Paulsen has shown that if MIN(X) = MAX(X) then X has dimension at most 4. Perhaps the formulation above in terms of classical Banach space theory will further reduce the class of candidates.

REFERENCES.

1. W. B. Arveson, *Subalgebras of C*-algebras I*, Acta Math., *123* (1969), 142-224.
2. D. P. Blecher, *Geometry of the tensor product of C*-algebras*, Math. Proc. Camb. Philos. Soc. *104* (1988), 119-127.
3. ——— *Commutativity in operator algebras*, (1989), Proc. Amer. Math. Soc. *109* (1990), 709-715.
4. ——— *The standard dual of an operator space*, (1989), to appear Pacific J. Math..
5. ——— *Tensor products of operator spaces II*, (1990), to appear Canad. J. of Math.
6. D. P. Blecher and V. I. Paulsen, *Tensor products of operator spaces*, (1989), to appear J. Funct. Anal.
7. ——— *Explicit construction of universal operator algebras and applications to polynomial factorization*, (1990), to appear Proc. Amer. Math. Soc.
8. D. P. Blecher, Z-J. Ruan and A. M. Sinclair, *A characterization of operator algebras*, (1988), J. Funct. Anal. *89* (1990), 188-201.
9. D. P. Blecher and R. R. Smith, *The dual of the Haagerup tensor product*, (1990), to appear J. London Math. Soc.
10. E. Christensen and A. M. Sinclair, *Representations of completely bounded multilinear operators*, J. Funct Anal. *72* (1987), 151-181.
11. E. G. Effros, *Advances in quantized functional analysis*, Proc. I.C.M. (1986).
12. E. G. Effros and A. Kishimoto, *Module maps and Hochschild-Johnson cohomology*, Indiana Math. J. *36* (1987), 257-276.

13. E. G. Effros and Z-J. Ruan, *On matricially normed spaces*, Pacific J. Math., *132* (1988), 243-264.
14. _______ *A new approach to operator spaces*, preprint (1989).
15. _______ *On approximation properties for operator spaces*, preprint (1989).
16. _______ *Self-duality for the Haagerup tensor product and Hilbert space factorization*, preprint (1990).
17. J. Gilbert and T. J. Leigh, *Factorization, tensor products, and bilinear forms in Banach space theory*, (Notes in Banach spaces, University of Texas Press, Austin and London, 1980, 182-305).
18. A. Grothendieck, *Resume de la theorie metrique des produits tensoriels topologiques*, Bull. Soc. Mat. Sao Paulo *8* (1956), 1-79.
19. U. Haagerup, *Injectivity and decompositions of completely bounded maps*, Lecture Notes in Math. No. 1132, Springer, Berlin (1983), 170-222.
20. _______ *The Grothendieck inequality for C^*-algebras*, Adv. in Math. *56* (1985), no. 2, 93-116.
21. V. I. Paulsen, *Completely Bounded Maps and Dilations*, (Pitman Research Notes in Math., Longman, London, 1986).
22. _______ *Representations of function algebras, abstract operator spaces, and Banach space geometry*, preprint (1990).
23. V. I. Paulsen and R. R. Smith, *Multilinear maps and tensor norms on operator systems*, J. Funct. Anal. *73* (1987), 258-276.
24. G. Pisier, *Grothendieck's theorem for non-commutative C^*-algebras with an appendix on Grothendieck's constants*, J. Funct. Anal. *29* (1978), 397-415.
25. _______ *Factorization of linear operators and the geometry of Banach space*, (C. B. M. S. Series No. 60, Amer. Math. Soc. Publ., 1986).
26. _______ *Factorization of operator valued analytic functions*, preprint (1989).
27. Z-J. Ruan, *Subspaces of C^*-algebras*, J. Funct. Anal. *76* (1988), 217-230.
28. R. Schatten, *A theory of cross spaces*, (Annals of Math. Studies 26, Princeton, 1950).
29. R. R. Smith, *Completely bounded module maps and the Haagerup tensor product*, preprint (1990).
30. M. Takesaki, *Theory of operator algebras I*, (Springer-Verlag, Berlin, 1979).

Operating Functions and Ultraseparating Function Spaces

EGGERT BRIEM Science Institute, University of Iceland, Reykjavik, Iceland

A *Banach function* space B of continuous functions on a compact Hausdorff space X is a subspace of $C(X)$, which separates the points of X and contains the constant functions, and which is a Banach space in some norm which dominates the sup-norm. We shall only consider subspaces of real valued functions. A function h defined on some open interval of the real line is said to *operate* on B if the composite function $h \circ b$ belongs to B for all b in B for which $h \circ b$ is defined. The Stone-Weierstrass Theorem may be stated in these terms as follows:

Let B be a uniformly closed subspace of $C(X)$ which separates the points of X and contains the constant functions. If the function $t \to t^2$ operates on B, then $B = C(X)$.

In [8] K. de Leeuw and Y. Katznelson showed that the function $t \to t^2$ can be replaced by any continuous function which is not affine i. e. not of the form $t \to at+b$.

These results do not hold for arbitrary Banach function spaces. The space B of continuously differentiable real valued functions on the interval $[0,1]$ is a Banach function space in the norm $||f|| = ||f||_\infty + ||f'||_\infty$ and the function $t \to t^2$ operates on B.

However, there are some positive results. The real part of a function algebra A equipped with the norm $||a|| = \inf\{||a+ib||_\infty : a+ib \in A\}$ is an example of a Banach function space on X. In [10] J. Wermer showed that if the function $t \to t^2$ operates on the real part of a function algebra A on X, then $A = C(X)$. The question is whether also in this case the function $t \to t^2$ can be replaced by any non-affine function. (It is known that a function which operates on the real part of a function algebra must be continuous). A powerful tool to answer this question, the notion of an ultraseparating function space (see definition below), was introduced by A. Bernard in [1], where he settled the problem with some extra conditions concerning the manner in which the non-affine function operates. Bernard also showed that Wermer's result remains true for Banach function algebras. The next step is due to S. J. Sidney. In

[9] he proved that if a function, non-affine on any subinterval, operates on the real part of a function algebra A on X, then $A = C(X)$. Finally, O. Hatori settled the question for the remaining case in [5]. Then in [6] Hatori generalized the result to ultraseparating Banach function algebras (of which function algebras with dense real parts are special instances). In [4], lecture notes by A. J. Ellis one can find most of the relevant background material.

In [1] A. Bernard proved that if B is a *normal* and ultraseparating Banach function space on X for which there is a continuous non-affine operating function then B is locally $C(K_x)$ for almost all x, i. e. for all but finitely many x in X there is a compact neighbourhood K_x of x such that $B|K_x$, the space obtained by restricting the functions in B to K_x, is all of $C(K_x)$. The *normality* hypothesis is that for every pair E and F of disjoint closed subsets of X and for each b in B there is a function a in B with $a = b$ on E and $a = 0$ on F. We show in this note that the normality hypothesis can be removed.

Let B be a Banach function space on X whith norm $||\cdot||$, and let $l^\infty(\mathbf{N}, B)$ denote the space of all $||\cdot||$-bounded sequences of functions from B. One can consider the space $l^\infty(\mathbf{N} \times B)$ in a natural way as a subspace of $C(\beta(\mathbf{N} \times X))$, where $\beta(\mathbf{N} \times X)$ denotes the Stone-Čech compactification of $\mathbf{N} \times X$. If $l^\infty(\mathbf{N}, B)$ separates the points of $\beta(\mathbf{N} \times X)$, then B is said to be *ultraseparating* on X. The real part of Dirichlet algebra on X, i. e. a function algebra on X whose real part is dense in $C(X)$, is an example of an ultraseparating Banach function space on X (see [1]). Dirichlet algebras arise naturally in connection with operating functions because if a non-affine function operates on a space of continuous functions on X, which separates the points of X and contains the constant functions, then that space must be dense in $C(X)$ by the result of de Leeuw and Katznelson.

The significance of the sequence space $l^\infty(\mathbf{N}, B)$ is seen in the following theorem due to Bernard [1]:

If B is a Banach functions space on X and if $l^\infty(\mathbf{N}, B)$ is dense in $C(\beta(\mathbf{N} \times X))$ then $B = C(X)$.

Our main tool will be a local version of Bernard's result together with a property of ultraseparating spaces expressed in lemma 2 below.

If $f \in C(X)$ we let $\{f\}$ denote the sequence $\{f_n\} \in l^\infty(\mathbf{N}, B)$, where $f_n = f$ for all n. Given $x \in X$ and $f_1, f_2, \cdots, f_k \in C(X)$ we let $\mathcal{F}_x = \mathcal{F}_x(f_1, f_2, \cdots, f_k)$ denote the subset of $\beta(\mathbf{N} \times X)$ given by

$$\mathcal{F}_x = \{\xi \in \beta(\mathbf{N} \times X) : \{f_i\}(\xi) = f_i(x), \text{ for } i = 1, 2, \cdots k\}.$$

The following lemma is a local version of Bernard's result. Although a proof has appeared in [2] we include a proof here for the sake of completeness. O. Hatori has a similar version of Bernard's theorem in [7].

LEMMA 1. *Let B be a Banach function space on a compact Hausdorff space X, let $x \in X$ and let $f_1, f_2, \cdots, f_k \in C(X)$. If $l^\infty(\mathbf{N}, B)|\mathcal{F}_x$ is dense in $C(\mathcal{F}_x)$, then there exists a compact neighbourhood K_x of x such that $B|K_x = C(K_x)$.*

PROOF: Let $K_n = \{y \in X : |f_i(y) - f_i(x)| < \frac{1}{n}, \text{ for } i = 1, \cdots, k\}$ for each $n \in \mathbf{N}$. We claim that there exists a number n_1 such that for all $f \in C(X)$ for which $||f||_\infty \leq 1$ there is a function $b \in B$ with $||b|| \leq n_1$ such that $|f - b| < \frac{1}{2}$ on K_{n_1}.

If not, then there is for each $n \in \mathbf{N}$ a function $g_n \in C(X)$ with $||g_n||_\infty \leq 1$ such that $||b|| > n$ for all $b \in B$ for which $|f_n - b| < \frac{1}{2}$ on K_n. By assumption there is a sequence $\{b_n\} \in l^\infty(\mathbf{N}, B)$ such that $|\{g_n\} - \{b_n\}| < \frac{1}{2}$ on $\mathcal{F}_x$ and hence on some neighbourhood $\mathcal{U}$ of $\mathcal{F}_x$. Since $\mathcal{U}$ is open the definition of $\mathcal{F}_x$ shows that there is a number n_1 such that

$$\{\xi \in \beta(\mathbf{N} \times X) : |\{f_i\}(\xi) - f_i(x)| \leq \frac{1}{n_1} \quad \text{for } i = 1, \cdots, k\} \subseteq \mathcal{U}.$$

It follows that $\mathbf{N} \times K_{n_1} \subseteq \mathcal{U}$ and hence $|\{g_n\} - \{b_n\}| \leq \frac{1}{2}$ on $\mathbf{N} \times K_{n_1}$. Since $\{b_n\}$ is a bounded sequence we have reached a contradiction, and thus the claim is true.

Let $f \in C(X)$ with $||f||_\infty \leq 1$. By induction we construct a sequence $\{v_n\}$ of elements from B such that $||v_n|| < n_1$ for all n and such that

$$|f - (v_0 + \frac{1}{2}v_1 + \cdots + \frac{1}{2^{n-1}}v_{n-1})| \leq \frac{1}{2^n} \quad \text{on} \quad K_{n_1}.$$

Let $v = \sum \frac{1}{2^n} v_n$. Then $v \in B$ and $v = f$ on K_{n_1}.

In [3] A. J. Ellis studies properties which are equivalent to the ultraseparating property. We use similar methods to obtain a weaker property:

LEMMA 2. *Let B be ultraseparating on X. Then for every $\epsilon > 0$ there is a constant M such that for every two points x and y in X there is a function $b \in B$ with $||b|| < M$ such that $b(x) > 1 - \epsilon$ and $b(y) < \epsilon$.*

PROOF: Let $\epsilon > 0$. If no such M exists then there are sequences $\{x_n\}$ and $\{y_n\}$ in X such that $||b|| > n$ if $b \in B$ with $b(x_n) > 1 - \epsilon$ and $b(y_n) < \epsilon$. Let $\{f_n\}$ be a sequence of continuous functions on X such that $||f_n||_\infty = 1$, $f_n(x_n) = 1$ and $f_n(y_n) = 0$. Since $l^\infty(\mathbf{N}, B)$ separates the points of $\beta(\mathbf{N} \times X)$ there are sequences $\{b_{jn}\}, \{c_{jn}\}$, $1 \leq j \leq k$ in $l^\infty(\mathbf{N}, B)$ such that

$$||\{f_n\} - (\{b_{1n}\} \wedge \{b_{2n}\} \wedge \cdots \wedge \{b_{kn}\} - \{c_{1n}\} \wedge \{c_{2n}\} \wedge \cdots \wedge \{c_{kn}\})||_\infty < \epsilon.$$

Since the sequences $\{b_{jn}\}, \{c_{jn}\}$, $1 \leq j \leq k$ are bounded, this inequality shows that there is a constant M' such that for each natural number n there are functions b_j, c_j, $1 \leq j \leq k$ (depending on n), in B such that $||b_j|| < M'$, $||c_j|| < M' for\, 1 \leq j \leq k$ and such that

$$||f_n - (b_1 \wedge b_2 \wedge \cdots \wedge b_k - c_1 \wedge c_2 \wedge \cdots \wedge c_k)||_\infty < \epsilon.$$

Let l and m be chosen such that

$$b_1(y) \wedge b_2(y) \cdots \wedge b_k(y) = b_l(y), \quad c_1(x) \wedge c_2(x) \wedge \cdots \wedge c_k(x) = c_m(x).$$

If $a = b_l - c_m$ then $a(x_n) > 1 - \epsilon$, $a(y_n) < \epsilon$ and $||a|| < 2M'$. Since we can find such a function a in B for each natural number n a contradiction has been reached.

We can now prove the main results of this note.

THEOREM. *Let B be an ultraseparating Banach function space on a compact Hausdorff space X and let h be a continuous non-affine function operating on B. Then there is a finite subset F of X such that for each $x \in X \backslash F$ there is a compact neighbourhood K_x of x such that $B|K_x = C(K_x)$.*

PROOF: We may suppose that h is not affine in any neighbourhood of 0. Let U be the set of points x in X for which there is a compact neighbourhood K_x of x such that $B|K_x = C(K_x)$ and let $F = X \backslash U$. Then F is a closed subset of X. We shall show that F has no limit points. Suppose on the contrary that x_0 is a limit point for F. Let $B(x_0) = \{b \in B : b(x_0) = 0\}$ and for each positive number r let $B_r(x_0) = \{b \in B(x_0) : ||b|| < r\}$. Now,

$$B_r(x_0) = \cup_n \{b \in B_r(x_0) : ||h \circ b|| \leq n\}.$$

Hence, by the Baire category theorem, there is an element $a_0 \in B_r(x_0)$ and positive numbers ϵ and M such that

$$||h \circ (a_0 + b)|| \leq M$$

for b in a dense subset of $B_\epsilon(x_0)$.

Using the lemma we pick a positive number M_1 such that for any two points x and y in X it is possible to find a function b in B with $||b|| < M_1$, $b(x) = 0$ and $b(y) = 1$.

Since x_0 is a limit point of F there is a point x_1 in F such that $|a_0(x_1)| < \frac{\epsilon}{2M_1}$. We now take a function a_1 in B with $||a_1|| < M_1$ such that $a_1(x_0) = 0$ and $a_1(x_1) = 1$ and let $c_0 = a_0 - a_0(x_1)a_1$. Then $c_0(x_0) = c_0(x_1) = 0$ and

$$||h \circ (c_0 + b)|| < M$$

for b in a dense subset of $B_{\epsilon/2}(x_0)$. Hence, $\{h \circ (c_0 + b_n)\} \in l^\infty(\mathbf{N}, B)$ for b_n in a dense subset of $B_{\epsilon/2}(x_0)$ and thus, since h is continuous,

$$\{h \circ (c_0 + b_n)\} \in \overline{l^\infty(\mathbf{N}, B)},$$

the closure of $l^\infty(\mathbf{N}, B)$ in the sup-norm on $\beta(\mathbf{N} \times X)$, if each b_n is in $B_{\epsilon/2}(x_0)$. Let c_1 and c_2 be functions in B such that $c_1(x_0) = 0, c_1(x_1) = 1$, $c_2(x_0) = 1$ and $c_2(x_1) = 0$, and let

$$\mathcal{F} = \mathcal{F}_{x_1}(c_0, c_1, c_2) = \{\xi \in \beta(\mathbf{N} \times X) : \{c_i\}(\xi) = c_i(x_1) \text{ for } i = 0, 1, 2\}.$$

Now, $h \circ \{c_0 + b_n\}|\mathcal{F} = h \circ \{b_n\}|\mathcal{F}$. It follows that

$$h \circ \{b_n\}|\mathcal{F} \in \overline{l^\infty(\mathbf{N}, B)|\mathcal{F}} \quad \text{if} \quad \{b_n\} \in l^\infty(\mathbf{N}, B) \quad \text{and} \quad b_n \in B_{\epsilon/2}(x_0).$$

Since $\{b_n - b_n(x_0)c_2\}|\mathcal{F} = \{b_n\}|\mathcal{F}$, and since $\{c_1\}|\mathcal{F} \equiv 1$, we can conclude that $l^\infty(\mathbf{N}, B(x_0))|\mathcal{F} = l^\infty(\mathbf{N}, B)|\mathcal{F}$ and that $l^\infty(\mathbf{N}, B(x_0))|\mathcal{F}$ contains the constant functions. The theorem of K. de Leeuw and Y. Katznelson now implies that $l^\infty(\mathbf{N}, B)|\mathcal{F}$ is dense in $C(\mathcal{F})$ and thus, by the lemma, that there is a compact neighbourhood K_{x_1} of x_1 such that $B|K_{x_1} = C(Kx_1)$, contradicting the fact that $x_1 \in F$. This finishes the proof of the theorem.

REMARK: Is it actually true that $B = C(X)$? We can prove that this is the case with some additional conditions on the operating function h. One such condition is that there is a constant M such that $|h(t)| \leq M|t|^{1+\epsilon}$ for t in some neighbourhood of 0. This will be discussed in a forthcoming paper.

REFERENCES

1. A. Bernard, Espaces des parties réelles des éléments d'une algebre de Banach de fonctions, *J. Funct. Anal.,* **10** (1972), 387-409.

2. E. Briem, Ultraseparating function spaces and operating functions for the real part of a function algebra, *Proc. Amer. Math. Soc.,* to appear.

3. A. J. Ellis, Separation and ultraseparation properties for continuous function spaces, *J. London Math. Soc.* (2), 29 (1984), 521-532.

4. A. J. Ellis, Topics in Banach spaces of continuous functions, *ISI Lecture Notes 2,* The Macmillan Co. of India Ltd., New Delhi (1978)

5. O. Hatori, Functions which operate on the real part of a uniform algebra, *Proc. Amer. Math. Soc.,***83** (1981), 565-568.

6. O. Hatori, Range transformations on a Banach function algebra, *Trans. Amer. Math. Soc.,***27** (1986), 629-643.

7. O. Hatori, Range transformations on a Banach function algebra II, *Pacific J. Math.,***138** (1989), 89-118.

8. K. de Leeuw and Y. Katznelson, Functions that operate on non-selfadjoint algebras, *J. Analyse Math.,* **11** (1963), 207-219.

9. S. J. Sidney, Functions which operate on the real part of a uniform algebra, *Pacific J. Math.,* 80 (1979), 265-272.

10. J. Wermer, The space of real parts of a function algebra, *Pacific J. Math.,* **13** (1963), 1423-1426.

Some Extremum Problems for Spaces of Weakly Continuous Functions

MICHAEL CAMBERN Department of Mathematics, University of California, Santa Barbara, CA 93106

Throughout this article the letters X and Y will denote compact Hausdorff spaces and E a Banach space. Given E, U_E will represent the closed unit ball of E. $C(X,E)$ denotes the space of continuous functions on X to E, and this space will be represented by $C(X)$ whenever E reduces to the scalar field. $C(X,(E,w))$ will stand for the continuous functions on X to E when the latter space is given its weak topology, and, if E is a dual space, $C(X,(E,w^*))$ denotes the analogous function space with E provided with its weak $*$ topology. Both $C(X,(E,w))$ and $C(X,(E,w^*))$ are, of course, given the supremum norm.

A major open problem involves finding an adequate description of the Banach duals $C(X,(E,w))^*$ and $C(X,(E,w^*))^*$. Not only would a useful description of these duals seem to be lacking, but also lacking are descriptions of subsets of the duals which are often focal points of interest – e.g. the extreme points of their unit balls. In fact the situation concerning extreme points is somewhat worse than has been stated: not only has one failed to obtain a description of the entire set of extreme points, but, moreover, it has been shown that certain obvious candidates for membership in this set may fail to belong. For $C(X,E)$ is a closed subspace of both $C(X,(E,w))$ and $C(X,(E,w^*))$, and the extreme points of $U_{C(X,E)^*}$ have long been known. They precisely constitute the set $\{L_{x,e^*} : x \in X$ and e^* extreme in $U_{E^*}\}$ where, for $F \in C(X,E)$, $L_{x,e^*}(F) = \langle F(x), e^* \rangle$, [5, p. 197]. The L_{x,e^*} obviously define elements of the unit balls of $C(X,(E,w))^*$ and $C(X,(E,w^*))^*$, and one might be tempted to hope that they would be extreme. E. Behrends, however, has shown that this need not be the case [1, Corollary 3.6].

Quite recently, though, Pei-Kee Lin [4] has shown that by switching attention from the extreme points to the exposed points of U_{E^*} one may obtain positive results. And his results apply not only to $C(X,(E,w))$ but also to certain subspaces of this space which are modules over a function algebra.

Thus let A be a closed subalgebra of $C(X)$ which separates the points of X and contains the constant functions. Then B will denote the Choquet boundary of A. $A(X,(E,w))$ stands for the subspace of $C(X,(E,w))$ consisting of all $F \in C(X,(E,w))$ such that $\langle F(\cdot), e^* \rangle \in A$ for all $e^* \in E^*$. We let S represent the surface of U_{E^*}, $S := \{e^* \in E^* : \| e^* \| = 1\}$ given its usual norm topology, and form the completely regular space $B \times S$. Then for $F \in A(X,(E,w))$ we define $\tilde{F}$ on $B \times S$ by $\tilde{F}((x,e^*)) = \langle F(x), e^* \rangle$. Clearly $\tilde{F}$ is a continuous function on $B \times S$, and it is also clear that $\sup_{(x,e^*) \in B \times S} | \tilde{F}((x,e^*)) | = \| F \|_\infty$. Each $\tilde{F}$ then has a unique continuous extension to a function $\check{F}$ defined on the Stone-Čech compactification Y of $B \times S$. Thus, if M denotes the subspace $M := \{\check{F} : F \in A(X,(E,w))\} \subseteq C(Y)$, then the map $F \to \check{F}$ gives an isometric representation of $A(X,(E,w))$ as a closed subspace of $C(Y)$.

Now let $\mathcal{E}$ denote the set of w^* strongly exposed points of the unit ball U_{E^*} of E^*. The significant fact realized by Lin [4], for the case $A(X,(E,w)) = H_E^\infty$, is that for each $(x,e^*) \in B \times \mathcal{E}$ the functional L_{x,e^*} defined on $A(X,(E,w))$ by $L_{x,e^*}(F) = \langle F(x), e^* \rangle = \check{F}((x,e^*)) = \int \check{F} d\mu_{(x,e^*)}$, $F \in A(X,(E,w))$, is an extreme point of the unit ball of $A(X,(E,w))^*$. This fact was established in [4] by a highly nontrivial argument using nets. We give here a measure theoretic proof. Our proof is formulated for the case in which E is a complex Banach space, but obvious modifications (in fact simplifications) may be made to handle the case in which E is real. The result is contained in the following:

THEOREM 1 (LIN): *If M is as above then for each $(x_o, e_o^*) \in B \times \mathcal{E}$ it follows that $\mu_{(x_o,e^*)}$ is an extreme point of U_{M^*}.*

PROOF: Suppose that $(x_o, e_o^*) \in B \times \mathcal{E}$ and that there exist $F_1^*, F_2^* \in U_{M^*}$ such that $\mu_{(x_o,e_o^*)} = \frac{1}{2}(F_1^* + F_2^*)$. For $i = 1,2$ take a Hahn-Banach extension of F_i^* to a regular Borel measure $\mu_i \in C(Y)^*$. We claim that if C is the "circular" subset of $B \times S$ given by $C = \{(x_o, ze_o^*) : z \in \mathbb{C} \text{ and } | z | = 1\}$ then $| \mu_i | (C) = 1$ for $i = 1,2$. For, assuming the contrary, there would exist an $\varepsilon > 0$ such that $(| \mu_1 | + | \mu_2 |)(C) < 2 - 2\varepsilon$. And thus by the regularity of the $| \mu_i |$ and the compactness of C, one could find a compact neighborhood K of C in Y such that $(| \mu_1 | + | \mu_2 |)(K) < 2 - 2\varepsilon$.

Now, for each $(x_o, ze_o^*) \in C$, $K \cap (B \times S)$ is a neighborhood of (x_o, ze_o^*) in $B \times S$ so that there exists a neighborhood V_z of x_o in B and a sphere $S_{ze_o^*}(2\varepsilon_z)$ centered at ze_o^* in S such that $V_z \times S_{ze_o^*}(2\varepsilon_z) \subseteq K$. Thus by the compactness of C there exist $V_{z_i} \times S_{z_ie_o^*}(\varepsilon_i)$, $1 \le i \le n$, such that if $V = \bigcap_{i=1}^{n} V_i$ then $C \subseteq V \times [\bigcup_{i=1}^{n} S_{z_ie_o^*}(\varepsilon_i)] \subseteq \bigcup_{i=1}^{n} V_i \times S_{z_ie_o^*}(\varepsilon_i)$.

Let $\varepsilon_o = \min_{1 \le i \le n} \varepsilon_i$ and let C_o be the "circular" set in S given by $C_o := \{ze_o^* : | z | = 1\}$. Suppose $(x, e^*) \in B \times S$ with $x \in V$ and $\| e^* - ze_o^* \| < \varepsilon_o$, for some $z \in \mathbb{C}$ with $| z | = 1$. Then $ze_o^* \in S_{z_ie_o^*}(\varepsilon_i)$ for some i so that $\| e^* - z_ie_o^* \| \le \| e^* - ze_o^* \| + \| ze_o^* - z_ie_o^* \| < 2\varepsilon_i$. Hence $(x, e^*) \in V \times S_{z_ie_o^*}(2\varepsilon_i) \subseteq V_{z_i} \times S_{z_ie_o^*}(2\varepsilon_i) \subseteq K$. That is, if N denotes the subset of $B \times S$ given by $N = V \times \{e^* \in S : d(e^*, C_o) < \varepsilon_o\}$ then $N \subseteq K$. Hence if $\overline{N}$ is the closure of N in Y we have $\overline{N} \subseteq K$ and thus $(| \mu_1 | + | \mu_2 |)(\overline{N}) < 2 - 2\varepsilon$.

Since e_o^* is w^* strongly exposed there exists an $e_o \in E$ with $\| e_o \| = 1$ and a number $\delta > 0$ such that $e^* \in S$ and $Re\langle e_o, e^* \rangle > 1 - \delta$ implies that $\| e^* - e_o^* \| < \varepsilon_o$. Thus suppose that e^* is any element of S with $| \langle e_o, e^* \rangle | > 1 - \delta$. Take $z \in \mathbb{C}$ with $| z | = 1$ such that $z\langle e_o, e^* \rangle = \langle e_o, ze^* \rangle = Re\langle e_o, ze^* \rangle = | \langle e_o, e^* \rangle | > 1 - \delta$. Then $\| ze^* - e_o^* \| = \| e^* - \overline{z}e_o^* \| < \varepsilon_o$. That is, $d(e^*, C_o) < \varepsilon_o$.

Next take $f \in A$ such that $1 = \| f \|_\infty = f(x_o)$ and $| f | < \delta\varepsilon/2$ on $B - V$, and suppose that $(x, e^*) \in [Y - \overline{N}] \cap (B \times S)$. Then either $x \notin V$ so that $| f(x) | < \delta\varepsilon/2$ or else $d(e^*, C_o) \geq \varepsilon_o$ so that $| \langle e_o, e^* \rangle | \leq 1 - \delta$. Thus if we define $F \in A(X, (E, w))$ by $F = f \cdot e_o$ and let D be the intersection of $Y - \overline{N}$ with the closure in Y of $\{(x, e^*) \in B \times S : | f(x) | < \delta\varepsilon/2\}$, then for $y \in D$ we have $| \check{F}(y) | \leq \delta\varepsilon/2$ and for $y \in G := [Y - \overline{N}] - D$ we have $| \check{F}(y) | \leq 1 - \delta$. Now $Y = \overline{N} \dot{\cup} D \dot{\cup} G$ and thus

$$
\begin{aligned}
1 = \check{F}(x_o, e_o^*) &= \int \check{F} d\mu_{(x_o, e_o^*)} = \frac{1}{2}[\int \check{F} d\mu_1 + \check{F} d\mu_2] \\
&= \frac{1}{2}[\int_{\overline{N}} \check{F} d(\mu_1 + \mu_2) + \int_D \check{F} d(\mu_1 + \mu_2) + \int_G \check{F} d(\mu_1 + \mu_2)] \\
&\leq \frac{1}{2}[(| \mu_1 | + | \mu_2 |)(\overline{N}) + \delta\varepsilon + (1 - \delta)(2 - (| \mu_1 | + | \mu_2 |)(\overline{N}))] \\
&< 1 - \delta\varepsilon/2 < 1
\end{aligned}
$$

which is absurd. Hence $| \mu_i | (C) = 1$, $i = 1, 2$ as claimed.

We thus have

$$1 = \langle e_o, e_o^* \rangle = \int \check{e}_o d\mu_{(x_o, e_o^*)} = \frac{1}{2}[\int_C \check{e}_o d\mu_1 + \int_C \check{e}_o d\mu_2]$$

so that $\int_C \check{e}_o d\mu_i = 1$ for $i = 1, 2$. (Here $\check{e}_o$ denotes the function that is constantly equal to e_o.) But for $(x_o, ze_o^*) \in C$, $\check{e}_o(x_o, ze_o^*)) = \langle e_o, ze_o^* \rangle = z$, so that if $\check{F}$ is any element of M whose value at the point (x_o, ze_o^*) of C is z, then $\int_C \check{F} d\mu_i = 1$ for $i = 1, 2$. Now if $\check{F} \in M$ and $\check{F}((x_o, e_o^*)) = \int \check{F} d\mu_{(x_o, e_o^*)} = \lambda$, then $\check{F}((x_o, ze_o^*)) = \lambda z$ so that $\int_C \check{F} d\mu_i = \lambda$, $i = 1, 2$ and hence $\mu_{(x_o, e_o^*)}$ is extreme as claimed.

Applying Lin's Theorem to the space $C(X, (E, w))$ we then obtain the following:

Corollary 1. If $x \in X$ and if e^* is a w^* strongly exposed point of U_{E^*} then the functional L_{x,e^*} defined for $F \in C(X, (E, w))$ by $L_{x,e^*}(F) = \langle F(x), e^* \rangle$ is an extreme point of the unit ball of $C(X, (E, w))^*$.

When E is reflexive, Corollary 1 obviously provides us with a set of extreme points of the unit ball of $C(X, (E, w^*))$. The fact that this set is reasonably large is contained in the following:

Corollary 2. Let E be a Banach space whose dual E^* has RNP. Then the convex hull of the set of all L_{x,e^*}, for $x \in X$ and e^* a w^* strongly exposed point of U_{E^*}, is weak $*$ dense in the unit ball of $C(X,(E,w))^*$.

PROOF: Suppose $F \in C(X,(E,w))$ and that we are given $\varepsilon > 0$. Take $x \in X$ such that $\| F(x) \| > \| F \|_\infty - \varepsilon$ and then take $e^* \in U_{E^*}$ such that $\langle F(x), e^* \rangle$ is real and greater than $\| F \|_\infty - \varepsilon$. Then since $co(\mathcal{E})$ is weak $*$ dense in U_{E^*} [2, p. 112] there exists a convex combination $\sum_{i=1}^n a_i e_i^*$ of elements $e_i^* \in \mathcal{E}$ such that

$$| \langle F(x), e^* \rangle - \sum_{i=1}^n a_i \langle F(x), e_i^* \rangle | < \varepsilon.$$

Thus for at least one value of i we must have $| \langle F(x), e_i^* \rangle | > \| F \|_\infty - 2\varepsilon$, and since $\varepsilon > 0$ is arbitrary we thus have $\| F \|_\infty = \sup_{(x,e^*) \in X \times \mathcal{E}} | L_{x,e^*}(F) |$. The conclusion then follows by the usual separation argument.

REFERENCES

1. E. Behrends, On the geometry of spaces of C_0K-valued operators, *Studia Math.*, 90: 135–151 (1988).
2. R. Bourgin, Geometric aspects of convex sets with the Radon-Nikodým property, Lecture Notes in Mathematics 993, Springer-Verlag, Berlin-Heidelberg-New York (1983).
3. M. Cambern and P. Greim, The bidual of $C(X,E)$, *Proc. Amer. Math. Soc.*, 85: 53–58, (1982).
4. P.-K. Lin, The isometries of $H^\infty(E)$, *Pacific J. Math.* (to appear).
5. I. Singer, Best approximations in normed linear spaces by elements of linear subspaces, Springer-Verlag, Berlin-Heidelberg-New York (1970).

C*-Algebras with the Dunford-Pettis Property

C.-H. CHU Goldsmiths' College, University of London, London S.E. 14, England.

B. IOCHUM Université de Provence and Centre de Physique Théorique, C.N.R.S., Luminy Case 907, F-13288 Marseille, Cedex 9, France.

S. WATANABE Department of Mathematics, Niigata University, Niigata 950-21, Japan.

It is a celebrated result of Grothendieck [9] that every weakly compact operator on $C(X)$ is completely continuous. He called this property of $C(X)$ the *Dunford-Pettis property*, referring to, of course, an earlier result of Dunford and Pettis [7] that all L_1- spaces enjoy the same property.

Naturally one would ask if $C(X)$-like spaces also have the Dunford-Pettis property. The case of uniform algebras has been studied, for instance, by Bourgain [2], Chaumat [3] and Delbaen [5]. On the other hand if one looks upon $C(X)$ as an arbitrary commutative C^*-algebra, one would ask another question, namely: does a noncommutative C^*-algebra also have the Dunford-Pettis property? The answer is 'no' because the C^*-algebra $K(H)$ of compact operators on an infinite dimensional Hilbert space H does not have this property. Which C^*-algebras have the Dunford-Pettis property? In [4], von Neumann algebras as well as separable C^*-algebras with the Dunford-Pettis property were characterized. We wish to point out in this exposition that the separability assumption in [4] is unnecessary. Therefore the second question above is completely settled. We should also point out that Hamana [10] has obtained many results similar to those of [4].

We refer to [11] for the basics of C^*-algebras and to [4] for unproved results. An excellent survey of results relating to the Dunford-Pettis property can be found in [6] from which we have derived much inspiration and pleasure. We note that Grothendieck [9] has also shown that a Banach space E has the Dunford-Pettis property if, and only if, whenever (x_n) and (f_n) are weakly null sequences in E and E^* respectively, then $\lim_{n\to\infty} f_n(x_n) = 0$. Therefore, if E^* has the Dunford-Pettis property, so does E, but not the converse. However, this property is not inherited by subspaces nor by quotients although it has been shown in [4] that it is inherited by C^*-subalgebras and also, that it is passed from a separable C^*-algebra to its quotient algebras. The latter was communicated to us by Professor G.K. Pedersen for which we are much indebted. We will show this result in the following lemma without the separability assumption. By [4], a C^*-algebra A has the Dunford-Pettis property if, and only if, given any weakly null sequence (a_n) in A, the sequence $(a_n^* a_n)$ is also weakly null in A.

THEOREM. *Let A be a C^*-algebra. The following are equivalent:*

(i) A *has the Dunford-Pettis property;*

(ii) A^* *has the Dunford-Pettis property;*

(iii) *For some closed two-sided ideal J of A, both J and A/J have the Dunford-Pettis property;*

(iv) *Every irreducible representation of A is finite-dimensional;*

(v) A^{**} *is a finite type I von Neumann algebra, that is,*

$$A^{**} = \ell_\infty - sum \sum_k \oplus \left(C(X_k) \otimes M_{n_k} \right),$$

where X_k is hyperstonean and M_{n_k} is the algebra of $n_k \times n_k$ complex matrices.

The equivalence of (*ii*) and (*v*) has been proved in [4; Theorem 7] and [10; Theorem 1]. Since the representations of A correspond to the direct summands of A^{**}, the implication (*v*) $\Rightarrow$ (*iv*) is clear while its converse has been established in [10; Lemma 5]. It has been shown in [4] that if A is separable, then (*i*) implies (*ii*). So we only need to show (*i*)$\Rightarrow$(*ii*) without separability assumption, and (*i*)$\Leftrightarrow$(*iii*). We prove the following lemma first.

LEMMA. *Let A be a C^*-algebra with the Dunford-Pettis property. Then any *-homomorphic image of A has the property.*

Proof. Let $\pi\colon A \mapsto \pi(A)$ be a *-homomorphism. Let (q_n) be a $\sigma\big(\pi(A), \pi(A)^*\big)$-null sequence in $\pi(A)$. It suffices to show that $(q_n^* q_n)$ is also weakly null in $\pi(A)$. Let $q_n = \pi(x_n)$ for $x_n \in A$ and let B be the C^*-algebra generated by $\{x_n\}$. Then B is separable and $\{q_n\} \subset \pi(B)$ is $\sigma(\pi(B), \pi(B)^*)$-null. By [4], (q_n) lifts to a $\sigma(B, B^*)$-null sequence (y_n) in B which is weakly null in A and by the Dunford-Pettis property of A, the sequence $(y_n^* y_n)$ is $\sigma(A, A^*)$-null which implies $q_n^* q_n = \pi(y_n^* y_n)$ tends to 0 weakly in $\pi(A)$.

Now we return to (*i*) $\Rightarrow$ (*ii*) in the theorem. For this, we show (*i*) $\Rightarrow$ (*iv*). First, A is a type I C^*-algebra. If not, A contains a subalgebra B and a closed two-sided ideal J in B such that B/J is isomorphic to the Fermion algebra $\otimes M_2$ [11; 6.7.4]. By the above lemma, B/J has the Dunford-Pettis property which implies that $\otimes M_2$ has the Dunford-Pettis property. This is absurd because $\otimes M_2$ contains a complemented copy of $K(\ell_2)$ [1]. So A is of type I. Thus, if A has an infinite-dimensional irreducible representation, then $K(\ell_2)$ shows up in a quotient of A which, by the lemma, implies that $K(\ell_2)$ has the Dunford-Pettis property. Impossible. So we have (*iv*).

Finally, the lemma gives (*i*) $\Rightarrow$ (*iii*). If (*iii*) holds, then by (*i*) $\Rightarrow$ (*ii*), both J^* and $(A/J)^*$ have the Dunford-Pettis property. This yields (*ii*) because A^* is the ℓ_1-sum of J^* and $(A/J)^*$.

In conclusion, we note that condition (*iv*) above implies readily that the ℓ_∞-sum

$\sum_k \oplus M_{n_k}$ ($n_k \uparrow \infty$) does not have the Dunford-Pettis property which seems in discord with [8; Theorem 1]. Moreover, as an application, we have the following improved version of the corollary in [8].

COROLLARY. *Let M be a von Neumann algebra. Then every C_0-semigroup of operators on M is uniformly continuous if, and only if, M is of the form ℓ_∞-sum $\sum_k \oplus \left(C(X_k) \otimes M_{n_k} \right)$ as above, with* $\sup n_k < \infty$.

References

1. J. Arazy, *Linear topological classification of C^*-algebras*, Math. Scand. 52 (1983), 89–111.
2. J. Bourgain, *New Banach space properties of the disc algebra and H^∞*, Acta Math. 152 (1984), 1–48.
3. J. Chaumat, *Une géneralisation d'un théorème de Dunford-Pettis*, Université de Paris XI, Orsay 1974.
4. C.-H. Chu and B. Iochum, *The Dunford-Pettis property in C^*-algebras*, Studia Math. 97 (1990), 59–64.
5. F. Delbaen, *The Dunford-Pettis property for certain uniform algebras*, Pacific J. Math. 65 (1976), 29–33.
6. J. Diestel, *A survey of results related to the Dunford-Pettis property*, Contemporary Math. vol. 2 (1980), 15–60.
7. N. Dunford and B.J. Pettis, *Linear operations on summable functions*, Trans. Amer. Math. Soc. 47 (1940), 323–392.
8. U. Groh, *Norm continuity of strongly continuous semigroups on W^*-algebras*, Semesterbericht Funktionalanalysis Tübingen 1983/84.
9. A. Grothendieck, *Sur les applications linéaires faiblement compactes d'espaces du type $C(K)$*, Cand. J. Math. 5 (1953), 129–173.
10. M. Hamana, *On linear topological properties of some C^*-algebras*, Tôhoku Math. J. 29 (1977), 157–163.
11. G.K. Pedersen, "C^*-algebras and their automorphism groups", Academic Press 1979.

Real Linear Isometries of Complex Function Spaces

A.J. ELLIS University of Hong Kong, Hong Kong

Let X be a compact Hausdorff space and M a *complex function space* on X, that is a closed linear subspace of $C_{\mathbb{C}}(X)$ containing constants and separating points of X. We will usually assume that X is the Šilov boundary for M.

Let $\sum$ denote the closed unit ball of M^* and let $S = \{\varphi \in \sum : \varphi(1) = 1\}$ be the *state space* of M, and $Z = co(S \cup -iS) \subset M^*$ the *complex state space* of M. Then $\sum$, S and Z are w^*-compact convex subsets of M^*, X is naturally embedded in S and identifies with $\overline{\partial S}$, the closure of the set of extreme points ∂S of S.

Whenever K is a compact convex set we denote by $A(K)$ (respectively $A_{\mathbb{C}}(K)$) the Banach space of continuous real-valued (respectively complex-valued) affine functions on K, with the supremum norm. It is well-known that there exists a real-linear homeomorphism θ between M and $A(Z)$, given by $(\theta f)(z) = re(z(f))$, $f \in M$, $z \in Z$, satisfying $\|\theta(u+iv)\| = \max\{\|u\|, \|v\|\}$.

If K_1, K_2 are compact convex sets then we will write $K_1 \cong K_2$ if there exists an affine homeomorphism from K_1 onto K_2.

If M is isometrically isomorphic over $\mathbb{C}$ to a uniform algebra M_1 then M need not be an algebra. Indeed, let M_1 be the disc algebra on the unit circle Γ and let $M = \{\bar{z}f(z) : f \in M_1\}$. In this case M and M_1 are isometrically isomorphic over $\mathbb{C}$ and $S \cong S_1 \cong M_1^+(\Gamma)$, the set of probability measures on Γ; however M is not an algebra, and it may be shown that Z and Z_1 are not affinely homeomorphic. On the other hand we can have $Z \cong Z_1$, for some uniform algebra M_1, with M not an algebra. In fact let M_1 be the disc algebra and $M = \{f : f(z) = u(z) + iv(-z), u + iv \in M_1\}$. These examples are discussed in [1].

The above considerations naturally lead to a question, whose affirmative answer is the main result of [1]:

THEOREM 1. *Let M be a complex function space and M_1, M_2 uniform algebras such that M is isometrically isomorphic over $\mathbb{C}$ to M_1 and $Z \cong Z_2$. Then M is an algebra.*

The condition in Theorem 1 that M and M_1 are isometrically isomorphic over $\mathbb{C}$ implies the existence of an affine homeomorphism $\varphi : \sum \to \sum_1$ which preserves the complex structure of the unit balls. Suppose now that we only know the existence of an affine homeomorphism φ between $\sum$ and $\sum_1$. Then it is elementary to check that $\varphi(0) = 0$ and that φ extends to a real-linear isometric isomorphism between the real-duals of M

and M_1. This mapping is the dual of a real-linear isometric isomorphism $T : M_1 \to M$, given by

$$\eta(Tf_1) = re(\varphi(\eta)f_1) - ire(\varphi(i\eta)f_1), \eta \in M^*, f_1 \in M_1 .$$

Conversely any real-linear isometric isomorphism $T : M_1 \to M$ induces such a mapping φ.

The question arises whether a function space M being real-linearly isometrically isomorphic to a uniform algebra implies that M is complex-linearly isometrically isomorphic to a uniform algebra. This question was answered affirmatively in [2]:

THEOREM 2. *Let M be a complex function space on X and let M_1 be either a uniform algebra on X_1 or a self-adjoint complex function space on X_1. Then if $T : M_1 \to M$ is a real-linear isometric isomorphism, and if $\lambda = T1$, there exists a homeomorphism $\sigma : X \to X_1$ and an open and closed subset E of X such that*

$$(Tf_1)(x) = \lambda(x)f_1(\sigma(x)), x \in E, f_1 \in M_1 ,$$
$$(Tf_1)(x) = \lambda(x)\overline{f_1}(\sigma(x)), x \in X \setminus E, f_1 \in M_1 .$$

COROLLARY. *If, in the notation of Theorem 2, M_1 is a uniform algebra then M is complex-linearly isometrically isomorphic to the uniform algebra $\overline{\lambda}M$. Consequently if $\sum \cong \sum_1$ and $Z \cong Z_2$ for uniform algebras M_1 and M_2, then M is an algebra.*

We will outline the proof of Theorem 2 since this will be directly relevant to later results.

As above, let $\varphi : \sum \to \sum_1$ be the affine homeomorphism associated with T. Now each extreme point η of $\sum$ has the form $\eta(f) = \lambda f(x)$, $f \in M$, for some λ in Γ and x in the Choquet boundary for M. Consequently, for such x we obtain $\varphi(x) = \ell(x)z$ and $\varphi(ix) = \ell'(x)z'$ for some $\ell(x)$, $\ell'(x)$ in Γ and z, z' in X_1, where we have identified x with the evaluation functional at x. Therefore $\lambda(x) = T1(x) = re\ell(x) - ire\ell'(x)$, and $Ti(x) = -im\ell(x) + i\, im\ell'(x)$. Since $\|\lambda\| = 1 = \|Ti\|$ it now follows that $|\lambda(x)| = 1$ for all such x, and hence for all x in X.

We now write $M_2 = \overline{\lambda}M$ and define $T_1 : M_1 \to M_2$ by $Tf_1 = \overline{\lambda}Tf_1$. Then T_1 is a real-linear isometric isomorphism with $T_1 = 1$, and so the associated map φ_1 maps S_2 onto S_1, and $-iS_2$ partly into $-iS_1$ and partly into iS_1. Defining $\varphi_2(x) = -i\varphi_1(ix)$ if $\varphi_1(ix) \in iS_1$ and $\varphi_2(x) = i\varphi_1(ix)$ if $\varphi_1(ix) \in -iS_1$, we obtain a homeomorphism $\varphi_2 : X \to X_1$ and an open and closed subset E of X such that for $f_1 = u_1 + iv_1 \in M_1$ we have

$$T_1f_1(x) = u_1(\varphi_1(x)) + iv_1(\varphi_2(x)), x \in E ,$$
$$T_1f_1(x) = u_1(\varphi_1(x)) - iv_1(\varphi_2(x)), x \in X \setminus E .$$

The result will now follow if we show that $\varphi_1 = \varphi_2$. To see this, take $\gamma = \alpha + i\beta \in \Gamma$ with $\alpha, \beta \neq 0$. Then γT_1 is a real-linear isometric isomorphism, so that if $x \in \partial S_2$ there is a $\lambda' \in \Gamma$ and $x_1 \in X$ with $re((\gamma T_1)f_1)(x) = re(\lambda' f_1(x_1))$, $f_1 \in M_1$. Putting $f_1 = 1$ and then $f_1 = i$ we obtain $\lambda' = \gamma$ on E, while $\lambda' = \overline{\gamma}$ on $X \setminus E$. Consequently we have

$$\alpha u_1(\varphi_1(x)) - \beta v_1(\varphi_2(x)) = \alpha u_1(x_1) - \beta v_1(x_1), x \in E\ ,$$

$$\alpha u_1(\varphi_1(x)) + \beta v_1(\varphi_2(x)) = \alpha u_1(x_1) + \beta v_1(x_1), x \in X \setminus E\ .$$

In the case where M_1 is a self-adjoint complex function space we may take $f_1 = u_1$, which leads to $u_1(\varphi_1(x)) = u_1(x_1)$, so that $\varphi_1(x) = x_1$; similarly $\varphi_2(x) = x_1$ and $\varphi_1 = \varphi_2$. In the case where M_1 is a uniform algebra, suppose that $x_1 \neq \varphi_1(x)$ and $x_1 \neq \varphi_2(x)$. Then we may choose $f_1 \in M_1$ with $f_1(x_1) = 1$, $f_1(\varphi_j(x)) = 0$ for $j = 1, 2$. This gives $\alpha = 0$, which is a contradiction. If $x_1 = \varphi_1(x)$ then we obtain $x_1 = \varphi_2(x)$ from the above equations; similarly if $x_1 = \varphi_2(x)$ we get $x_1 = \varphi_1(x)$. In either case we have $\varphi_1(x) = \varphi_2(x)$, which completes the proof.

The following example shows that the result of Theorem 2 fails to hold for general complex function spaces M_1, and neither does it hold for the stated spaces if the isometric property of T is dropped.

EXAMPLE: Let $\triangle$ be the closed unit disc in $\mathbb{C}$ and put $M = M_1 = \{f \in C_{\mathbb{C}}(\triangle) : f(z) = \alpha z + \beta, \alpha, \beta \in \mathbb{C}\}$, so that M is a complex function space on $\triangle$ with Šilov boundary Γ. Note that for f of the form given we have $\|f\| = |\alpha| + |\beta|$. Hence if we write $T : M \to M$, where $(Tf)(z) = \overline{\alpha} z + \beta$, then T is a real-linear isometric isomorphism on M with $T1 = 1$. If the result of Theorem 2 holds in this case then, since Γ is connected, either we would have $Tf = f \circ \sigma$ on Γ or $Tf = \overline{f} \circ \sigma$; in the first case T would be complex-linear while in the second case T would be conjugate-complex-linear, giving a contradiction.

If we associate $f \in M$ with $(\alpha, \beta) \in \mathbb{C}^2$, then we see that M is $\mathbb{C}$-linearly homeomorphic to the uniform algebra M_1 on two points; in this case we have $\|(\alpha, \beta)\| = \max\{|\alpha|, |\beta|\}$. Therefore $T : M_1 \to M$ has the form $T(\alpha, \beta) = (\overline{\alpha}, \beta)$ and is a real-linear homeomorphism which does not split up in the form required in Theorem 2.

We will now show that if the hypotheses on M_1 in Theorem 2 are retained then, in substance, that result holds for general complex Banach spaces M.

THEOREM 3. *Let M be a complex Banach space and let M_1 be either a uniform algebra or a self-adjoint complex function space. Then if $T : M_1 \to M$ is a real-linear isometric isomorphism, M and M_1 have M-summand decompositions $M = M' \oplus M''$, $M_1 = M_1' \oplus M_1''$ such that $T : M_1' \to M'$ is complex-linear while $T : M_1'' \to M''$ is conjugate-complex-linear.*

PROOF: We first assume that M is a complex function space on X, not necessary containing constants. Then, as before, T induces an affine homeomorphism $\varphi : \sum \to \sum_1$

and the previous argument shows that if $\lambda = T1$ then $|\lambda(x)| = 1$ for all x in the Choquet boundary Y for M.

We may assume that X is the closure of Y, and define the space $M_2 = \{\overline{\lambda} f : f \in M\}$. M_2 is a space of continuous functions on X which contains the constants, but will not generally separate the points of X. Consequently M_2 is a complex function space on the quotient space $\tilde{X}$ corresponding to the quotient map $q : X \to \tilde{X}$, where $q(x) = q(y)$ if and only if $h(x) = h(y)$ for all h in M_2. The map q is the restriction to $\sum$ of the dual of the isometric mapping $Q : M_2 \to M$ given by $Qf_2 = \lambda f_2$, and hence q maps Y onto the Choquet boundary of M_2; consequently $\tilde{X}$ is the Šilov boundary for M_2.

If we define $T_1 : M_1 \to M_2$ by $T_1 f_1 = \overline{\lambda} T f_1$, $f_1 \in M_1$, then T_1 is a real-linear isometric isomorphism with $T_1 = 1$. By Theorem 2 there exists an open and closed subset E of $\tilde{X}$ and a homeomorphism $\sigma : \tilde{X} \to X_1$ such that

$$T_1 f_1(\tilde{x}) = f_1(\sigma(\tilde{x})), \tilde{x} \in E, \; T_1 f_1(\tilde{x}) = \overline{f}_1(\sigma(\tilde{x})), \tilde{x} \in \tilde{X} \setminus E \,.$$

We put $\tau = \sigma \circ q : X \to X_1$, so that τ is a continuous surjection. If $D = q^{-1}(E)$ then we have

$$\begin{aligned} T f_1(x) = \lambda(x) T_1 f_1(\tilde{x}) &= \lambda(x) f_1(\tau(x)), x \in D \,, \\ &= \lambda(x) \overline{f_1}(\tau(x)), x \in X \setminus D \,. \end{aligned}$$

For f_1 in M_1 we have $i(T_1 f_1) \in M_2$, and so there exists some $g \in M_1$ with $g \circ \sigma = i(f_1 \circ \sigma)$ on E, while $\overline{g} \circ \sigma = i(\overline{f}_1 \circ \sigma)$ on $\tilde{X} \setminus E$; that is $g = i f_1$ on $\sigma(E)$, while $g = -i f_1$ on $X_1 \setminus \sigma(E)$. Therefore we see that $f_1 \chi_{\sigma(E)}$ and $f_1 \cdot \chi_{X_1 \setminus \sigma(E)}$ belong to M_1.

Now if $f \in M$ we have $f = T f_1$ for some $f_1 \in M_1$ and hence $T(f_1 \cdot \chi_{\sigma(E)}) = f \cdot \chi_D \in M$ and $T(f_1 \cdot \chi_{X_1 \setminus \sigma(E)}) = f \cdot \chi_{X \setminus D} \in M$. Writing $M' = M \cdot \chi_D$, $M'' = M \cdot \chi_{X \setminus D}$ and $M_1' = M_1 \cdot \chi_{\sigma(E)}$, $M_1'' = M_1 \cdot \chi_{X_1 \setminus \sigma(E)}$ we see that $M = M' \oplus M''$, $M_1 = M_1' \oplus M_1''$, where the sums form M-summands. Moreover T maps M_1' onto M' and M_1'' onto M'', with $T|M_1'$ being complex-linear while $T|M_1''$ is conjugate-complex-linear.

Returning to the general case, let M be any complex Banach space identified as a space of continuous complex-valued affine functions on $\sum$, restricted to $X = \partial \sum$. The previous reasoning will now give the general result. We note however in this case that $q(x) = q(y)$ (or $\tau(x) = \tau(y)$) if and only if $y = \alpha x$ for some $\alpha \in \Gamma$. Indeed, $q(x) = q(y)$ implies $\overline{\lambda}(x) f(x) = \overline{\lambda}(y) f(y)$, that is $f(\overline{\lambda}(x)x) = f(\overline{\lambda}(y)y)$, for all f in M, and this implies that $\overline{\lambda}(x)x = \overline{\lambda}(y)y$. Conversely, $x = \alpha y$ for α in Γ implies, since $\lambda \in M$, $\overline{\lambda}(x) f(x) = \overline{\alpha}\, \overline{\lambda}(y) \alpha f(y) = \overline{\lambda}(y) f(y)$, that is $q(x) = q(y)$.

COROLLARY 1. *Let M be a complex Banach space which is real-linearly isometrically isomorphic to a complex Lindenstrauss space M_1. Then M is a complex Lindenstrauss space.*

PROOF: First assume that the unit ball of M_1 contains an extreme point. In this case M_1 may be identified with a self-adjoint complex function space (cf. Olsen [4, Cor.20]). Applying Theorem 3 to the real-linear isometric isomorphism $T : M_1 \to M$ we obtain, with the notation of the proof of that theorem, $Tf_1 = \lambda(f_1 \circ \tau)$ on D while $Tf_1 = \lambda(\overline{f_1} \circ \tau)$ on $X \setminus D$, for f_1 in M_1. If we now define $T'f_1 = \lambda(f_1 \circ \tau)$ then, since $\lambda(f_1 \circ \tau) = T(f_1 \cdot \chi_{\sigma(E)}) + T(\overline{f_1} \cdot \chi_{X_1 \setminus \sigma(E)})$ and since M_1 is self-adjoint, it follows that T' is a complex-linear isometric isomorphism between M_1 and M. Consequently M is a complex Lindenstrauss space.

In the general case we note that any real-linear isometric isomorphism between complex Banach spaces induces a real-linear isometric isomorphism between their dual spaces. Now M_1^{**} is a complex Lindenstrauss space whose unit ball contains an extreme point, and so the first part of the proof shows that M^{**} is a complex Lindenstrauss space. Consequently M is a complex Lindenstrauss space.

The proof of Theorem 3, togther with the Mazur-Ulam theorem shows that if a complex Banach space M is isometric to a self-adjoint complex function space M_1, then M is complex-linearly isometrically isomorphic to M_1. In order to deduce a corresponding result in the case where M_1 is a uniform algebra we would need to know when a uniform algebra is complex-linearly isometrically isomorphic to its complex conjugate algebra. This latter property for a uniform algebra is equivalent to a geometric property of its complex state space.

THEOREM 4. *Let M be a uniform algebra. Then M and $\overline{M}$ are isometrically isomorphic over $\mathbb{C}$ if and only if there exists an affine homeomorphism $\varphi : Z \to Z$ such that $\varphi(S) = -iS$ and $\varphi(-iS) = S$.*

PROOF: We write M_1 for $\overline{M}$. Then, by Ellis and So [3, Th.5], M and M_1 will be isometrically isomorphic over $\mathbb{C}$ if and only if there exists an affine homeomorphism $\psi : Z_1 \to Z$ such that $\psi(S_1) = S$ and $\psi(-iS_1) = -iS$.

The real-linear isometric isomorphism $T : M \to M_1$, given by $Tf = i\overline{f}$, $f \in M$, induces an affine homeomorphism $\Phi : \sum_1 \to \sum$. If x is a Choquet boundary point for M_1 then arguing as before, there exist y, y' in X and ℓ, ℓ' in Γ such that

$$i\overline{f}(x) = Tf(x) = re(\ell f(y)) - ire(\ell' f(y')), f \in M\ .$$

Choosing $f = 1$ we obtain $i = re\ell - ire\ell'$, while choosing $f = i$ we obtain $1 = re(i\ell) - ire(i\ell')$; it follows that $\ell = -i$ and $\ell' = -1$. It is now straightforward to verify that $y = y' = x$ and that Φ maps S_1 onto $-iS$ and $-iS_1$ onto S.

Given a map φ with the properties stated in the Theorem, we define $\psi = \varphi \circ \Phi : Z_1 \to Z$ so that ψ has the required properties. If conversely M and M_1 are isometrically isomorphic over $\mathbb{C}$ and ψ is the associated map, then $\varphi = \psi \circ \Phi^{-1} : Z \to Z$ is an affine homeomorphism satisfying $\varphi(S) = -iS$ and $\varphi(-iS) = S$.

A special case relating to Theorem 4 is the uniform algebra $M = P(E)$, the uniform limits of polynomials on E, where E is a compact simply connected subset of the complex plane. If we write $E^* = \{\bar{z} : z \in E\}$ then it is straightforward to verify that $\overline{M}$ is isometrically isomorphic over $\mathbb{C}$ to $P(E^*)$. Consequently M and $\overline{M}$ are isometrically isomorphic over $\mathbb{C}$ if and only if $P(E)$ and $P(E^*)$ have the same property.

Now $P(E)$ and $P(E^*)$ are isometrically isomorphic over $\mathbb{C}$ if and only if there exsits a homeomorphism $\tau : \delta E^* \to \delta E$ (where δE denotes the boundary of E), such that the functions $g \in P(E^*)$ are those satisfying $g(\bar{z}) = f(\tau(\bar{z}))$, $z \in \delta E$, for some $f \in P(E)$. It may now be verified that τ extends to a function in $P(E^*)$, and that τ^{-1} extends to a function in $P(E)$. Consequently τ is a conformal equivalence between E^* and E. Conversely, any such conformal equivalence induces a complex-linear isometric isomorphism between $P(E)$ and $P(E^*)$. If, in particular, E is the closure of a simply-connected region with simple boundary points then M and $\overline{M}$ are isometrically isomorphic over $\mathbb{C}$.

However $P(E)$ and $P(E^*)$ are not conformally equivalent in general. In fact let E be the closed unit disc Δ with the additional line segments attached at $1, -1$ and i as shown.

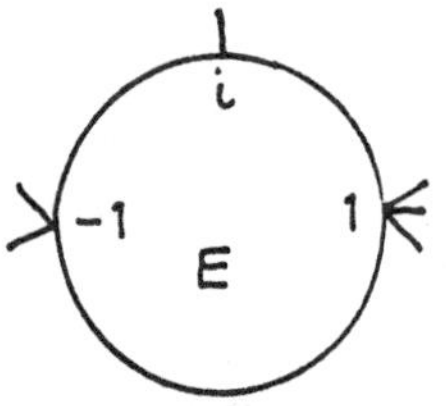

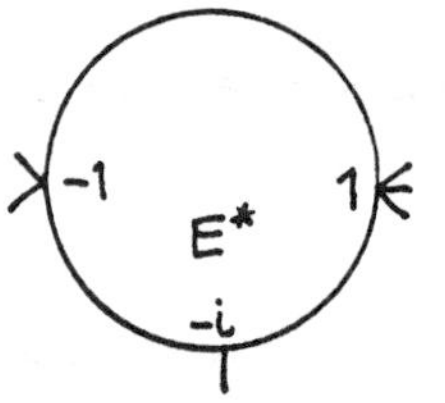

If $\tau : E \to E^*$ is a conformal equivalence then τ must map Δ° onto Δ°, and hence Γ onto Γ. Consequently τ is a bilinear function. For evident topological reasons we must also have $\tau(1) = 1$, $\tau(-1) = -1$ and $\tau(i) = -i$. The only bilinear function satisfying these equations is $\tau(z) = \frac{1}{z}$, which does not map Δ onto Δ. Consequently $P(E)$ and $P(E^*)$ are not $\mathbb{C}$-linearly isometrically isomorphic.

I am grateful to Professor Y.T. Siu and Dr. P.W. Wong for this example.

We note also that if we take $M = P(E \cup E')$, where E' is a translate of E^* which is disjoint from E, then $M = P(E) \oplus P(E^*)$ so that $\overline{M}$ is $\mathbb{C}$-linearly isometrically isomorphic to $P(E^*) \oplus P(E)$, and hence to M. However $M|E$, i.e. $P(E)$, is not $\mathbb{C}$-linearly isometrically isomorphic to $\overline{M}|E$. This example therefore shows that if a uniform algebra is $\mathbb{C}$-linearly isometrically isomorphic to its complex conjugate then its restriction to a component of the maximal ideal space need not have the same property.

Taking these remarks, together with the comments following Corollary 1 we obtain the following result.

THEOREM 5. *Let X and Y be isometric complex Banach spaces, such that Y is either a self-adjoint complex function space or a uniform algebra M with the property that M^1 is $\mathbb{C}$-linearly isometrically isomorphic to $\overline{M^1}$ for each M-summand M^1 of M. Then X and Y are $\mathbb{C}$-linearly isometrically isomorphic.*

We conclude with some remarks concerning the proof of Theorem 2 in relation to the hypotheses required by M_1 in Theorem 2 and 3. Taking $\alpha = \beta = (\sqrt{2})^{-1}$ the equations we obtain are

$$u_1(\varphi_1(x)) - v_1(\varphi_2(x)) = u_1(x_1) - v_1(x_1), x \in E ,$$
$$u_1(\varphi_1(x)) + v_1(\varphi_2(x)) = u_1(x_1) + v_1(x_1), x \in X \setminus E .$$

It is not difficult to deduce that we have

$$\varphi_1(x) + i\varphi_2(x) = (1+i)x_1, x \in E \quad \text{and}$$
$$\varphi_1(x) - i\varphi_2(x) = (1-i)x_1, x \in X \setminus E .$$

In order to deduce that $\varphi_1(x) = \varphi_2(x)$ it will be sufficient for M_1 to satisfy the condition that whenever $y \in \partial S_1$ the set $co\{y, -iy\}$ is a face of Z_1, and the set $co\{y, iy\}$ is a face of $co(S_1 \cup iS_1)$.

These geometric properties of Z_1 and $co(S_1 \cup iS_1)$ hold for the case when M_1 is self-adjoint since S_1 is a split face of Z_1 and of $co(S_1 \cup iS_1)$ in that case, and hold for the case when M_1 is a uniform algebra since each $y \in \partial S_1$ is a split face of Z_1 and of $co(S_1 \cup iS_1)$ in that case.

References

1. A.J. Ellis, 'Algebraic structure in complex function spaces', Proc. Amer. Math. Soc. 107(1989), 621-626.
2. A.J. Ellis, 'Real characterizations of function algebras amongst function spaces', Bull. London Math. Soc. (to appear).
3. A.J. Ellis and W.S. So, 'Isometries and the complex state spaces of uniform algebras', Math. Z. 195(1987), 119-125.
4. G.H. Olsen, 'On the classification of complex Lindenstrauss spaces', Math. Scand, 35(1974), 237-258.

Contractive Projections onto Subsets of L^p-Spaces

PER ENFLO Department of Mathematics & Computer Science, Kent State University, Kent, Ohio

1. Introduction.

In this paper we will characterize the subsets of (real) $L^p = L^p(0,1)$, $1 < p < \infty$, which are ranges of contractive, possibly non-linear, projections, i.e. projections that satisfy $\|Px - Py\| \leq \|x - y\|$ for all x and y. We call these sets *contractive* sets. Similar problems — both linear and non-linear — have earlier been studies in a number of papers.,

A complete characterization of Banach spaces which are ranges of contractive linear projections in any space containing them was given by Nachbin [10], Goodner [7], and Kelley [9] for the real case and by Hasumi [8] for the complex case. In [1] Ando characterized the subspaces of L^p which are ranges of contractive, linear projections.

The non-linear version of this problem has been studied in Beauzamy-Maurey [4], Beauzamy [2] Westphal [8], Davis-Enflo [5] and Enflo [6] (cf. also Beauzamy-Enflo [3]). In the study of the non-linear problem the concepts "minimal point" and "optimal set" play an important role, so we will briefly recall their definitions and some of their properties here.

We say that a point x in a metric space is minimal with respect to M if $d(y,m) \leq d(x,m)$ for all $m \in M$ implies $x = y$. The set of minimal points with respect to M is denoted $\min M$. Obviously $M \subset \min M$. We say that M is optimal if $\min M = M$.

If M is a contractive subset of a Banach space then M is obviously optimal. It is still an open problem whether, in a strictly convex Banach space, an optimal set is always contractive. In many important cases this is true. In [4] it is proved that if B is a reflexive,

---

Supported in part by NSF Grant #440702

strictly convex and smooth Banach space, then an optimal subspace is always contractive. And in [2] it is proved that — under the same assumptions on B — an optimal subset of B with non-empty interior is always contractive. In [5] it is proved that optimal subsets of ℓ^p, $1 < p < \infty$, are contractive. It is part of the main theorem of this paper that optimal subsets of L^p, $1 < p < \infty$, are contractive.

In [5] contractive subsets of ℓ^p, $1 < p < \infty$, are characterized. Part of this characterization is in [5] proved only for bounded subsets. The techniques used in [5] do not carry over to L^p. With the techniques used in this paper we also get the main result of [5], with the restriction of boundedness removed.

In [6] contractive convex subsets of L^1 are characterized. The characterization of contractive convex subsets of L^1 uses contractive cones and is similar to the characterization of contractive subsets of L^p. However, for the characterization of contractive subsets of L^p a somewhat more general class of cones is used.

The proof technique for one direction of the characterization in [6] can be modified to work in L^p. For the other direction, though, the technique in [6] only works in L^1 and so a new technique is introduced in this paper, and is carried out in the Lemmas 7-12 below.

2. The Main Result.

We recall that we work in real L^p. To characterize the contractive subsets of L^p we first define "the cone $C_{f,\mathcal{T},x}$". Let $f \in L^p$ and consider a measurable subset $E \subset [0,1]$, $m(E) > 0$, s.t. $f > 0$ on E or $f < 0$ on E. Put $F = [0,1] \backslash E$. We will below, when we talk about subsets of $[0,1]$, assume that they are measurable.

If g is a real valued function on F, then G_g is the following subset of R^2:

$$\{(t,y) | t \in F, \; y \in [0, g(t)]\}$$

where is is not necessary that $g(t) \geq 0$, we put $[a,b] = [b,a]$. Now, $\mathcal{T}$ is a map from subsets of E to $G_g : s$ where

i) $g(t)$ is defined for almost all $t \in F$

ii) sign $g(t) =$ sign $f(t)$ for almost all $t \in F$

and iii) $|g(t)| \leq |f(t)|$ for almost all $t \in F$

We assume the following:

1. $\mathcal{T}(E_1 \cup E_2) = \mathcal{T}(E_1) \cup \mathcal{T}(E_2)$

2. $m(E_1) = 0 \Longrightarrow m(\mathcal{T}(E_1)) = 0$

3. $\mathcal{T}$ is left continuous, that is, if E_n is an increasing sequence of subsets of E, then $\mathcal{T}(\bigcup_n E_n) = \bigcup_n(\mathcal{T}(E_n))$.

We let $g_{\mathcal{T}(E_1)}$ denote the function s.t. $\mathcal{T}(E_1) = G_g$. We let $C_{f,\mathcal{T},0}$ be the following cone: $h \in C_{f,\mathcal{T},0}$ if and only if $h \in L^p$ and for every $E_1 \subset E$

$$\operatorname*{ess\,inf}_{t\in E_1} \frac{h(t)}{f(t)} \le \operatorname*{ess\,inf}_{t\in F} \frac{h(t)}{g_{\mathcal{T}(E_1)}(t)}$$

(for division by 0 we use the convention $\frac{a}{0} = \begin{cases} +\infty & \text{if } \ a > 0 \\ & \text{if } \ a = 0 \ \ \text{not specified} \\ -\infty & \text{if } \ a < 0 \end{cases}$)

We put $C_{f,\mathcal{T},x} = x + C_{f,\mathcal{T},0}$. We let $C_{E,+,0}(C_{E,-,0})$ be the cone of functions in L^p that are $\ge 0(\le 0$ on E. And we put $C_{E,+,x}(C_{E,-,x}) = x + C_{E,+,0}(C_{E,-,0})$. We can now state our main result.

Theorem 1. *For subsets M of L^p, $1 < p < \infty$, $p \neq 2$, the following are equivalent*

1. *M is optimal*

2. *M is contractive*

3. *M is a non-empty intersection of a family of cones $C_{f_\alpha,\mathcal{T}_\alpha,x_\alpha}$, $C_{E_\beta,+,x_\beta}$, $C_{E_\gamma,-,x_\gamma}$.*

Proof: In this section we will prove 3) $\to$ 2). In the next section we will prove 1) $\to$ 3). 2) $\to$ 1) is trivial.

The proof that 3) $\to$ 2) is given by the Lemmas 1-5 and Proposition 1 below. It is similar to the corresponding part in [6] but since we work here with more general cones, the details have to be worked out somewhat differently. We start with

Lemma 1. $C_{f,\mathcal{T},0}$ is a closed, convex cone.

Proof: It is obvious that $C_{f,\mathcal{T},0}$ is a cone. To show that it is closed assume that $h_n \in C_{f,\mathcal{T},0}$ and $h_n \to h$ but $\operatorname*{ess\,inf}_{t\in E_1} \frac{h(t)}{f(t)} > \operatorname*{ess\,inf}_{t\in F} \frac{h(t)}{g_{\mathcal{T}(E_1)}(t)} + \delta$ for some

$$E_1 \subset E \text{ and } \delta > 0. \tag{1}$$

There is a sequence $E_1^{(n)} \subset E_1$ s.t. $m(E_1 \backslash E_1^{(n)}) \to 0$ and s.t.

$$\operatorname*{ess\,inf}_{t\in E_1^{(n)}} \frac{h_n(t)}{f(t)} \longrightarrow \operatorname*{ess\,inf}_{t\in E_1} \frac{h(t)}{f(t)}. \tag{2}$$

By left continuity of $\mathcal{T}$

$$m(\mathcal{T}(E_1)\backslash \mathcal{T}(E_1^{(n)})) \to 0$$

and so, if n is large enough

$$\operatorname*{ess\,inf}_{t\in E_1^{(n)}} \frac{h_n(t)}{f(t)} \le \operatorname*{ess\,inf}_{t\in F} \frac{h_n(t)}{g_{\mathcal{T}(E_1^{(n)})}(t)} \le \operatorname*{ess\,inf}_{t\in F} \frac{h_n(t)}{g_{\mathcal{T}(E_1)}(t)} + \frac{\delta}{3} < \operatorname*{ess\,inf}_{t\in F} \frac{h(t)}{g_{\mathcal{T}(E_1)}(t)} + \frac{2\delta}{3}.$$

This contradicts (1) and (2) and so we have proved that $C_{f,\mathcal{T},0}$ is closed. To show that $C_{f,\mathcal{T},0}$ is convex we argue again by contradiction. WLOG we assume $f \ge 0$ on $[0,1]$. Assume $h_1 \in C_{f,\mathcal{T},0}$ and $h_2 \in C_{f,\mathcal{T},0}$ but that there is an $E_1 \subset E$ and a $\delta > 0$ s.t.

$$\operatorname*{ess\,inf}_{t\in E_1} \frac{h_1(t)+h_2(t)}{f(t)} > \operatorname*{ess\,inf}_{t\in F} \frac{h_1(t)+h_2(t)}{g_{\mathcal{T}(E_1)}(t)} + \delta \tag{3}$$

Take a subset

$$F_1 \subset F \text{ s.t. } \frac{h_1(t)}{g_{\mathcal{T}(E_1)}(t)} \text{ and } \frac{h_2(t)}{g_{\mathcal{T}(E_1)}(t)}$$

vary less than $\frac{\delta}{10}$ on F_1 and s.t. $\frac{h_1(t)+h_2(t)}{g_{\mathcal{T}(E_1)}(t)} < \operatorname*{ess\,inf}_{u\in F} \frac{h_1(u)+h_2(u)}{g_{\mathcal{T}(E_1)}(u)} + \frac{\delta}{10}$ for all $t \in F_1$ and s.t. $\{g_{\mathcal{T}(E_1)}(t) \,|\, t \in F_1\}$ is contained in an interval

$$[K, 2K] \tag{4}$$

Given $\delta' > 0$, by an exhaustion argument, there is a maximal subset $E_2 \subset E_1$, $m(E_1 \backslash E_2) > 0$ s.t. for all $t \in F_1$

$$g_{\mathcal{T}(E_1)}(t) - g_{\mathcal{T}(E_2)}(t) \ge \delta' \tag{5}$$

Then, for every $E_3 \subset E_1 \backslash E_2$, $m(E_3) > 0$ we have

$$\operatorname*{ess\,inf}_{t\in F} g_{\mathcal{T}(E_1)}(t) - g_{\mathcal{T}(E_3)}(t) < \delta' \tag{6}$$

Now consider an $E_3 \subset E_1 \backslash E_2$, $m(E_3) > 0$ s.t.

$$\frac{h_1(t)}{f(t)} < \operatorname*{ess\,inf}_{u \in E_1 \backslash E_2} \frac{h_1(u)}{f(u)} + \frac{\delta}{10} \text{ for all } t \in E_3$$

and an $E_4 \subset E_3$, $m(E_4) > 0$ s.t.

$$\frac{h_2(t)}{f(t)} < \operatorname*{ess\,inf}_{u \in E_3} \frac{h_2(u)}{f(u)} + \frac{\delta}{10} \text{ for all } t \in E_4 \tag{7}$$

This gives

$$\begin{aligned} &\operatorname*{ess\,inf}_{t\in E_1} \tfrac{h_1(t)+h_2(t)}{f(t)} \le \operatorname*{ess\,inf}_{t\in E_4} \tfrac{h_1(t)+h_2(t)}{f(t)} < \\ &\le \operatorname*{ess\,inf}_{T\in E_4} \tfrac{h_1(t)}{f(t)} + \operatorname*{ess\,inf}_{T\in E_4} \tfrac{h_2(t)}{f(t)} + \tfrac{2\delta}{10} \le \\ &\le \operatorname*{ess\,inf}_{t\in F} \tfrac{h_1(t)}{g_{T(E_4)}(t)} + \operatorname*{ess\,inf}_{t\in F} \tfrac{h_2(t)}{g_{T(E_4)}(t)} + \tfrac{2\delta}{10} \le \\ &\le \text{ (by(4) and (6))} \operatorname*{ess\,inf}_{t\in F_1} \tfrac{h_1(t)}{g_{T(E_1)}(t)} + \\ &\operatorname*{ess\,inf}_{t\in F_1} \tfrac{h_2(t)}{g_{T(E_1)}(t)} + \tfrac{2\delta}{10} + \tfrac{2\delta}{10} + \delta''(\delta') \le \\ &\operatorname*{ess\,inf}_{t\in F_1} \tfrac{h_1(t)+h_2(t)}{g_{T(E_1)}(t)} + \tfrac{4\delta}{10} + \delta''(\delta') \le \\ &\le \text{(by (4))} \operatorname*{ess\,inf}_{t\in F} \tfrac{h_1(t)+h_2(t)}{g_{T(E_1)}(t)} + \tfrac{5\delta}{10} + \delta''(\delta') \end{aligned} \tag{8}$$

Here δ'' depends only on δ', K and the sizes of h_1 and h_g on F_1 and $\delta''(\delta') \to 0$ as $\delta' \to 0$. Thus (8) contradicts (3) and the Lemma is proved.

In the Lemmas 2-5 below we will construct a contractive projection on to $C_{f,\mathcal{T},0}$. In order to do that we first define "elementary projections" — $R : s$ — in L^p.

Let E' and F' be disjoint sets of positive measure on (0,1) and let $r \geq 0$. If $g \in L^p(0,1)$ we put $R(E', F', r)(g) = R' \circ S(g)$ where R' and S are the following maps:

1. $(Sg)(t) = g(t)$ if $t \notin E' \cup F'$.

2. $(Sg)(t) = \frac{\int_{E'} g(u)du}{m(E')}$ if $t \in E'$ and $(Sg)(t) = \frac{\int_{F'} g(u)du}{m(F')}$ if $t \in F'$.

3. $R'(h)(t) = h(t)$ if $t \notin E' \cup F'$.

Now, on the two-dimensional space of functions h that are constant on E' and constant on F', R' acts as follows: R' is the unique contractive projection, that maps outside points to boundary points, onto the half-space defined by

$$h|_{E'} \leq r h|_{F'}$$

We observe that if $r \geq 0$, then R' will decrease the value on E' and increase the value on F'. If $r < 0$, $R(E', F', r)(g)$ is defined by first multiplying g by -1 on F', then applying $R(E', F', |r|)$ and then multiply the result by -1 on F'.

Now let (h_m) be a sequence in L^p s.t. $\{h_m\}$ is dense in L^p and $\{h_m\} \cap C_{f,T,0}$ is dense in $C_{f,T,0}$.

We can assume WLOG that the union of the sets where some h_m takes a value of the form $\frac{k}{2^n}$ (k and n integers) has measure 0. Now we define f_m on [0,1] by $f_m(t) = [f(t) \cdot 2^m] \cdot 2^{-m}$ where as usual $[x]$ denotes the integer part of x. Then $f_m \uparrow f$ uniformly and f_m takes values of the form $\frac{k}{2^m}$, k integer.

We now make a sequence of splittings of E and F in the following way: In the n:th stage E is divided into pairwise disjoint subsets $E_{n1}, E_{n2} \ldots$ which are the minimal sets in the algebra generated by the following sets by taking finite intersections and finite unions and complements:

1. The subsets of E, where f_m, $1 \leq m \leq n$, is constant
2. The subsets of E, where h_m, $1 \leq m \leq n$, takes values between $\frac{k}{2^n}$ and $\frac{k+1}{2^n}$, k integer.

In the n:th stage F is divided into pairwise disjoint subsets $F_{n1}, F_{n2}, \ldots$ which are the minimal sets in the algebra generated as above by the following sets:

1. The subsets of F where f_m, $1 \leq m \leq n$, is constant
2. The subsets of F where h_m, $1 \leq m \leq n$, takes values between $\frac{k}{2^n}$ and $\frac{k+1}{2^n}$, k integer
3. The subsets of F where $g_{T(E_{ni})}$, $i = 1, 2, \ldots$ takes values $\geq \frac{k}{2^n}$ and $< \frac{k+1}{2^n}$, k integer.

It is obvious that the $E_{n+1,i}$: s and $F_{n+1,j}$: s are subsets of E_{ni} : s and F_{nj} : s. We will now define projections Q_n which will be first approximations to projections on $C_{f,T,0}$. In order to define Q_n we enumerate all pairs (E_{ni}, F_{nj}) as (E_{ni_L}, F_{nj_L}), $L = 1, 2, \ldots$. We now assume $f > 0$ on E. The case $f < 0$ on E is done similarly.

We define r_{nij} by

$$f_n|_{E_{ni}} = \frac{r_{nij}}{2^n} \cdot [2^n \cdot g_{T(E_{ni})}]|_{F_{nj}} \tag{9}$$

Since on F_{nj} $g_{T(E_{ni})}$ takes values $\geq \frac{k}{2^n}$ and $< \frac{k+1}{2^n}$, the right hand side of (9) is well-defined.

If the right hand side is 0 we put $r_{nij} = +\infty$. We now put

$$\begin{aligned} Q_n h = \lim_{L\to\infty} R(E_{ni_L}, F_{nj_L}, r_{n i_L j_L})\circ \\ \circ R(E_{ni_{L-1}}, F_{ni_{L-1}}, r_{n i_{L-1} j_{L-1}})\circ\cdots\circ \\ \circ R(E_{ni_1}, F_{nj_1}, r_{n i_1 j_1})(h) \end{aligned} \tag{10}$$

Every R is a contractive projection, that decreases values on E. On F it decreases values where f is negative and increases values where f is positive. Thus the right hand side of (10) exists in L^p and so Q_n is well-defined. We now have

Lemma 2. $\lim_{s\to\infty} Q_{m+s}\circ Q_{m+s-1}\circ\cdots\circ Q_m(h)$ exists for every m and every $h\in L^p$.

Proof: We see that

$Q_{m+s}\circ Q_{m+s-1}\circ\cdots\circ Q_m(h)$ decreases with S on E. On F it increases with s where f is positive and decreases with s where f is negative. Since it is obviously bounded in L^p the lemma follows. We put

$$P_m h = \lim_{s\to\infty} Q_{m+s}\circ Q_{m+s-1}\circ\cdots\circ Q_m(h) \tag{11}$$

We now have

Lemma 3. $P_m h \in C_{f,T,0}$ for every $h\in L^p$.

Proof: WLOG we can assume $f\geq 0$ on $[0,1]$. Since, by Lemma 1, $C_{f,T,0}$ is closed we can assume $h = h_n$ for some n. We now assume $P_m h_n \notin C_{f,T,0}$ and argue by contradiction.

If $P_m h_n \notin C_{f,T,0}$ then there is an $E_1\subset E$ and $\delta > 0$ s.t.

$$\operatorname*{ess\,inf}_{t\in E_1} \frac{P_m h_n(t)}{f(t)} > \operatorname*{ess\,inf}_{t\in F} \frac{P_m h_n(t)}{g_{T(E_1)}(t)} + \delta \tag{12}$$

Now

$$\begin{aligned} \operatorname*{ess\,inf}_{t\in E_1} \tfrac{P_m h_n(t)}{f(t)} &= \lim_{r\to\infty}\left(\operatorname*{ess\,inf}_{t\in E_1} \frac{P_m h_n(t)}{f_r(t)}\right) = \\ &= \lim_{r\to\infty}\left(\operatorname*{ess\,inf}_{t\in E_1}(\lim_{s\to\infty} \frac{Q_{m+s}\circ Q_{m+s-1}\circ\cdots\circ Q_m h_n(t)}{f_r(t)})\right) \end{aligned} \tag{13}$$

Since $\frac{Q_{m+s}\circ Q_{m+s-1}\circ\cdots\circ Q_m h_n(t)}{f_r(t)}$ decreases with s on E, (13) gives the following: Given n, r and s, there is an $N = \max(m+s, r, n)$ and a set, $E_1' \supset E_1$, in the algebra of sets from the

N^{th} splitting of E s.t.

$$\begin{aligned} &\operatorname*{ess\,inf}_{t\in E_1'} \tfrac{Q_N \circ Q_{N-1} \circ \cdots \circ Q_m h_n(t)}{f_r(t)} > \\ &> \operatorname*{ess\,inf}_{t\in F} \tfrac{P_m h_n(t)}{g_{T(E_1')}}(t) + \tfrac{\delta}{2} \geq \\ &\operatorname*{ess\,inf}_{t\in F} \tfrac{Q_N \circ Q_{N-1} \circ \cdots \circ Q_m h_n(t)}{g_{T(E_1')}(t)} + \tfrac{\delta}{2} \end{aligned} \tag{14}$$

The last inequality of (14) holds since the right hand side increases with N. Now let F_{Nj} be a subset of F where

$$\begin{aligned} &\operatorname*{ess\,inf}_{t\in F_{N_j}} \tfrac{Q_N \circ Q_{N-1} \circ \cdots \circ Q_m h_n(t)}{g_{T(E_1')}(t)} = \\ &\operatorname*{ess\,inf}_{t\in F} \tfrac{Q_N \circ Q_{N-1} \circ \cdots \circ Q_m h_n(t)}{g_{T(E_1')}(t)} = \end{aligned} \tag{15}$$

And let E_{Ni} be a subset of E_1' s.t.

$$g_{T(E_1')}(t) < g_{T(E_{Ni})}(t) + \frac{1}{2^N} \text{ on } F_{Nj}. \tag{16}$$

Then if N is large enough we get by (14), (15) and (16)

$$\frac{Q_N \circ Q_{N-1} \circ \cdots \circ Q_m h_n}{fr}|_{E_{Ni}} > \sup_{t\in F_{Nj}} \frac{Q_N \circ Q_{N-1} \circ \cdots \circ Q_m h_n(t)}{g_{T(E_{Nj})}(t)} + \frac{\delta}{4} \tag{17}$$

Now consider the pair (E_{Ni}, F_{Nj}). When forming Q_N one uses the elementary projection

$$R(E_{Ni}, F_{Nj}, r_{Nij}) \tag{18}$$

and composing with other projections will only decrease values on E and increase values on F. Thus (17) contradicts (9) and (18) and so the lemma is proved.

We will define the contractive projection onto $C_{f,T,0}$ as a weak limit of P_m : s. In order to show that we get the identify map on $C_{f,T,0}$ we need the following

Lemma 4. Let $h_n \in C_{f,T,0}$. Then, given $\epsilon > 0$, there is an $h \in C_{f,T,0}$ s.t. $\|h - h_n\| < \epsilon$ and s.t. h has the following properties:

1. There is an N s.t. h is constant on each E_{Ni} and F_{Nj} in the N:th partition.
2. $h \leq h_n$ on E—where we have assumed $f > 0$ $h \geq h_n$ on the subset of F where $f \geq 0$ and $h \leq h_n$ on the subset of F where $f \leq 0$.

Proof: Since h_n is used to define the Nth subdivision, if $N \geq m$, h_n will vary not more than $\frac{1}{2^N}$ on each E_{Ni} and F_{Nj}.

Now put $h = \operatorname{ess\,inf} h_n$ on each E_{Ni} and $h = \operatorname{ess\,sup} h_n$ or $\operatorname{ess\,inf} h_n$ on each F_{Nj} depending on the sign of f. This h obviously has the required properties.

Lemma 5. $C_{f,\mathcal{T},0}$ is the range of a contractive projection.

Proof: Since for every $h \in L^p$ $(P^m h)$ is a bounded sequence we can by a standard diagonal procedure, extract a subsequence P_{mj} which converges weakly for each h. Let $Ph = \lim_{j\to\infty} P_{m_j} h$ where lim is in the weak topology. Obviously P is contractive and since by Lemma 1, $C_{f,\mathcal{T},0}$ is closed and convex $Ph \in C_{f,\mathcal{T},0}$ for every $h \in L^p$. We finally show that P is the identity map on $C_{f,\mathcal{T},0}$. For this it is obviously enough to consider $h_n : s$. Now given h_n we find h as in Lemma 5. We now WLOG assume $f \geq 0$ on F. And, then for $\delta > 0$ we consider the function $h^1 = h - \delta$ on E and $h + \delta$ on F. It is easy to check that then $Q_n h^1 = h^1$ for all sufficiently large n and so $Ph^1 = h^1$. But since $\|h_n - h^1\|$ is arbitrarily small we get that $P =$ identity on $C_{f,\mathcal{T},0}$. To complete 3) $\to$ 2) in Theorem 1 we now use the following

Proposition 1. Let B be a reflexive and strictly convex Banach space. Then a non-empty intersection of a family of contractive sets in B is contractive.

Proof: Let $\{M_\alpha | \alpha \in I\}$ be the family of contractive sets and let $\{P_\alpha | \alpha \in I\}$ be the corresponding projections. Now consider the smallest family $\mathcal{F}$ of maps on B, which contains the identify and is closed under the following operations:

1. Composition to the left with a P_α
2. Taking weak limits of maps
3. Taking convex combinations of maps

Introduce an order $<<$ in $\mathcal{F}$ by putting $S_1 << S_2$ if $\|S_1 y - m\| \leq \|S_2 y - m\|$ for all $y \in B$ and all $m \in \bigcap_\alpha M_\alpha$. By Zorn's lemma there is a minimal element S in $\mathcal{F}$. We prove that S is a contractive projection onto

$$\bigcap_\alpha M_\alpha. \tag{19}$$

To prove (19) we observe that $P_\alpha \circ S << S$ for every α. Thus, if for some α, $P_\alpha \circ S \neq S$, then $\frac{1}{2}(S + P_\alpha \circ S) << S$ but since B is strictly convex the reverse relation does not hold. Thus $P_\alpha \circ S = S$ for all α. Thus $Sy \in \bigcap_\alpha M_\alpha$ for all $y \in B$. Moreover, since every P_α is the identity on $\bigcap_\alpha M_\alpha$, this is true for every element in $\mathcal{F}$. Thus $Sm = m$ for all $m \in \bigcap_\alpha M_\alpha$ and so S is a contractive projection onto $\bigcap_\alpha M_\alpha$. This completes the proof of the proposition.

It is easy to see that every $C_{E,+,x}$ and $C_{E,-,x}$ is contractive, the contractive projection begin just a truncation. This, together with Lemma 5 and Proposition 1 finishes the proof that 3) implies 2) in Theorem 1.

3. Proof that 1) implies 3) in Theorem 1.

In this section we will prove that an optimal subset of L^p is an intersection of cones of type $C_{f,T,x}$, $C_{E,+,x}$ and $C_{E,-,x}$. By the following Lemma 6 we reduce the problem to optimal cones. We first make a definition. Let K be a closed, convex set in a Banach space and let x be a support point of K. That is $x \in K$ and there is a supporting hyperplane at x, i.e. a functional f s.t. $f(y) \geq f(x)$ for all $y \in K$. Consider all $z \in B$ s.t. $x + tz \in K$ for some $t > 0$. then the cone generated by K at the support point x is the following set:

$$\overline{\{x + uz | u \geq 0, \text{ there is a } t > 0 \text{ s.t. } x + tz \in K\}}$$

We now have

Lemma 6. Assume that K is an optimal subset of L^p. Let K_x be the cone generated by K at a support point x. Then K_x is optimal.

Proof: Assume that K_x is not optimal and that $x+y$ is minimal with respect to K_x, $x+y \notin K_x$. Then, given $\delta > 0$, there is y_1 with $\|y_1 - y\| < \delta$ and a finite subset $M \subset K_x$ s.t. $x + y_1$ is minimal with respect to M [See [3], Th. 1 sec. 2]. Let $M = \{x + z_i | i = 1, 2, \ldots, n\}$

Since L^p is uniformly convex it is easy to see that we have y_2, with $\|y_2 - y_1\|, < \delta$ s.t. $x + y_2$ is minimal with respect to $\{x + z_i^1 | i = 1, 2, \ldots, n\}$ where for every i there is $t > 0$ s.t. $x + tz_i^1 \in K$. Then if $t > 0$ is small enough $x + ty_2$ is minimal with respect to $\{x + tz_i^1 | i = 1, 2, \ldots, n\}$. But if δ is small enough $x + ty_2 \notin K$ and this contradicts that K

is optimal. This proves the lemma.

Since $K = \bigcap_x K_x$ where x runs through the support points of K, it is enough to prove

$$1) \to 3) \text{ for optimal cones} \tag{20}$$

The next lemma is a simple approximation principle. However, it will lead to the important lemmas 9 and 10 which are basic tools for finding the set maps that give the $C_{f,\mathcal{T},0}$-cones that give the optimal cone, that we will consider.

Lemma 7. Assume $f_i \in L^p$, $1 \leq i \leq n$ and $g_i \in L_p$, $1 \leq i \leq n$. Assume $f_i(t) \leq g_i(t)$ for all $t \in [0,1]$ for all i. Assume $a_i \geq 0$, $\sum_{i=1}^n a_i = 1$. Assume that m_F is the minimal point of $\{f_i | 1 \leq i \leq n\}$ s.t.

$$\sum_{i=1}^n a_i sign(f_i - m_F)|f_i - m_F|^{p/q} = 0 \tag{21}$$

and that m_G is the minimal point of $\{g_i | 1 \leq i \leq n\}$ such that

$$\sum_{i=1}^n a_i sign(g_i - m_G)|g_i - m_G|^{p/q} = 0 \tag{22}$$

Then $m_F(t) \leq m_G(t)$ for all $t \in [0,1]$.

Proof: We observe that

$$a_i sign(f_i(t) - m_F(t))|f_i(t) - m_F(t)|^{p/q} \tag{23}$$

is an increasing function of $f_i(t)$ for fixed $m_F(t)$ and a decreasing function of $m_F(t)$ for fixed $f_i(t)$. From this the lemma immediately follows.

We introduce the notion

$$mx\{t_1, t_2\} = \begin{cases} t_1 & \text{if } |t_1| > |t_2| \\ t_2 & \text{if } |t_2| > |t_1| \\ \{t_1 t_2\} & \text{if } |t_1| = |t_2| \end{cases}$$

and

$$mn\{t_1, t_2\} = \begin{cases} t_1 & \text{if } |t_1| \leq |t_2| & \text{and } sign t_1 = sign t_2 \\ t_2 & \text{if } |t_2| \leq |t_1| & \text{and } sign t_1 = sign t_2 \\ 0 & \text{if } sign t_1 \neq sign t_2 & \end{cases}$$

The notation extends in an obvious way to more than 2 numbers. When we below write $mx\{f_1, f_2\}$ where f_1 and f_2 are functions we mean the whole set of measurable functions where for all those points t where $|f_1(t)| = |f_2(t)|$ we can choose either $f_1(t)$ or $f_2(t)$. If $\mathbf{x}$ and $\mathbf{y}$ are vectors in R^n $mx\{\mathbf{x}, \mathbf{y}\}$ denotes the set of vectors where the ith coordinate is $mx\{x_i, y_i\}$ and if $|x_i| = |y_i|$ either choice x_i or y_i is possible. The meaning of $mn\{f_1, f_2\}$ and $mn\{\mathbf{x}, \mathbf{y}\}$ is obvious. We have

Lemma 8. Assume that an optimal cone — with 0 as vertex — in ℓ_n^p contains the vectors $\mathbf{y}$ and $\mathbf{z}$. Then it also contains the vector $mn\{\mathbf{y}, \mathbf{z}\}$ and the set of vectors $mx\{\mathbf{y}, \mathbf{z}\}$.

Proof: We know that an optimal cone in ℓ_n^p is an intersection of half-spaces $ax_i + bx_j \leq 0$. [See [5]].

So to prove the lemma it is enough to prove that if $ay_i + by_j \leq 0$ and $az_i + bz_j \leq 0$ then also a $mx\{y_i, z_i\} + bmx\{y_j, z_j\} \leq 0$ and a $mn\{y_i, z_i\}+$ b $mn\{y_j, z_j\} \leq 0$. To do this we check the following cases: $a \geq 0$ and $b \geq 0$, $a \geq 0$ and $b \leq 0$, $a \leq 0$ and $b \leq 0$. We leave this routine check to the reader. Lemma 7 and Lemma 8 give

Lemma 9. Assume that an optimal cone in L^p — with 0 as vertex — contains the functions f_1 and f_2. Then it also contains the function $mn\{f_1, f_2\}$ and the set of functions $mx\{f_1, f_2\}$

Proof: The smallest optimal cone containing a given cone is obtained by repeating the operation of taking minimal hull a countable number of times and then taking the closure [See [4]]. Assume first that f_1 and f_2 are bounded.

Consider two sets of simple functions $\{g_1, g_2\}$ and $\{h_1, h_2\}$ s.t. $g_1 \leq f_1 \leq h_1$, and $g_2 \leq f_2 \leq h_2$. Now, we can think of g_1, g_2, h_1, h_2 as being in a finite-dimensional ℓ^p space. To get the optimal hulls of the 2-dimensional cones $\{a_1 g_1, a_2 g_2\}$ and $\{a_1 h_1, a_2 h_2\}$, $\alpha_1, \alpha \geq 0$, we repeat the min-operation a countable number of times. We say that $\min^n M$ is the set obtained from repeating the min-operation on M n times. Now by induction on n and Lemma 7 for every point $m_G^{(n)}$ in $\min^{(n)}\{\alpha_1 g_1, \alpha_2 g_2\}$ and the corresponding point $m_H^{(n)}$ in $\min^{(n)}\{\alpha_1 h_1, \alpha_2 h_2\}$ — "corresponding" meaning that the same linear combinations and same a_i : s are used at every stage — there is a point $m_F^{(n)}$ in $\min^{(n)}\{\alpha_1 f_1, \alpha_2 f_2\}$ s.t.

$$m_G^{(n)} \leq m_F^{(n)} \leq m_H^{(n)} \tag{24}$$

Now by Lemma 8 the result follows. The case of unbounded f_1 and f_2 follows immediately by approximation. The lemma is proved.

In the next lemma we assume that $E \subset [0,1]$, that $E \cup F = [0,1]$ and $E \cap F = \emptyset$.

Lemma 10. Assume that f is a support point of an optimal cone C — with 0 as vertex. Assume that the following holds: If $h|_E = f|_E$ and $h \in C$ then

$$\|h|_F\| \geq \|f|_F\|.$$

Then for every subset $F_1 \subset F$ the following holds: If $h^1|_E = f|_E$ and $h^1 \in C$ then

$$\|h^1|_{F_1}\| \geq \|f|_{F_1}\|.$$

Proof: Assume $\|h^1|_{F_1}\| < \|f|_{F_1}\|$. By Lemma 9 $mn\{h^1, f\} \in C$. But then $\|mn\{h^1, f\}|_F\| < \|f|_F\|$ which contradicts the assumption on f. This proves the lemma.

For the following lemmas we need to make some definitions. Assume that C is an optimal cone with 0 as vertex — and that $g \notin C$. We say that $E^1 \subset [0,1]$ is of type $(\alpha, 1, w, g)$ if there is an $h \in C$ s.t. $\|h\| \leq w\|g\|$ and s.t. $\alpha g(t) \leq h(t) \leq g(t)$ for those $t \in E^1$ where $g(t) \geq 0$ and $\alpha g(t) \geq h(t) \geq g(t)$ for those $t \in E^1$ where $g(t) \leq 0$. We have the simple

Lemma 11. Let E_n^1 be of type $(\alpha, 1, w, g)$ for every n and $E_n^1 \subset E_{n+1}^1$ for every n. Then $\bigcup_n E_n^1$ is of type $(\alpha, 1, w, g)$.

Proof: For every N consider an h_N that works for $\bigcup_{n=1}^{N} E_n^1$ and then take a weak limit.

From Lemma 11 we get that every set of type $(\alpha, 1, w, g)$ is contained in a maximal set of type

$$(\alpha, 1, w, g) \tag{25}$$

We also observe that, given $g \notin C$ and w there is an $\epsilon > 0$ s.t. [0,1] is not of type

$$(1 - \epsilon, 1, w, g) \tag{26}$$

We now can prove

Lemma 12. Let C be an optimal cone — with 0 as vertex — in L^p. Let $g \notin C$. Then there is a cone $C_{f,T,0}$ (or $C_{E,+,0}$ or $C_{E,-,0}$) s.t. $C \subset C_{f,T,0}$ (or $C_{E,+,0}$ or $C_{E,-,0}$) and $g \notin C_{f,T,0}$ (or $C_{E,+,0}$ or $C_{E,-,0}$).

Proof: By (25) and (26) we now consider an $\epsilon > 0$ and an $E^1 \subset [0,1]$ $m([0,1]\backslash E^1) > 0$ s.t. E^1 is a maximal set of type $(1-\epsilon, 1, 2, g)$. Now consider an $h_o \in C$ s.t. $h_o(t)$ is in the closed interval $[(1-\epsilon)g(t), g(t)]$ for every

$$t \in E^1 \tag{27}$$

With $f^1 = [0,1]\backslash E^1$, assume the

$$\|h_o|_{F^1}\| \tag{28}$$

is minimal given (27).

We consider the set F_1^1 where $\frac{h_o(t)}{g(t)} > 1$ and the set F_2^1 where $\frac{h_o(t)}{g(t)} > 1 - \epsilon$. We now will consider some different cases

Case 1. $m(F_1^1) > 0$, Then there is $F_1'' \subset F_1'$ with $m(F_1'') > 0$ and a $\delta > 0$ s.t. $\frac{h_o(t)}{g(t)} > 1 + \delta$ on F_1''.

We now consider $E_+^1 = \{t | t \in E^1, g(t) > 0\}$

$$E_o^1 = \{t | t \in E^1, g(t) = 0\}$$

and

$$E_-^1 = \{t | t \in E^1, g(t) < 0\}.$$

We define h_{o+} by $(1-\epsilon)g \leq h_{o+} \leq g$ on E_+^1 and, given that condition, h_{o+} has minimal norm on F_1''. h_{o-} is defined by $(1-\epsilon)g \geq h_{o-} \geq g$ on E_-^1 and, given that condition, h_{o-} has minimal norm on F_1''. Now, by applying Lemma 9 we easily see that a.e. on F_1'' we either have $\frac{h_{o+}(t)}{g(t)} \geq 1 + \delta$ or $\frac{h_{o-}(t)}{g(t)} \geq 1 + \delta$. We assume WLOG that $\frac{h_{o+}(t)}{g(t)} > 1 + \delta$. on $F_1''' \subset F_1''$ where $m(F_1''') > 0$. Now let $f = h_{o+}$ on E_+^1 and given this assume that f has minimal norm on $[0,1]$. By Lemma 10 we have $f = h_{o+}$ on F_1'''. Now, put $E = E_+^1$ and $F = [0,1]\backslash E$. For every subset $E_1 \subset E$ consider the function g_{E_1} in C that is $\equiv f$ on E_1 and, given that, has minimal norm on F. In an obvious way this defines a map $\mathcal{T}$ from subsets of E to G_g : s, where for every $E_1 \subset E$ $g_{E_1} = g_{\mathcal{T}(E_1)}$.

We now show $\mathcal{T}(E_1 \cup E_2) = \mathcal{T}(E_1) \cup \mathcal{T}(E_2)$.

Let $f_1 = f$ on E_1 and $f_2 = f$ on E_2 and, given these conditions, let f_1 and f_2 have minimal norm on F. We can assume $|f_i| \le |f|$ and a.e. on E either $f_i(t) = 0$ or $\text{sign} f_i(t) = \text{sign} f(t)$ $i = 1, 2$. Otherwise, instead of f_i, consider

$$mn(f_i, f), \ i = 1, 2 \tag{29}$$

We now show that a.e. on F, either

$$f_1(t) \text{ or } f_2(t) \text{ is } 0 \text{ or } \text{sign} f_1(t) = \text{sign} f_2(t) \tag{30}$$

To prove (30), assume on the contrary that $\text{sign} f_1(t) \neq \text{sign} f_2(t)$ for all $t \in F^{(1)}$, $F^{(1)} \subset F$, $m(F^{(1)}) > 0$. WLOG, we assume

$$|f_1| \ge |f_2| > 0 \text{ on } F^{(1)} \tag{31}$$

Consider now $mx\{f_1, f_2\}$ where by (29) we can choose

$$mx\{f_1, f_2\} = f \text{ on } E_1 \bigcup E_2 \tag{32}$$

We can also assume

$$mx\{f_1, f_2\} = f_1 \text{ on } F^{(1)} \tag{33}$$

We now consider $mn\{f_2, mx\{f_1, f_2\}\} \in C$. This is equal to

$$f_2 \text{ on } E_2 \text{ and is equal to } 0 \text{ on } F^{(1)} \tag{34}$$

But, since f_2 has minimal norm on F_1 it has, by Lemma 10, minimal norm on $F^{(1)}$. By (31)

$$\|f_2|_{F^{(1)}}\| > 0. \tag{35}$$

Thus (34) contradicts (35) and so (30) is proved. From (30) and Lemma 10 it follows that $T(E_1 \bigcup E_2) \supset T(E_i)$, $i = 1, 2$. And by considering $mx\{f_1, f_2\}$ we see that $T(E_1 \bigcup E_2) = T(E_1) \bigcup T(E_2)$.

Since C is closed we immediately get that T is left continuous.

We also see that, by its definition, $C \subset C_{f,T,0}$ and by considering E and F_1''' we see that $g \notin C_{f,T,0}$.

Case 2. $m(F_1') = 0$. Then $\frac{h_o(t)}{g(t)} < 1 - \epsilon$ a.e. on F'. Now some cases can occur

Case 2A. For every $h \in C$ $sign\ h(t) \neq sign g(t)$ a.e. on F'. Then either $C \subset C_{F',+,0}$ (or $C_{F',-,0}$) but $g \notin C_{F',+,0}$) (or $g \notin C_{F',-,0}$) and so the lemma is proved.

Case 2B. There is a subset F_2'' of $F_2' = F'$ of type $(1-\epsilon, 1, 2, g)$, $m(F_2'') > 0$. By Lemma 11 we extend this to a maximal subset F_2^o of type $(1-\epsilon, 1, 2, g)$. Since E' is also a maximal subset of type $(1-\epsilon, 1, 2, g)$ we know $E' \not\subset F_2^o$. Now exactly as we did for E' we consider F_{2+}^o, F_{2o}^o and F_{2-}^o and we consider h_{o+}^1 and h_{o-}^1. Now for h_{o+}^1 or h_{o-}^1, say h_{o+}^1, Case 1 must occur. Otherwise we could consider $mx\{h_{o+}, h_{o-}, h_{o+}^1\}$ which would contradict that E^1 is maximal of type $(1-\epsilon, 1, 2, g)$. Then we put $E = F_{2+}^o$ and as above in Case 1 we find f and T s.t. $g \notin C_{f,T,0}$ but $C \subset C_{f,T,0}$. This finishes the proof of Lemma 12.

From Lemma 12 it immediately follows that an optimal cone C — with 0 as vertex — is equal to the intersection of all cones $C_{f,T,0}, C_{E,+,0}$ and $C_{E,-,0}$ that contain C. And so 1) → 3) in Theorem 1, and so the proof of Theorem 1 is complete.

Remark. By doing the argument of this paper in ℓ^p, we get the following improvement of the main result in [5].

Theorem: For subsets $M \subset \ell^p, 1 < p < \infty, p \neq 2$ the following are equivalent.

1. M is optimal
2. M is contractive
3. M is an intersection of half-spaces of the form $ax_i + bx_j \leq c$.

References

1. Ando, T.: Contractive projections in L_p-spaces, Pacific J. Math. 17(1966), 391-405.

2. Beauzamy, B.: Projections contractantes dans les espaces de Banach, Bull. Sc. Math. 2° série 102(1978), 43-47.

3. Beauzamy, B., Enflo, P.: Theoremes des point fixe et d'approximation, Ark. Mat. Vol 23 no. 1 (1985), 19-34.

4. Beauzamy, B., Maurey, B.: Points minimaux et ensembles optimaux dans les espaces de Banach, J. Funct. Anal. 24(1977), 107-139.

5. Davis W., Enflo P: Contracive projections on ℓ_p spaces, London Math. Soc. Lecture Note Series 137 (1989) p. 151-161.

6. Enflo P.: Contractive projections onto subsets of $L^1(0,1)$, London Math. Soc. Lecture Note Series 137(1989), 162-184.

7. Goodner, D.B.: Projections in normed linear spaces, Trans. Amer. Math. Soc. 69 (1950), 89-108.

8. Hasumi, M.: The extension property of complex Banach spaces, Tohoku Math, J.(2) 10(1958), 135-142.

9. Kelley J.: Banach spaces with the extension property, Trans. Amer. Math. Soc. 72 (1952) 323-326.

10. Nachbin L.: A. theorem of the Hahn-Banach type for linear transformations, Trans. Amer. Math. S 68 (1950), 28-46.

11. Westphal U.: Cosuns in $\ell^p(n)$, J. of Approximation Theory.

The Largest Coefficient in Products of Polynomials

PER ENFLO Department of Mathematics & Computer Science, Kent State University, Kent, Ohio

1. Introduction

It is a classical result by Gelfond (see [4]) that if P and Q are polynomials of degrees m and n, then

$$|PQ|_\infty \geq e^{-(m+n)}|P|_\infty|Q|_\infty$$

where

$$|\sum_{j=0}^{N} a_j z^j|_\infty = \max_{0\leq j\leq N} |a_j| \tag{1}$$

In this paper we will prove inequalities of the same type as (1) but with more general assumptions on P and Q. These more general assumptions will, in Theorems 1 and 2, be, that one of the factors, say Q, has some concentration up to degree n and the other factor P has one coefficient which is large compared to the sum of the moduli of the coefficients. A motivation for considering these assumptions comes from an operator theoretic viewpoint. P is then thought of as a polynomial in the shift operator operating on an element Q. Under certain conditions, large norms of products is connected to cyclic vectors of the shift operator and small norms of products to invariant subspaces of the shift operator (see Enflo [3]). We want to know how P operates not only on polynomials of degree n but also on neighborhoods of these polynomials. In every norm under which the coefficient functionals are continuous the polynomials near to a polynomial of degree n have some concentration up to degree n,

Supported in part by NSF grant. #440702

although they may have terms of higher degree. In Theorem 3 we assume that both P and Q have some concentration up to degrees m and n. Similar assumptions on P and Q as in this paper were used in Beauzamy-Enflo [1], (See also [2]). Theorem 2 below is in fact, an improvement of Theorem 1 in [1]. The methods used in this paper combine results from [1] with new techniques.

We first introduce some notation. If $P(z) = \sum_{-\infty}^{\infty} a_j z^j$ we put $\|P\|_1 = \int_0^{2\pi} |P(e^{i\theta})| \frac{d\theta}{2\pi}$, $\|P\|_2 = (\int_0^{2\pi} |P(e^{i\theta})|^2 \frac{d\theta}{2\pi})^{1/2}$, $|P|_1 = \sum |a_j|$, and more generally $|P|_p = (\sum |a_j|^p)^{1/p}$ for $1 \leq p < \infty$, $|P|_\infty = \sup |a_j|$. We have $|P|_\infty \leq \|P\|_1 \leq \|P\|_2 = |P|_2 \leq |P|_1$. Also, $|P|_p$ decreases with p. If $Q(z) = \sum_{j \geq 0} a_j z^j$ we put $Q|^k = \sum_{j=0}^{k} a_j z^j$ and $Q|_k = a_k z^k$. We put $Q|_E = \sum_{j \in E} a_j z^j$.

2. The Results.

We first prove

Theorem 1. *Assume* $|P|_\infty \geq \delta |P|_1$ *and* $|Q|^n|_p \geq \delta' |Q|_p$, $1 \leq p < 2$. *Then there is a* $d = d(\delta, \delta', p) > 0$ *such that*

$$|PQ|_\infty \geq d^n |P|_1 |Q|_p.$$

Remark. In this result, the largest coefficient of PQ decays exponentially with n like in Gelfond's result. However, the d in d^n depends on δ, δ' and p. We do not know whether there exists a number $\lambda = \lambda(\delta, \delta', p)$ and an absolute constant A such that $|PQ|_\infty \geq \lambda A^n |P|_1 |Q|_p$.

Proof of Theorem 1. We start with the simple

Lemma 1. Consider $S(z) = \sum_{-\infty}^{\infty} a_j z^j$ with $|S|_p = 1$, $1 \leq p < 2$. then, given $\epsilon > 0$, we have

$$\Big| \sum_{|a_j| \leq \epsilon} a_j z^j \Big|_2 \leq \epsilon^{1-\frac{p}{2}}$$

Proof: We have $\sum_{|a_j| \leq \epsilon} |a_j|^p \leq 1$ and since $|a_j|^2 \leq \epsilon^{2-p} |a_j|^p$ if $|a^j| \leq \epsilon$ we get $\sum_{|a_j| \leq \epsilon} |a^j|^2 \leq \epsilon^{2-p}$

and

$$\left(\sum_{|a_j|\le\epsilon}|a_j|^2\right)^{1/2}\le\epsilon^{1-\frac{p}{2}}.$$

We now use Corollary 9 in [1] which states: Assume $\|P\|_2>\delta_1\|P\|_\infty$ and $\|Q|^n\|_2\ge\delta_1'\|Q\|_2$. Then with $\alpha=\delta_1{\delta_1'}^{8/\delta_1^2}/(e^{5/2}9^{n+1})^{4/\delta_1^2}$ one has $|PQ|_1\ge\alpha\|P\|_2\|Q\|_2$. Now, if $|P|_\infty\ge\delta|P|_1$, then $\|P\|_2\ge|P|_\infty\ge\delta|P|_1\ge\delta\|P\|_\infty$. And if $|Q|^n|_p\ge\delta'|Q|_p$, then

$$|Q|^n|_2\ge n^{\frac{1}{2}-\frac{1}{p}}|Q|^n|_p\ge\delta' n^{\frac{1}{2}-\frac{1}{p}}|Q|^p\ge$$

$$\ge\delta' n^{\frac{1}{2}-\frac{1}{p}}|Q|_2.$$

So if P and Q fulfill the conditions of Theorem 1, they fulfill the conditions of Corollary 9 in [2] with $\delta_1=\delta$ and $\delta_1'=\delta' n^{\frac{1}{2}-\frac{1}{p}}$. Since we also have

$$|Q|_p\ge\frac{1}{\delta'}|Q|^n|_p\le n^{\frac{1}{p}-\frac{1}{2}}\cdot\frac{1}{\delta'}|Q|^n|_2\le$$

$$\le n^{\frac{1}{p}-\frac{1}{2}}\cdot\frac{1}{\delta'}|Q|_2$$

We get the following

Lemma 2. Under the assumptions of Theorem 1, there is a $d_1=d_1(\delta,\delta',p)$ s.t.

$$|PQ|_2\ge d_1^m|P|_1|Q|_p.$$

We now assume P and Q as in the theorem with $|P|_1=|Q|_p=1$. Put $Q(z)=\sum_{j\ge0}a_jz^j$. We use Lemma 1 on Q with $\epsilon=d^{\frac{2n}{1-p/2}}$ where d_1 is as in Lemma 2. We assume *WLOG* $d_1\le\frac{1}{10}$. We put $Q_0(z)=\sum_{|a_j|\ge\epsilon}a_jz^j$. Then by Lemma 1

$$|Q-Q_0|_2\le d_1^{2n}$$

and so

$$|PQ-PQ_0|_2\le d_1^{2n} \tag{2}$$

We also have

$$\text{card}\,\{j|\ \ |a_j|\ge\epsilon\}\le\frac{1}{\epsilon^p}=\frac{1}{d_1^{\frac{2pn}{1-p/2}}} \tag{3}$$

(3) gives $|Q_0|_1 \le \frac{1}{d_1^{\frac{2pn}{1-p/2}}} = D^n$ with

$$D = \frac{1}{d_1^{\frac{2p}{1-p/2}}} \tag{4}$$

Now we use Lemma 1 on $P(z) = \sum_{-\infty}^{\infty} b_j z^j$ with $p = 1$ and $\epsilon_1 = d^{\frac{2n}{1-1/2}}/D^{\frac{n}{1-1/2}}$ and put $P_0(z) = \sum_{|b_j| \ge \epsilon_1} b_j z^j$. We get $|P - P_0|_2 \le d_1^{2n}/D^n$ and so by (4)

$$|PQ_0 - P_0Q_0|_2 \le d_1^{2n} \tag{5}$$

We also get

$$\text{card}\,\{j|\ \ |b_j| \ge \epsilon_1\} = \frac{D^{2n}}{d_1^{4n}} \tag{6}$$

Now (2) and (5) give

$$|PQ - P_0Q_0|_2 \le 2d_1^{2n} \tag{7}$$

Thus by Lemma 2

$$|P_0Q_0|_2 \ge d_1^n - 2d_1^{2n} > \frac{3}{4}d_1^n \tag{8}$$

Put $PQ(z) = \sum_j c_j z^j$ and $P_0Q_0(z) = \sum_j c'_j z^j$. Put $E = \{j|c'_j \ne 0\}$. We get from (3) and (6)

$$\text{card}\, E \le \frac{1}{d_1^{\frac{2pn}{1-p/2}}} \cdot \frac{D^{2n}}{d_1^{4n}} = D_1^{2n} \tag{9}$$

where D_1 is defined by (9).

We get from (7) and (8) $|\sum_{j\in E} c_j z^j|_2 > \frac{3}{4}d_1^n - 2d_1^{2n} > \frac{1}{2}d_1^n$. By (9) this gives $\max_{j\in E}|c_j| > \frac{1}{2}d_1^n/D_1^n$. With $d = \frac{d_1}{2D_1}$ this gives the theorem.

The proof of Theorem 1 breaks down for $p = 2$, since then Lemma 1 breaks down. We now consider the case $p = 2$. This is treated also in [1] (Theorem 1 in that paper). A careful analysis of the argument [1] shows that it gives a $C = C(\delta, \delta') > 0$ such that $|PQ|_\infty \ge \frac{1}{C^{(C^{(C^n)})}}|P|_1|Q|_2$. In this paper we will improve this to "double exponential", we have

Theorem 2. *Assume $|P|_\infty \ge \delta|P|_1$ and $|Q|^n|_2 \ge \delta'|Q|_2$. Then there is a $C = C(\delta, \delta')$ such that $|PQ|_\infty \ge \frac{1}{C^{(C^n)}}|P|_1|Q|_2$.*

Proof. Put supp $(\sum_{-\infty}^{\infty} a_j z^j) = \{j | a_j \neq 0\}$. We have the following

Lemma 3. Put $S(z) = \sum_{-\infty}^{\infty} a_j z^j$ and $T(z) = \sum_{-\infty}^{\infty} b_j z^j$. Assume $|ST|_2 = A|S|_1|T|_2$, $0 \leq A \leq 1$. Then there is an m such that

$$|ST|_{\text{supp } x^m T(z)}|_2 \geq A^2|S|_1|T|_2 > A^2|S|_1|T|_2.$$

Remark. This lemma provides a tool to estimate the size of the coefficients in ST, since it says that if ST is large, then it is large already on a translate of T.

Proof of Lemma 3. We assume $|S|_1 = |T|_2 = 1$. We put $< \sum_{-\infty}^{\infty} c_j z^j, \sum_{-\infty}^{\infty} d_j z^j >= \sum_{-\infty}^{\infty} c_j \bar{d}_j$. We get $A^2 =< ST, ST >=< ST, \sum_{-\infty}^{\infty} a_j z^j T >= \sum_j \bar{a}_j < ST, z^j T >$. Since $\sum |a_j| = 1$, this gives that there is an m for which $| < ST, z^m T > | \geq A^2$. This gives

$$|ST|_{\text{supp } z^m T(z)}|_2 = |ST|_{\text{supp } z^m T(z)}|_2 |z^m T|_2$$

$$\geq | < ST|_{\text{supp } z^m T(z)}, z^m T > | = | < ST, z^m T > | \geq A^2.$$

This proves the lemma.

We also need the following lemma, which is - like Lemma 2 - a simple consequence of Corollary 9 in [1].

Lemma 4. Under the assumptions of Theorem 2, there is a $d_2 = d_2(\delta, \delta') > 0$, $1 > d_2$, such that $|PQ|_2 \geq \frac{1}{\delta'} d_2^n |P|_1 |Q|_2$.

We now put $P(z) = \sum_{-\infty}^{\infty} a_j z^j$ and $Q(z) = \sum_{j \geq 0} b_j z^j$. We put $Q_0 = Q|^n$, $E_0 = \text{supp } Q_0$ and $M_0 = \text{ card } \{\text{supp } Q_0\} \leq n + 1$. We assume $|P|_1 = 1$ $|Q_0|_2 = \delta'$ which gives $|Q|_2 \leq 1$. Then P and Q_0 satisfy the assumptions of Theorem 2 and so by Lemma 4 $|PQ_0|_2 \geq d_2^n$. Thus by Lemma 3 there is an m such that

$$|PQ_0|_{\text{supp } z^m Q_0}|_2 \geq d_2^{2n} \tag{10}$$

Now 2 cases can occur

Case 0)

$$|PQ|_{\text{supp } z^m Q_0}|_2 \geq \frac{1}{2} d_2^{2n}.$$

Since card $\{\text{supp } z^m Q_0\} \leq n+1$ the theorem is proved in this case

Case 1)

$$|PQ|_{\text{supp } z^m Q_0}|_2 < \frac{1}{2} d_2^{2n}. \tag{11}$$

If Case 1 occurs we observe the following. Put $Q_1' = \sum b_j z^j$ where the summation is extended over those j in supp $(Q - Q_0)$ for which $|b_j| < \frac{1}{4}\frac{d_2^{2n}}{\sqrt{M_0}}$. Then $|PQ_1'|_\infty \leq |P|_1 |Q_1'|_\infty \leq \frac{1}{4}\frac{d_2^{2n}}{\sqrt{M_0}}$ and so

$$|PQ_1'|_{\text{supp } z^m Q_0}|_2 \leq \sqrt{M_0} \cdot \frac{1}{4}\frac{d_2^{2n}}{\sqrt{M_0}} = \frac{1}{4} d_2^{2n} \tag{12}$$

Now put $Q_1 = \sum_{j \in E_1} b_j z^j$ where E_1 consists of those j in supp $(Q - Q_0)$ for which

$$|b_j| \geq \frac{1}{4}\frac{d_2^{2n}}{\sqrt{M_0}} \tag{13}$$

(10),(11) and (12) give $|PQ_1|_2 > \frac{1}{4} d_2^{2n}$ and so

$$|Q_1|_2 > \frac{1}{4} d_2^{2n} \tag{14}$$

Since $|Q_1|_2 < 1$ (13) gives

$$\text{card } \{\text{supp } Q_1\} \leq \frac{16 M_0}{d_2^{4n}}$$

Put $M_1 =$ card$\{\text{supp } (Q_0 + Q_1)\}$. We get

$$M_1 \leq M_0 \left(1 + \frac{16}{d_2^{4n}}\right) \tag{15}$$

We now assume that it is defined that the case $\overbrace{11\cdots 1}^{r\ 1:s}$ occurs. We assume that we then have defined $Q_0, Q_1, \cdots, Q_r$ where $Q_i = \sum_{j \in E_j} b_j z^j$ for $0 \leq i \leq r$, $E_i \cap E_j = \phi$ if $i \neq j$, with

$$|Q_i|_2 > \frac{1}{4} d_2^{2n} \text{ for } 0 \leq i \leq r \tag{14'}$$

We put card$\{\text{supp } (Q_0 + Q_1 + \cdots + Q_r)\} = M_r$

We have that P and $(Q_0 + Q_1 + \cdots + Q_r)$ satisfy the assumptions of Theorem 2 and so by Lemma 4 $|P(Q_0 + Q_1 + \cdots + Q_r)|_2 \geq d_2^n$. Thus by Lemma 3 there is an m such that

$$|P(Q_0 + Q_1 + \cdots + Q_r)|_{\text{supp } z^m (Q_0 + Q_1 + \cdots + Q_r)}|_2 \geq d_2^n. \tag{10''}$$

Now 2 cases can occur

Case $\overbrace{11\cdots 10}^{r\,1:s}$

$$|PQ|_{\text{supp } z^m(Q_0+Q_1+\cdots+Q_r)|_2} \geq \frac{1}{2}d_2^{2n}.$$

Since card{supp $z^m(Q_0 + Q_1 + \cdots + Q_r)\} = M_r$ the estimates on r and M_r given below will show that the theorem is proved in this case. (16)

Case $\overbrace{11\cdots 11}^{r+1\,1:s}$

$$|PQ|_{\text{supp } z^m(Q_0+Q_1+\cdots+Q_r)|_2} < \frac{1}{2}d_2^{n}. \tag{11''}$$

In this case we observe the following. Put $Q'_{r+1} = \sum b_j z^j$ where the summation is extended over those j in supp $(Q - (Q_0 + Q_1 + \cdots + Q_r))$ for which $|b_j| < \frac{1}{4}\frac{d_2^{2n}}{\sqrt{M_r}}$. Then $|PQ'_{r+1}|_\infty \leq |P|_1|Q'_{r+1}|_\infty \leq \frac{1}{4}\frac{d_2^{2n}}{\sqrt{M_r}}$ and so

$$|PQ'_{r+1}| \text{ supp } z^m(Q_0 + Q_1 + \cdots + Q_r)|_2 \leq \sqrt{M_r}\cdot\frac{1}{4}\frac{d_2^{2n}}{\sqrt{M_r}} = \frac{1}{4}d_2^{2n} \tag{12''}$$

Now put $Q_{r+1} = \sum_{j\in E_{r+1}} b_j z^j$ where E_{r+1} consists of those j in supp $(Q-(Q_0+Q_1+\cdots+Q_r))$ for which

$$|b_j| \geq \frac{1}{4}\frac{d_2^{2n}}{\sqrt{M_r}} \tag{13''}$$

(10"),(11") and (12") give $|PQ_{r+1}|_2 > \frac{1}{4}d_2^{2n}$ and so

$$|Q_{r+1}|_2 > \frac{1}{4}d_2^{2n} \tag{14''}$$

Since $|Q_{r+1}|_2 < 1$ (13") gives card{supp $(Q_{r+1}\} \leq \frac{16M_r}{d_2^{4n}}$.

Put card {supp $(Q_0 + Q_1 + \cdots + Q_{r+1}\} = M_{r+1}$. We get

$$M_{r+1} \leq M_r(1 + \frac{16}{d_2^{4n}}) \tag{15''}$$

We can now easily complete the proof of Theorem 2. We have $1 \geq |Q|_2 \geq |Q_0 + Q_1 + \cdots + Q_r|_2 \geq \sqrt{r+1}\cdot\frac{1}{4}d_2^{2n}$, where the last inequality follows from (14') and the fact that the Q_i's are disjointly supported. This gives $r \leq \frac{16}{d_2^{4n}} - 1$ and so for some $r \leq \frac{16}{d^{4n}}$ in case $\overbrace{11\cdots 10}^{r\,1:s}$ must occur. (17)

We get from (15) and (15") that

$$M_r \leq (n+1)\left(1 + \frac{16}{d_2^{4n}}\right)^r \tag{18}$$

Now Theorem 2 follows immediately from (16), (17) and (18).

We finally consider the situation when both P and Q have concentration on degrees up to m and n. The simple argument below gives exponential decay in m and n. Put $P = \sum_{j\geq 0} a_j z^j$ and $Q = \sum_{j\geq 0} b_j z^j$.

Theorem 3. Assume $|P|^m|_\infty \geq \delta |P|_\infty$ and $|Q|^n|_\infty \geq \delta' |Q|_\infty$. Then there is a $C = C(\delta, \delta')$ such that

$$|PQ|_\infty \geq \frac{C}{256^{3(m+n)}} |P|_\infty |Q|_\infty.$$

Proof. We consider

$$P'(z) = P(\frac{z}{2}) \text{ and } Q'(z) = Q(\frac{z}{2}).$$

We assume $|P|^m|_\infty = |Q|^n|_\infty = 1$ and this gives $|P'|^m|_2 \geq \frac{1}{2^m}$ and $|Q'|^n|_2 \geq \frac{1}{2^n}$. Moreover we get $|P'|_2 \leq \frac{2}{\delta}$ and $|Q'|_2 \leq \frac{2}{\delta'}$. It is now a simple consequence of Corollary 7 in [1] that

$$|P'Q'|_2 \geq \frac{1}{4} \frac{1}{2^m} \cdot \frac{1}{2^n} \frac{\delta^6 (\delta')^6}{(2^{m+1})^6 (2^{n+1})^6 \cdot e^{15} \cdot 9^{3m+3n+6}} \tag{19}$$

With $P(z) = \sum a_j z^j$, $Q(z) = \sum b_j z^j$ and $PQ(z) = \sum c_j z^j$, we get by the assumptions above $|a_j| \leq \frac{1}{\delta}$ $|b_j| \leq \delta^1$, and so with $c_j = \sum_{m=0}^{j} a_{j-m} b_m$ we have $|c_j| \leq \frac{j+1}{\delta\delta'}$. Thus

$$|P'Q' - P'Q'|^r|_2 = |\sum_{j=r+1}^{\infty} \frac{c_j^2}{4^j}|^{1/2} \leq$$

$$\leq \frac{1}{\delta\delta'} \left(\sum_{r+1}^{\infty} \frac{(j+1)^2}{4^j} \right)^{1/2} \leq \frac{r}{\delta\delta'} \cdot \frac{1}{2^{r-1}} \text{ if}$$

$$r \geq 10. \tag{20}$$

(19) and (20) give that there is a number $s = s(\delta, \delta')$ such that if $r = geq24(m+n) + s(\delta, \delta')$ then $|(P'Q')|^r|_2 > \frac{1}{2}||(P'Q')|_2$. Now (19) combined with the fact $|PQ|^r|_\infty \geq |(P'Q')|^r_\infty$ gives the result.

References

1. B. Beauzamy et P. Enflo, *Estimations de produits de polynômes*, J. of Number Theory 21, 390-412, (1985).

2. B. Beauzamy, E. Bombieri, P. Enflo, H. Montgomery, *Products of polynomials in many variables*, J. of Number Theory, Vol. 36 No. 2 (1990), 219-245.

3. P. Enflo, *On the invariant subspace problem in Banach spaces*, Acta Math. 158 (1987) 213-313.

4. M. Waldschmidt, *Nombres Transcendants*, Lecture Notes in Math. Vol. 402, Springer-Verlag, Berlin/New York 1974.

New Results on Regular and Irregular Sampling Based on Wiener Amalgams

HANS G. FEICHTINGER, University of Vienna, Department of Mathematics Strudlhofgasse 4, A-1090 Wien, AUSTRIA

1 INTRODUCTION

This paper should be seen as a companion to the article [F7] "WIENER AMALGAMS OVER EUCLIDEAN SPACES AND SOME OF THEIR APPLICATIONS" published in this issue. We are going to show how Wiener amalgam spaces (and not just ordinary amalgam spaces) can be used to derive results of interest in sampling theory. In this sense it represents a continuation of a series of papers [F8], [FG4–6] on the theoretical background of the irregular sampling problem for band-limited functions.

In view of the close connection between the two articles we will keep the same notations and have decided to avoid duplication in the list of references. Thus references which are *not* listed in the bibliography of the present note will be found in the previous note.

2 VARIATIONS ON THE SAMPLING THEOREM, L^p-ERROR ANALYSIS

One of the corner-stones of digital signal analysis is the so-called sampling theorem, according to which a band-limited signal can be completely reconstructed from the sampling values taken at any sufficient fine lattice. In fact, the critical rate, also known as the Nyquist rate, is inversely proportional to the bandwidth of the signal ($(2\cdot$maximal frequency$)^{-1}$ in the usual engineering terminology). Usually this result is presented as a Hilbert space result. Using Plancherel's theorem and Poisson's formula it can be verified that the classical Shannon sampling theorem is indeed equivalent to the Fourier series expansion of periodic functions, a result which is the prototype of the concept of general orthogonal expansion in a Hilbert space (cf. [Br], [Pa2], [LO], and [Bu]; [Je] or [Ma1] for surveys).

We will describe the setting in a function space terminology and show how statements about the sampling series (and later results on irregular sampling) can be derived by arguments based on the use of Wiener amalgam spaces. Most of the results in this section are new and cover limiting cases or variants of results given in a series of papers on the irregular sampling problem for band-limited functions [FG4-6].

Definition 1. A tempered distribution $\sigma \in \mathcal{S}(\mathbb{R}^m)$ is called *band-limited* with *spectrum* Ω if the (generalized) Fourier transform $\hat{\sigma}$ vanishes on the complement of the closed, bounded set $\Omega \subseteq \mathbb{R}^m$. For integrable functions f the statement spec$(f) \subseteq \Omega$ simply means that $\hat{f}(s) = 0$ for $s \notin \Omega$.

It is a standard result due to Paley and Wiener, and, in its most general version, to Schwartz, that band-limited tempered distributions are represented by analytic, hence continuous and differentiable (ordinary) functions. Therefore single function values are well defined for band-limited functions in L^p-spaces, for $1 \leq p \leq \infty$. However, we shall not base our arguments on this fact, but use the following result instead (this is a special case of Thm.5 of [F1] and holds for lca. groups).

Lemma 1. Let a compact subset Ω of $\mathbb{R}^m$ and $\alpha \geq 0$ be given. Then there exists some constant C_Ω (only depending on Ω and α) such that for all $p \geq 1$

$$\|f\,|W(C^0, L^p_w)\| \leq C_\Omega \cdot \|f\|_{L^p_w} \tag{1}$$

for $f \in L^p_w(\mathbb{R}^m)$ with spec$(f) \subseteq \Omega$, and all weights with $L^1_\alpha * L^p_w \subseteq L^p_w$.

Proof. We choose an arbitrary $h \in \mathcal{S}(\mathbb{R}^m) \subseteq W(C^0, L^1_\alpha)$ satisfying $\hat{h}(\omega) \equiv 1$ on Ω. Then $f = h * f$, and by the convolution relations for amalgams

$$\|f\,|W(C^0, L^p_w)\| \leq C_1 \|h\,|W(C^0, L^1_\alpha)\| \cdot \|f\,|W(L^1, L^p_w)\| \leq C_\Omega \cdot \|f\|_{L^p_w} .$$

□

Although the following result is true (by almost the same argument) for lca. groups, we present it for simplicity in the setting of $\mathbb{R}^m$. The product of multi-indices such as $\boldsymbol{an}$ is to be understood as $(a_1 n_1, \ldots, a_m n_m)$.

Theorem 2 (*(Weighted L^p-version of the Classical Sampling Theorem)*. Given a compact set $\Omega \subseteq \mathbb{R}^m$ and a band-limited function $h \in L^1_s(\mathbb{R})$ satisfying $\hat{h}(t) = 1$ on Ω there exists $\boldsymbol{c} = (c_1, \ldots c_m)$, $c_i > 0$ (depending only on h) such that for any $\boldsymbol{a} \leq \boldsymbol{c}$ (coordinatewise) one has:

Any band-limited function $f \in L^p_w(\mathbb{R}^m)$ with spec$(f) = $ supp$(\hat{f}) \subseteq \Omega$ can be reconstructed from the sampling values over the lattice $(\boldsymbol{an})_{n \in Z^m}$ by means of the *cardinal series*

$$f = \sum_{n \in Z^m} \boldsymbol{a}^{-1} f(\boldsymbol{an}) T_{\boldsymbol{an}} h \tag{2}$$

Unconditional convergence of the series takes place in the $W(C^0, L^p_w)$-norm, hence in L^p_w as well as uniformly over compact subsets of $\mathbb{R}^m$ for $1 \leq p < \infty$.

Proof. We shall use the symbol III for the so-called 'shah-distribution' III $=$ III$_1$, given by

$$\text{III}_{\boldsymbol{a}} := \sum_{n \in Z^m} \delta_{\boldsymbol{an}} \ .$$

It is clear that $\text{III}_{\boldsymbol{a}} \in W(M, L^\infty)$ for each $\boldsymbol{a}$, actually $\boldsymbol{a} \cdot \text{III}_{\boldsymbol{a}} := \prod_{j=1}^m a_j \cdot \text{III}_{\boldsymbol{a}}$ is uniformly bounded in $W(M, L^\infty)$. Since f is continuous by Lemma 1, $f \cdot \text{III}_{\boldsymbol{a}}$ is well defined as a discrete Radon measure, but the pointwise multiplier result for amalgams gives more: $f \cdot \boldsymbol{a}\text{III}_{\boldsymbol{a}} \in$

$W(C^0, L^p_w) \cdot W(M, L^\infty) \subseteq W(M, L^p_w)$, and even uniform bounded in $W(M, L^p_w)$ with respect to $\boldsymbol{a}$.

Interpretation of (2) in the distribution theoretic sense shows that we have to verify

$$(\boldsymbol{c}f \cdot \text{III}_c) * h = f \ . \tag{3}$$

Given Poisson's formula in the form $(\boldsymbol{a}\text{III}_a)^\wedge = \text{III}_b$ for $\boldsymbol{b} = \boldsymbol{a}^{-1}$ we may rewrite (applying the usual rules for the Fourier transform on $\mathcal{S}'(I\!R^m)$) that this conditions is equivalent to $(\hat{f} * \text{III}_b) \cdot \hat{h} = \hat{f}$. Drawing a picture of the compactly supported function $\hat{f}$ and its β-periodic extension $\hat{f} * \text{III}_b$ the reader will immediately verify that the given plateau-condition allows to find $\boldsymbol{c}$ such that $\boldsymbol{b} = \boldsymbol{c}^{-1}$ is large enough for the formula to hold (Ω has to fit into a rectangle of the form $[c_1^{-1}, \dots, c_m^{-1}]$).

In order to check convergence let us observe that Lemma 1 implies convergence of $\boldsymbol{a} \cdot \sum_{|n|\le k} f(an)\delta_{an}$ to $\boldsymbol{a}f \cdot \text{III}_a$ for $k \to \infty$ in the norm of $W(M, L^p_w)$ for any $p < \infty$. Applying $W(M, L^p_w) * W(C^0, L^1_s) \subseteq W(C^0, L^p_w)$ we derive that

$$\boldsymbol{a} \cdot \sum_{|n|\le k} f(an)\, \delta_{an} * h = \boldsymbol{a} \cdot \sum_{n=-k}^{k} f(an)\, T_{an} h$$

is convergent in $W(C^0, L^p_w)$, and therefore in L^p_w as well as locally uniform. □

Remark 1. Using invertible linear transformations of $I\!R^m$ the above results can be easily transformed into a result on more general lattices and no significant change in the arguments is required. That this is the most general approach to the sampling problem based on a Poisson-type formula can be seen from a recent result of Cordoba [Co].

We want to show next how amalgams can be used to describe the aliasing error in L^p-norms, i.e. the consequences of applying the above formula to $f \in L^p$ which are not band-limited. Obviously the part of $\hat{f}$ exceeding Ω will be responsible for the error, but a simple L^p-estimate of that part is certainly not sufficient. After all, the sampling points are just a set of measure zero in $I\!R^m$. A sufficient extra condition on $\hat{f}$ is integrability, which allows one to obtain *uniform* estimates for the aliasing error (cf. [BSS], Theorem 3.8). We show that $W(C^0, L^{p'})$-estimates can be obtained under a slightly stronger $W(L^p, L^1)$ assumption on $\hat{f}$.

Lemma 3. (*aliasing error estimate using amalgams*)

Assume $\hat{f} \in W(L^p, L^1)$ for some $p \in [1, 2]$. Then $f \in W(C^0, L^{p'})$, and the aliasing error can be estimated as follows. For any $\hat{h}(t) \equiv 1$ on Ω and $0 \le \hat{h}(t) \le 1$ on $I\!R^m$ there exists some $C_2 > 0$ such that

$$\|f - (\boldsymbol{a}f \cdot \text{III}_a) * h \,|W(C^0, L^{p'})\| \le C_2 \|\hat{f} - \hat{f} \cdot \mathbf{1}_\Omega \,|W(L^p, L^1)\| \ . \tag{4}$$

In particular, the aliasing error tends to zero for $\boldsymbol{a} \to (0, \dots, 0)$.

Proof. The first statement is a simple consequence of the generalized HY-theorem, by which $\mathcal{F}$ maps $W(L^p, L^1) \subseteq W(\mathcal{F}L^{p'}, L^1)$ to $W(\mathcal{F}L^1, L^{p'}) \subseteq W(C^0, L^{p'})$ (cf. [F2], see [F3] for weighted versions). In order to estimate the aliasing error we split f into a good and a bad part by setting $f_\Omega := \mathcal{F}^{-1}(f \cdot \mathbf{1}_\Omega)$ and $f_r := f - f_\Omega$. Then of course $f_\Omega = (\boldsymbol{a}f_\Omega \text{III}_a) * h$, and therefore the aliasing error can be estimated by

$$\begin{aligned} &\|f - (\boldsymbol{a}f \cdot \text{III}_a) * h \,|W(C^0, L^{p'})\| = \|f_r - (\boldsymbol{a}f_r \text{III}_a) * h \,|W(C^0, L^{p'})\| \le \\ &\le \ \|f_r \,|W(C^0, L^{p'})\| + \|(\boldsymbol{a}f_r \text{III}_a) * h \,|W(C^0, L^{p'})\| \le \\ &\le \ \|f_r|W(C^0, L^{p'})\| + C \cdot \|\boldsymbol{a}\text{III}_a \,|W(M, L^\infty)\| \, \|h \,|W(C^0, L^1)\| \cdot \|f_r \,|W(C^0, L^{p'})\| \ . \end{aligned}$$

Since the family $a\mathrm{III}_a$ is uniformly bounded in $W(M, L^\infty)$ we obtain

$$\|f - (af \cdot \mathrm{III}_a) * h \,|W(C^0, L^{p'})\| \le C_1 \cdot \|f_r \,|W(C^0, L^{p'})\| \le C_2 \cdot \|\hat{f} - \hat{f} \cdot \mathbf{1}_\Omega \,|W(L^p, L^1)\| \ .$$

From this it is clear that the aliasing error tends to zero as Ω increases to $\mathbb{R}^m$, and even the speed of convergence can be controlled by the decay of the norm of the high frequency tails, measured in the norm of $W(L^p, L^1)$. □

In Theorem 2 we have restricted ourselves to the use of band-limited functions $h \in L^1_s$ in the reconstruction process, because we wanted to have the result for the full range of values $p \ge 1$ and weights. In fact, the use of the traditional sinc-function, the inverse Fourier transform of the rectangular function, which is *not* in L^1, has to be excluded from the discussion for this reason. On the other hand, the sinc-function (or its multidimensional analog, obtained by pointwise products), belongs to L^p, for any $p > 1$. Thus there is some hope that estimates involving the sinc-function can be obtained in this case. It turns out, however, that the convolution estimate based on the fact that $\sum_{n\in Z^m} f(an)\,\delta_{an} \in W(M, L^p_w)$, and therefore $\in W(C^0, L^r)$ for any $r > 1$, is too weak to ensure L^p_w (or even $W(C^0, L^p_w)$ convergence) of the sampling series, even for the trivial weight w. The problem even gets worse if one wants to study jitter errors, because then the usual argument for the L^2-case (it is based on orthogonal series expansions) breaks down too. We will show how the generalized HY-theorem can be used to establish appropriate estimates. We write $\mathbf{sinc}(\boldsymbol{x}) := \mathrm{sinc}(x_1)\cdots\mathrm{sinc}(x_m),\ \boldsymbol{x} \in \mathbb{R}^m$.

But first a very useful corollary to the generalized HY inequality.

Lemma 4. The *ideal low pass filter*, i.e. convolution by **sinc**, defines a bounded multiplier from $W(M, L^p)$ into $W(C^0, L^p)$. In particular, for any convergent sequence $(\mu_n)_{n\ge1}$ in $W(M, L^p)$, with limit μ_0, the sequence $\mu_n * \mathbf{sinc}$ is convergent in $W(C^0, L^p)$.

Proof. We will apply the generalized HY inequality twice. First we observe that $\mu \in W(M, L^p) \subseteq W(\mathcal{F}L^\infty, L^p)$ implies that $\hat{\mu} \in W(\mathcal{F}L^p, L^\infty)$. Since $\mathrm{rect} = \mathcal{F}(\mathbf{sinc})$ is known to be a bounded pointwise multiplier for $\mathcal{F}L^p$ for $1 < p < \infty$ (cf. [Pe], Chap.7 or [St], Chap.4), hence $\hat{\mu}\cdot\mathrm{rect} \subseteq \mathrm{W}(\mathcal{F}\mathrm{L^p}, \mathrm{L^1})$, thus $\mu * \mathbf{sinc} \in \mathcal{F}^{-1}(W(\mathcal{F}L^p, L^1)) \subseteq W(\mathcal{F}L^1, L^p) \subseteq W(C^0, L^p)$, and the required norm estimates hold as well. □

The following result is a partial extensions of Theorem 5 in [Go] to the *irregular* case (cf. Thm.14 below).

Corollary 5. Let $X = ((x_i)_{i\in I})$ be a relatively separated family in $\mathbb{R}^m$ and $\Omega \subseteq \mathbb{R}^m$ be compact.

Then

$$S_X f := \sum_{i\in I} f(x_i) T_{x_i} \mathbf{sinc}$$

is unconditionally convergent in $W(C^0, L^p)$, and there exists $C_2 = C(X, \Omega) > 0$ such that

$$\|S_X f \,|W(C^0, L^p)\| \le C_2 \cdot \|f\|_p \tag{5}$$

for any $p \in (1, \infty)$ and any $f \in L^p(\mathbb{R}^m)$ with $\mathrm{spec}(f) \subseteq \Omega$.

Proof. If X is relatively separated then $\delta_X := \sum_{i\in I} \delta_{x_i} \in W(M, L^\infty)$, and by Lemma 1

$$\sum_{i\in I} f(x_i)\delta_{x_i} = f \cdot \delta_X \in W(C^0, L^p) \cdot W(M, L^\infty) \subseteq W(M, L^p), \tag{6}$$

for any band-limited $f \in L^p(\mathbb{R}^m)$ and the previous Lemma applies. □

In various situations, especially in the discussion of the irregular sampling problem with highly irregular sampling sets (which might have arbitrary high density at some places), the following modification is of interest.

Lemma 6. Let $\Psi = (\psi_i)_{i\in I}$ be a family of measurable functions which satisfy $0 \le \psi_i(x) \le 1$, $\mathrm{supp}(\psi_i) \subseteq B_\delta(x_i)$ for all $i \in I$ and $\sum_{i\in I} \psi_i(x) \le C_\Psi < \infty$ for $x \in \mathbb{R}^m$. If $c_i := \|\psi_i\|_1$ then the discrete measure $\mu_\Psi := \sum_{i\in I} c_i \delta_{x_i}$ belongs to $W(M, L^\infty)$, and $\|\mu \,|W(M, L^\infty)\| \le C_\delta \cdot C_\Psi$ for some $C_\delta > 0$.

Proof. We shall make use of the duality $W(M, L^\infty) = W(C^0, L^1)'$. Thus we only have to obtain an estimate for $f \in \mathcal{K}(\mathbb{R}^m)$ (using the positivity of c_i)

$$|\mu_\Psi(f)| \le \sum_{i\in I} c_i |f(x_i)| = \|\sum_{i\in I} f(x_i)\psi_i\|_1 \tag{7}$$

Fixing $h \in \mathcal{K}(\mathbb{R}^m)$ with $h(x) \equiv 1$ on $B_\delta(0)$ we check that $\sum_{i\in I} f(x_i)\psi_i(x) \le \|T_x h \cdot f\|_\infty$, and consequently we complete the proof by

$$|\mu_\Psi(f)| \le \|T_x h \cdot f\|_1 = \|F_h\|_1 = \|f \,|W(C^0, L^1)\| \ .$$

□

Corollary 7. For Ψ, δ as above and any compact set $\Omega \subseteq \mathbb{R}^m$ there exists $C = C(X, \delta, \Omega) > 0$ such that $S_X f := \sum_{i\in I} f(x_i)\, c_i \cdot T_{x_i}\mathbf{sinc}$ is unconditionally convergent in $W(C^0, L^p)$, and satisfies

$$\|S_X f \,|W(C^0, L^p)\| \le C \cdot \|f\|_p \tag{8}$$

for any $p \in (1, \infty)$ and any $f \in L^p(\mathbb{R}^m)$ with $\mathrm{spec}(f) \subseteq \Omega$.

Proof. This corollary follows from Lemma 6 by means of Lemma 4.

In the discussion of the sampling theorem various kinds of error analysis are of interest. In some of the classical papers (cf. [Pal,2]) uniform error estimates for L^2-data were considered as sufficient for practical purposes). As we have seen, one may expect $W(C^0, L^p)$ estimates in many cases. That these can be obtained has been shown for a variety of situations in [FG3]. Some limiting cases and situations mostly not covered in [FG3] are discussed in the sequel.

Theorem 8 (*Jitter error estimate for p-norms*). Let X be relatively separated , Ω bounded in $\mathbb{R}^m$ and $p, 1 < p < \infty$ be given. Then for every $\varepsilon > 0$ there exists a $\delta > 0$ such that for any family $\tilde{X} = (\tilde{x}_i)_{i\in I}$ satisfying $|x_i - \tilde{x}_i| \le \delta$ the jitter error is small in the $W(C^0, L^p)$-sense, i.e.

$$\|\sum_{i\in I} (f(x_i) - f(\tilde{x}_i))\, T_{x_i}\mathbf{sinc} \,|W(C^0, L^p)\| \le \varepsilon \cdot \|f\|_p \tag{9}$$

Proof. Without loss of generality we may assume that $f \in \{f \in L^p(\mathbb{R}^m),\ \|f\|_p \le 1 \text{ and } \mathrm{spec}(f) \subseteq \Omega\}$, a set which is known to be equicontinuous in $L^p(\mathbb{R}^m)$. Next we observe that

$$|f(x_i) - f(\tilde{x}_i)| \le \mathrm{osc}_\delta f(x_i) \text{ for } i \in I\ . \tag{10}$$

By Theorem 3 of [F7] $\mathrm{osc}_\delta f \in W(C^0, L^p)$ (with arbitrary small norm for sufficiently small δ), thus it is possible to find to any given $\eta > 0$ some $\delta > 0$ such that

$$\|\sum_{i\in I} (f(x_i) - f(\tilde{x}_i))\delta_{x_i} \,|W(M, L^p_w)\| \le \eta \ .$$

An application of Lemma 4 of [F7] concludes the proof. □

Let et us look at a delicate point in the above argument: We did *not* go to absolute values of the sinc-function in the proof. However, we used the fact that smaller (by the absolute value) complex coefficients in the series allow better norm estimates of the corresponding discrete measures in $W(M, L^p)$.

The above result has an important corollary.

Corollary 9. Let Ω be a bounded subset of $\mathbb{R}^m$ and $1 < p < \infty$ be given. Then for any sufficiently small $\boldsymbol{a} > 0$ there exists some $\delta = \delta(\boldsymbol{a}, \Omega, p)$ such that the operator $f \mapsto S_X f$ is invertible over $L^{p,\Omega}(\mathbb{R}^m)$ if $|x_n - \boldsymbol{a}n| \leq \delta$ for all $n \in \boldsymbol{Z}^m$. In particular, it is possible to recover f from the irregular sampling values $(f(x_n))_{n \in Z^m}$ by applying the inverse operator S_X^{-1} to $S_X f$ (it can be obtained by Neumann's series).

Remark 2. The above result can also be described alternatively as an iterative algorithm (involving iterative application of $Id - S_X$), which is convergent in the sense of the $W(C^0, L^p)$-norm. Actually, we have just shown that the family $\{T_{x_i}\mathbf{sinc}\}$ is a *Banach frame* in the sense of [Gr] in the Banach space

$$L^{p,\Omega}(\mathbb{R}^m) := \{ f \in L^p(\mathbb{R}^m), \ \mathrm{spec}(f) \subseteq \Omega \} \ .$$

This means that the mapping $f \mapsto \langle f, T_{x_i}\mathbf{sinc}\rangle = f(x_i)$ is a mapping from $L^{p,\Omega}$ to the sequence space ℓ^p, that $(\sum_{i \in I} |f(x_i)|^p)^{1/p}$ defines an equivalent norm on $L^{p,\Omega}$, and that there is a bounded operator $U : \ell^p \to L^{p,\Omega}$, with $U \circ S_X = Id$ on $L^{p,\Omega}$.

The jitter error discussed above is the traditional one. Plotkin (cf. [PRS1,2]) also mentioned a *jitter error of second kind*, arising in the synthesis process, i.e. from the considerations of sums $\sum_{i \in I} \lambda_i T\tilde{x}_i h$ instead of sums $\sum_{i \in I} \lambda_i T_{x_i} h$. It is an open question whether it is possible to give a satisfactory estimate for this jitter error in the L^p-norm for the case of the **sinc**-function (another argument for the use of better decaying kernels), but we can at least prove the following (which at least guarantees uniform estimates).

Proposition 10 (*Jitter error of the second kind*). Let $h \in W(C^0, L^1)$, $p \in [1, \infty)$, and a relatively separated family X be given. Then for any $\varepsilon > 0$ there exists $\delta > 0$ such that

$$\|\sum_{i \in I} \lambda_i T\tilde{x}_i h - \sum_{i \in I} \lambda_i T_{x_i} h \,|W(C^0, L^p)\| \leq \varepsilon \cdot \|\sum_{i \in I} \lambda_i \delta_{x_i} \,|W(M, L^p)\| \tag{11}$$

$$\text{whenever } \ |x_i - \tilde{x}_i| \leq \delta \text{ for } i \in I \ .$$

In the limiting case of $h = \mathbf{sinc}$ we can obtain an estimate in $W(C^0, L^r)$ for any r, $p < r < \infty$, and in particular the uniform jitter error will be small.

Note. The typical application of this result involves $h \in L^1(\mathbb{R}^m)$, which is band-limited, e.g. a classical de la Vallée Poussin kernel (cf. Lemma4).

Proof. In this case we are forced to use absolute values involving h.
The pointwise estimate $|T\tilde{x}_i h - T_{x_i} h| \leq T_{x_i}(\mathrm{osc}_\delta h)$, gives via Lemma 4

$$\|\sum_{i \in I} \lambda_i T\tilde{x}_i h - \sum_{i \in I} \lambda_i T_{x_i} h \,|W(C^0, L^p)\| \leq C \cdot \|\sum_{i \in I} \lambda_i \delta_{x_i} \,|W(M, L^p)\| \cdot \|\mathrm{osc}_\delta h \,|W(C^0, L^1)\| \ .$$

By Theorem 3 $\mathrm{osc}_\delta h$ will be small in $W(C^0, L^1)$ for sufficiently small δ. The choice $h = \mathbf{sinc}$ is not covered by this statement, but since the **sinc**-function belongs only to $L^s(\mathbb{R}^m)$ for any $s > 1$, hence to $W(C^0, L^s)$ by Lemma 4 (actually the norms tend to infinity for $s \to 1$). Therefore the following estimate is possible (but requires smaller and smaller δ for $\varepsilon > 0$, as s goes to 1):

$$\|\sum_{i \in I} \lambda_i T_{\tilde{x}_i}\mathbf{sinc} - \sum_{i \in I} \lambda_i T_{x_i}\mathbf{sinc} \,|W(C^0, L^r)\| \leq$$

$$\leq C \cdot \| \sum_{i \in I} \lambda_i \delta_{x_i} | W(M, L^p) \| \cdot \| \mathrm{osc}_\delta \mathbf{sinc} \, | W(C^0, L^s) \| \quad \text{for } 1/r = 1 - (1/p + 1/s) \quad .$$

Uniform convergence follows in each case, or by the choice $s = p'$. □

Corollary 11 (*uniform jitter estimate for sinc-functions*). For any compact Ω, relatively separated X, and p, with $1 < p < \infty$, the uniform total jitter is small if $\tilde{X}$ is close to X, i.e. for any $f \in L^p(\mathbb{R}^m)$ with $\mathrm{spec}(f) \subseteq \Omega$

$$\| \sum_{i \in I} f(x_i) T_{x_i} \mathbf{sinc} - f(\tilde{x}_i) T_{y_i} \mathbf{sinc} \|_\infty \leq \varepsilon \cdot \|f\|_p \tag{12}$$

as long as $|x_i - y_i| \leq \gamma$ and $|\tilde{x}_i - x_i| \leq \gamma$ for some $\gamma \leq \gamma_0 = \gamma_0(\varepsilon)$. If **sinc** is replaced above by some function $h \in W(C^0, L^1)$ the estimate even holds true in the sense of $W(C^0, L^p)$ and for $p \geq 1$.

There is, however, no hope to extend the jitter error estimate for the sinc-function to the case $p = \infty$, as can be shown by the following one-dimensional counterexample. It also gives an answer to the following problem: Given an irregular sampling family $X = (x_n)_{n=1}^\infty$, is it possible that the series $\sum_{n=1}^\infty T_{x_n}$ sinc can have a zero, in other words, is it possible that for some choice of x_n one has $\sum_{n=1}^\infty \mathrm{sinc}(x_n - x_0) = 0$ for some x_0? This series may be considered as a low pass filtered version of the discrete measure $\delta_X := \sum_{n=1}^\infty \delta_{x_n}$ (using the rectangular filter), and is proposed as a correction term in the reconstruction procedure suggested by Plotkin (cf.[PRS1] and [PRS2]).

Proposition 12. Given any lattice constant $\boldsymbol{a} > 0$ and any $\delta > 0$ (the allowed jitter constant). Then for any $r \in \mathbb{R}$ (or $r = \infty$) it is possible to find a jitter-sequence j_n with $|j_n| \leq \delta$, such that

$$\sum_{n=1}^{\infty} \mathrm{sinc}(\boldsymbol{a}n - j_n) = r \quad ,$$

i.e. the 'jittered series' can take arbitrary values at zero, or may be divergent, even if the jitter error is uniformly small (it would be easy to use summation over $\boldsymbol{Z}$ as well).

Proof. The argument is based on the fact that the sinc-function, defined by $\mathrm{sinc}(x) := \sin(\pi x)/(\pi x)$ for $x \neq 0$, decays only like $1/x$, and that the harmonic series $\sum_{n=1}^\infty 1/n$ is known to be divergent. Without loss of generality we assume that $\boldsymbol{a} = (1, \ldots, 1)$, and that $r > 0$. By means of dilations and choosing some j_n's negative, our argument covers the remaining cases.

Setting $M := \{1 + 2k, \ k \in \mathbb{N}\}$ we plan to keep $j_n = 0$ for $n \notin M$, and to choose j_n in a suitable way for $n \in M$. We will assume for simplicity that $\delta \leq 1/8$.

The main estimate concerns an estimate for the derivative of the sinc-function over the intervals I_k, defined by $I_k := [2k + 3/8, 2k + 5/8]$, $k \geq 2$. One obtains (since $y > 4$, hence $\pi y > 6$ for $y \in I_k, k \geq 2$):

$$\mathrm{sinc}'(y) = \frac{\cos(\pi y)}{y} - \frac{\sin(\pi y)}{\pi y^2} \leq \frac{(\cos(\pi y) + 1/6)}{y} \quad . \tag{13}$$

Noting furthermore that $\cos(z) \leq -\frac{1}{2y} < -\frac{2}{3}$ if $|(2k+1)\pi - z| < \frac{\pi}{4}$ we conclude that

$$\mathrm{sinc}'(y) < \frac{-2/3 + 1/6}{y} = -\frac{1}{2y} \leq -\frac{1}{4k} \quad \text{for } y \in I_k \quad .$$

The mean value theorem implies that for any $u \in I\!R$

$$\operatorname{sinc}(u - j_n) = \operatorname{sinc}(u) - j_n \operatorname{sinc}'(\xi_n) \quad \text{for some } \xi_n \in (n - j_n, n) \ , \tag{14}$$

hence

$$\operatorname{sinc}(2k + 1 - \delta) \geq \frac{\delta}{4k} \quad \text{for } k \in I\!N \ . \tag{15}$$

Since the partial sums of the harmonic series are unbounded there exists $k_1 \in I\!N$ such that

$$s := \sum_{k=0}^{k_1 - 1} \operatorname{sinc}(2k + 1 - \delta) \leq r \ , \quad \text{but} \quad s + \operatorname{sinc}(2k_1 + 1 - \delta) > r \ .$$

Using the fact that sinc is strictly decreasing and continuous on $[2k_1 + 1 - \delta, 2k_1 + 1]$ we can find some δ_1 with $0 \leq \delta_1 \leq \delta$ such that $\operatorname{sinc}(2k_1 + 1 - \delta_1) = r - s$. Setting $j_n = \delta$ for $n = 2k + 1$, $2 \leq k \leq k_1$, $j_{k_1} := \delta_1$ and $j_n = 0$ otherwise, we find that $\sum_{n=1}^{\infty} \operatorname{sinc}(n - j_n) = r$, i.e. the jittered sampling series can take any prescribed value. □

As an immediate corollary we obtain the counterexample, showing that the low-pass filtered version of an almost regular lattice may have zeros.

Since $\operatorname{sinc}(0) = 1$ it is sufficient to choose j_n in such a way that $\sum_{n=1}^{\infty} \operatorname{sinc}(n - j_n) = -1$ (note that $\operatorname{sinc}(k) = 0$ for any $k \in \boldsymbol{Z}$, $k \neq 0$). Note that it would be possible to ask for *no* jitter error for any term with $n \leq n_0$ (some given number), and still having this disastrous phenomenon. It is also possible to show that there is a uniform error estimate if the sequence of jitter errors is itself square summable, i.e. if $\sum_{n=1}^{\infty} j_n^2 < \infty$. Results in this direction will be discussed elsewhere.

Remark 2. The above arguments also can be used to obtain an L^p-error estimate for the jitter error for reconstruction from sampling values together with derivates as described in [BDo]. One has only to observe that the derivative of a band-limited function in L^p is itself a band-limited function in L^p.

In a discussion of band-limited functions on $I\!R$ ([Cl],[CC]) the following question came up: Given a band-limited, square integrable function and a smooth deformation of the real line $\varphi : I\!R \to I\!R$, what can be said about $f \circ \varphi$? Since band-limited functions are differentiable it makes sense to discuss only differentiable functions φ which are strictly increasing, i.e. satisfying $0 < \varphi'(t) < \infty$ for all $t \in I\!R$. According to the conjecture stated in [Cl] the composition mapping $f \mapsto f \circ \varphi$ should always produce some non-band-limited functions from band-limited ones, except the special case where φ is an affine mapping of the form $\varphi(z) = \alpha z + \beta$ for $\alpha > 0$ and $\beta \in I\!R$. Along with this question, however, it has to be checked under which circumstances the composition mapping preserves square integrability. Using Wiener amalgam spaces we can give a partial answer.

Proposition 13 (*Preservation of p-integrability under composition*). Assume that for some positive $a > 0$ and $K \geq 0$ we have $\varphi(t) \in [a(t - K), a(t + K)]$ for all $t \in I\!R$ (i.e. that the graph of φ is contained in a strip in $I\!R^2$), then for some $C = C(\Omega, K, a) > 0$

$$\|f \circ \varphi\|_p \leq C \cdot \|f\|_p \quad \forall f \in L^p \text{ with } \operatorname{spec}(f) \subseteq \Omega \ . \tag{16}$$

Note that this statement is *not* valid without the hypothesis of band-limitedness on f. Let us demonstrate this by an L^2-example. Assume that for some point $t_n \in I\!R$ with $|\varphi'(t_n)| \leq \frac{1}{2n^2}$, then $|\varphi'(t)| \leq 1/n^2$ for some interval $[a_n, b_n]$. Setting $\alpha_n := \varphi^{-1}(a_n)$ and

$\beta_n := \varphi^{-1}(n_n)$ it follows from the mean value theorem, that $|a_n - b_n| \geq n^2|\alpha_n - \beta_n|$, and that the indicator function $f := \mathbf{1}_{[\alpha_n,\beta_n]}$ satisfies $\|f \circ \varphi\|_2 \geq n \cdot \|f\|_2$.

Proof. Choosing some function $k \in \mathcal{K}(\mathbb{R}^m)$ such that $k(x) \equiv 1$ on $[-aK, aK]$ we have

$$|f \circ \varphi(t)| \leq \sup_{z \in [a(t-K), a(t+K)]} |f(z)| \leq \|T_{at}k \cdot f\|_\infty = F_k(at) \, .$$

Since we know that band-limited functions belong to $W(C^0, L^p)$, with $\|F_k\|_p \leq C_\Omega \|f\|_p$, we end up with the estimate (using the dilation invariance of L^p)

$$\|f \circ \varphi\|_p \leq a^{m(1-1/p)} \cdot \|F_k\|_p \leq C \cdot \|f\|_p \; . \tag{17}$$

□

The above proposition gives a sufficient condition to preserve square integrability of band-limited L^p-functions. Improving on the necessary conditions on φ (in order to preserve band-limitedness) as given in [Cl] we mention the following result: Any function φ, which has an unbounded derivative will produce functions which are *not* band-limited!

In fact, band-limited functions satisfy (what is usually called Bernstein's inequality, cf. [FG2], Prop.3.4 for an L^p-version)

$$\sup_{t \in \mathbb{R}} |f'(t)| \leq C \sup_{t \in \mathbb{R}} |f(t)| \; .$$

Functions φ with unbounded $\gamma'(t)$ (arbitrary steep parts in the graph of φ) apparently destroy this property, i.e. $f \circ \varphi$ will not satisfy the last estimate for such functions φ and cannot be band-limited for that reason.

3 GEOMETRIC CONDITIONS ON THE SAMPLING SETS

One of the basic estimates in irregular sampling theory is the following (we state the L^p-version here), taking a discrete sampling family $X = (x_i)_{i \in I}$.

For any bounded set Ω and p, $1 \leq p < \infty$, there is some $C_\Omega > 0$ such that

$$\left(\sum_{i \in I} |f(t_i)|^p \right)^{1/p} \leq C_\Omega \|f\|_p \; \forall f \in L^p(\mathbb{R}^m) \text{ with } \operatorname{spec}(f) \subseteq \Omega \; . \tag{18}$$

This estimate has been first shown for $f \in L^2(\mathbb{R}^m)$ and regular lattices by S. Nikolskij, and is known as Nikolskij's inequality. In a more classical setting such estimates have been proved by Plancherel and Polya (cf. [PP]). Recently an irregular version (for sets in the plane which arise as products of irregular sets in each coordinate) has been proposed by Butzer and Hinsen (cf. [BH1,2]). For certain irregular sets these results were proved by Duffin and Shaeffer in [DS] in the one-dimensional case (cf. [DS], or [Y]).

It is clear, that X must not be too dense that such an inequality holds, even if we have very smooth and nice functions f. Certainly there must be no accumulation points to the family X. Since the L^p-spaces are translation invariant the correct necessary condition is the following one.

Definition 2. A discrete set $X = (x_i)_{i \in I}$ is called *relatively separated* if there exists an upper bound on the local density of X in $\mathbb{R}^m$ in the following sense: For some $r_0 > 0$ there is a uniform bound on

$$d(y) := \#\{i \,|\, x_i \in B_{r_0}(y)\}, \text{ i.e. } d_r(X) := \sup_{y \in \mathbb{R}^m} d(y) < \infty \; . \tag{19}$$

It is then clear (any ball can be covered by a finite family of balls of any given size) that for any ball $B \subseteq \mathbb{R}^m$ the following is true.

The number of elements in any translate $x + B$ is uniformly bounded.

(I) For some $C_B > 0$ we have $\#\{i \,|\, t_i \in (x+B)\} \leq C_B \;\; \forall x \in \mathbb{R}^m$.

(II) For any $r > 0$ the family $(B_r(x_i)_{i\in I})$ of balls of radius r is of bounded height $h(r)$, i.e. there is a maximal number $h(r)$ of balls $B_r(x_i)$ covering any given point. Actually, somewhat more holds.

(II′) Given any compact subset $K \subseteq \mathbb{R}^m$ there is a uniform bound on the balls intersecting $y + K$, independently of y , i.e.

$$\sup_{y\in\mathbb{R}^m} \#\{i \,|\, (y+K) \cap B_r(x_i) \neq \emptyset\} < \infty \,. \tag{20}$$

Theorem 14. A discrete family $X = ((x_i)_{i\in I})$ in $\mathbb{R}^m$ is relatively separated if one (hence all) of the following conditions are satisfied: (I), (II) or

(III) The measure $\sum_{i\in I} \delta_{x_i}$ belongs to $W(M, L^\infty)$, i.e. is *translation bounded* in the sense of [AL].

(IV) For some (any) $1 \leq p < \infty$ there is a constant $C > 0$ such that

$$(\sum_{i\in I} |f(x_i)|^p)^{1/p} \leq C \cdot \|f \,|W(C^0, L^p)\| \;\; \forall f \in W(C^0, L^p) \tag{21}$$

(or only for all functions $f = T_x k$, for some non-zero $k \in \mathcal{K}(\mathbb{R}^m)$).

(V) For any $\delta > 0$ the family is a finite union of *separated* families $X^k = (x_i^k)_{i\in I}$, each satisfying $|x_i^k - x_j^k| \geq \delta > 0$ for $i \neq j$.

(VI) X is a finite union of sets which are subsets of sequences each of which is *uniformly dense* in the sense of Duffin and Shaeffer:

$$|x_n - \alpha n| \leq L \;\; \forall n \in \mathbf{Z}^m (\text{ for some } \alpha > 0 \text{ and } L > 0) \,. \tag{22}$$

Proof. Taking for granted that the concept of relative separation does not depend on the choice of the radius r_0 we obtain the stronger version of (II) by choosing some r_1 such that $K \subseteq B_{r_1}(x_0)$, observing then that $(y+K) \cap B_r(x_i) \neq \emptyset$ only if $x_i \in B_{r_2}(y)$, for $r_2 = r_1 + r$, showing thus that (20) is equivalent with (19). In order to show the equivalence with (III) choose some $k \in \mathcal{K}(\mathbb{R}^m)$ such that $0 \leq k(y) \leq 1$, $k(x) \equiv 1$ on $B_s(0)$ and $\mathrm{supp}(k) \subseteq B_r(0)$. Then we have for $\mu := \sum_{i\in I} \delta_{x_i}$

$$M_k(x) := \|(T_x k) \cdot \mu\|_M = \sum_{i\in I} |k(x_i - x)| \leq \#\{i | x_i \in B_r(x)\} \;\; , \tag{23}$$

but on the other hand $M_k(x) \geq \#\{i | x_i \in B_s(x)\}$.

To check that a relatively separated family X satisfies (IV) we use

$$\mu \cdot f \in W(M, L^\infty) \cdot W(C^0, L^p) \subseteq W(M, L^p) \,.$$

Assume conversely, that X is *not* relatively separated. Then for each $r > 0$ there are points x_n in $\mathbb{R}^m$ such that $\#\{i | x_i \in B_r(x)\} \geq n$. Let now $k \in \mathcal{K}(\mathbb{R}^m)$ be any function with $\min_{y \in B_r(0)}(k(y)) \geq \eta > 0$. Then the sequence $T_{x_n}k$ is bounded in $W(C^0, L^p)$ for any $p \geq 1$, but $(\sum_{i \in I} |T_{x_n}k(x_i)|^p)^{1/p} \geq \eta \cdot n^{1/p}$ for each $n \geq 1$, in contradiction to (IV).

(V): It is left as an exercise to the reader to verify that a relatively separated set satisfies (V). Conversely, let X be relatively separated. We cover $\mathbb{R}^m$ by (almost) disjoint cubes of side length $\geq \delta > 0$. Then in each of these cubes together with all its 2^m neighbors there are a maximal number n_1 of points available. This can be easily used to split X into at most n_1 δ-separated subsequences (cf. [FGr]). Conversely any separated set is obviously relatively separated, and the same is true for finite unions.

(VI): We observe first that property (18) has the following features: If this property holds for any subset Y of set X satisfying (18), and it also holds for finite unions of sets X_i, each of them satisfying (18). Since a set satisfying (22) is relatively separated, the sets described in (VI) are relatively separated by the argument just given. The converse requires only slight modifications of arguments used for (V). □

The above result also sheds light on the so-called Parseval relationship for non-uniform sampling appearing in a note by Marvasti and Chuande [MC] which has just appeared. Their proof makes implicit use of the definition of a 'sampling set' in the sense of Duffin and Schaeffer [DS], which implies that we have an estimate of the form (21) for $p = 2$, which means that X is relatively separated by Theorem 6. Actually, the argument used in [MC] is not valid without relative separation. In fact, for general frequencies $(\lambda_n)_{n=1}^{\infty}$ the convergence of $\sum_{n=1}^{\infty} a_n e^{-i\pi\lambda_n}$ for sequences $(a_n)_{n=1}^{\infty}$ in ℓ^2 is only guaranteed in the sense of some mean (cf. [Be] for details), and *not* locally in L^2 (it is not difficult to set up simple counter-examples which are divergent over some interval). Given this restriction we can give a proof of Parseval's relationship for non-uniform sampling in several variables.

Theorem 15. Let $(x_n)_{n=1}^{\infty}$ be a relatively separated sampling sequence in $\mathbb{R}^m$, and M_{lp} be the Fourier transform of $S_X f := \sum_{n=1}^{\infty} f(x_n) \cdot T_{x_n} g$, where $g \in L^2$ satisfies $\hat{g}(t) \equiv 1$ on Ω. Then M_{lp} belongs locally to L^2, and the following relation holds for any $f \in L^2$ with $\mathrm{spec}(f) \subseteq \Omega$

$$\sum_{n=1}^{\infty} |f(x_n)|^2 = \int_{\mathbb{R}^m} \hat{f}(s) \cdot \overline{M}_{lp}(s)\, ds \quad . \tag{24}$$

Proof. We have already discussed convergence of the series on the left side. On the other hand Corollary 5 shows that $S_X f \in W(M, L^2) \subseteq W(\mathcal{F}L^\infty, L^2)$ and thus by the generalized HY inequality $M_{lp} \in W(L^2, L^\infty) \subseteq L^2_{loc}(\mathbb{R}^m)$. Since $\hat{f}$ is a compactly supported L^2-function converence of the right hand integral follows. These observations allows us to use the duality pairing $\langle \cdot, \cdot \rangle$ (this time considered as the natural extension of the Hilbert space duality in the argument below) in varying pairs in order to obtain.

$$\sum_{n=1}^{\infty} |f(x_n)|^2 = \langle |f|^2, \delta_X \rangle = \langle f, \delta_X \cdot f \rangle = \langle f * g^*, \delta_X \cdot f \rangle = \tag{25}$$

$$= \langle f, (\delta_X \cdot f) * g \rangle = \langle f, S_X f \rangle = \langle \hat{f}, M_{lp} \rangle = \int_{\mathbb{R}^m} \hat{f}(s) \overline{M}_{lp}(s)\, ds$$

the third step following from the fact that f has spectrum in Ω and that convolution by $g^* := \mathcal{F}^{-1}(\hat{g}^-)$ acts therefore trivial. The last step uses Plancherel's theorem. □

Remark 4. Observe that this result is not only true in several dimensions but is *not* restricted to particular sampling schemes arising as product sets of one-dimensional irregular sampling sets, as in the two-dimensional result given in [MC].

4 PRODUCT CONVOLUTION OPERATORS AND SIGNAL RECOVERY

Estimates for certain product-convolution operators are at the heart of a reconstruction method suggested by Donoho and Stark [DS1,2]. The situation discussed there is the following one: A function f is given with several parts being missing. This missing information is complemented by some a priori information on its Fourier transform. The well known Papoulis–Gerchberg algorithm covers the case where the spectrum is known to be contained in some bounded spectral set, i.e. covers the case of band-limited functions. Given only a small part of the function it is then possible to recover the full function by an iterative procedure. However, the method may be very instable and sensitive to noise. It also does not cover the case of a possibly unbounded spectrum. The results of Donoho and Stark [DS1,2] show that it is possible to solve the problem in certain cases, e.g. if the spectrum is unbounded, but has finite measure. Of course, there has to be a trade-off between the size of the set T of missing values, and the set Ω on which the Fourier transform $\hat{f}$ of f is concentrated (or unknown). It turns out that one has stable reconstruction by means of an iterative algorithm if $|T|\,|\Omega| < 1$, i.e. if the product of the (Lebesgue) measures of these sets is small enough. Although only the one-dimensional case is treated explicitly in [DS1,2] their arguments extend to m dimensions, and even to locally compact abelian groups, as pointed out by Smith [Sm].

The key estimate for the recovery (cf. [DS1]), concerns the operator

$$f \mapsto PQf \;\text{ with }\; Q : f \mapsto \mathcal{F}^{-1}(\mathbf{1}_\Omega \hat{f}), \;\; P : f \mapsto \mathbf{1}_T f \;, \tag{26}$$

where $\mathbf{1}_\Omega$ denotes the indicator function of the set $\Omega \subseteq \mathbb{R}^m$ and $T \subseteq \mathbb{R}^m$ is some subset of $\mathbb{R}^m$. Clearly, another way of considering this operator is to see it as a convolution product, with convolution by $\mathcal{F}^{-1}(\mathbf{1}_\Omega)$, followed by pointwise multiplication with $\mathbf{1}_T$. It is shown that $\|P \circ Q\| \leq |W|\,|T|$, the operator norm being for L^p, $1 \leq p \leq 2$. It is easy to verify this result (due to K. Smith [Sm]). By Hausdorff-Young $\|h_\Omega\|_{p'} \leq \|\mathbf{1}_\Omega\|_p = |\Omega|^{1/p}$, and by Hölder's inequality $L^p * L^{p'} \subseteq C^0$, thus

$$\|PQf\|_p \leq \|P(Qf)\|_p \leq \|\mathbf{1}_T\|_p \|Qf\|_\infty \leq (|T|\,|\Omega|)^{1/p}\,\|f\|_p \;\; \forall f \in L^p(\mathbb{R}^m) \;. \tag{27}$$

It is clear that under the given circumstances $Q \circ (Id - P) \circ Q$ is invertible as an operator on $L^{p,\Omega}$. However, if $f \in L^{p,\Omega}$ is given over $\mathbb{R}^m \backslash T$, we exactly know $(Id - P) \circ Q$, and therefore, applying the inverse operator, we are able to recover f. Of course, the inversion is carried out by means of Neumann's series and can thus be formulated as an iterative procedure (cf. [DS1], section 4).

The other way of looking at their result was to decompose the mapping $f \mapsto PQf$ into 4 different mappings, which in principle could go through arbitrary Banach spaces of functions or distributions (not only through L^p-spaces).

With $B = L^p$ (on any lca. group) the following is natural (and gives the same result as mentioned above). Consider the sequence of mappings

$$f \mapsto \mathcal{F}f \mapsto \mathbf{1}_\Omega \cdot \mathcal{F}f \mapsto \mathcal{F}^{-1}(\mathbf{1}_\Omega \cdot \mathcal{F}f) \mapsto \mathbf{1}_T \cdot \mathcal{F}^{-1}(\mathbf{1}_\Omega \mathcal{F}f) \tag{28}$$

The composed operator is treated as a composition of operators each of which is either a (inverse) Fourier transform or a pointwise multiplier of some indicator function. Using the fact that L^p is (contractively) embedded into the pointwise multiplier algebra from $L^{p'}$ into L^1, we see that the optimal way of looking at the above sequence as operators between the spaces L^p, $L^{p'}$, L^1, L^∞ and L^p (in this order), and in each case the norm of the multiplication operator is just the $|\cdot|^{1/p}$ (a power of the volume of the underlying set).

It is now evident, that the above chain (28) can run through various other spaces. The general idea behind such an approach is of course to describe situations, which are not covered by the above estimates, but still allow (maybe under some extra conditions on f) to apply the signal recovery algorithm. As a typical result in this direction obtained by using amalgam spaces we discuss a theorem concerning L^2-functions. Furthermore we fix some r such that $r \geq 2$. Then for $1/p := 1/2 + 1/r$

$$\|f\mathbf{1}_\Omega\,|W(L^2,\ell^p)\| \leq \|\hat{f}\,|W(L^2,\ell^2)\| \cdot \|\mathbf{1}_\Omega\,|W(L^\infty,\ell^r)\| \quad , \tag{29}$$

and by the Hausdorff-Young theorem for Wiener amalgams an estimate for Qf:

$$\|Qf\,|W(L^{p'},\ell^2)\| \leq C \cdot \|\hat{f}\mathbf{1}_\Omega\,|W(L^2,\ell^p)\| \leq C \cdot \|f\|_2\,\|\mathbf{1}_\Omega\,|W(L^\infty,\ell^r)\| \quad . \tag{30}$$

Applying now the pointwise multiplier rule for amalgams one has

$$\|PQf\|_2 = \|PQf\,|W(L^2,\ell^2)\| \leq \|Qf\,|W(L^{p'},\ell^2)\| \cdot \|\mathbf{1}_T\,|W(L^r,\ell^\infty)\| \leq$$

$$\leq C \cdot \|f\|_2\,\|\mathbf{1}_\Omega\,|W(L^\infty,\ell^r)\| \cdot \|\mathbf{1}_T\,|W(L^r,\ell^\infty)\| \quad .$$

We have shown that at the expense of a more sensitive measurement of W ($\|\mathbf{1}_\Omega\,|W(L^\infty,\ell^2)\|$ instead of $\|\mathbf{1}_\Omega\|_2 = |W|^{1/2}$) we can replace $\|\mathbf{1}_T\|_2 = |T|^{1/2}$ by the much less sensitive measure $\|\mathbf{1}_T\,|W(L^2,\ell^\infty)\| = \sup_{x\in\mathbb{R}^m}|T\cap(x+Q)|$, where Q is the unit cube in $\mathbb{R}^m$ (which may be considered as a *local density measure*). This result can be used as follows.

Assume we know that the set W consists of few disjoint intervals (or cubes), far apart from each other (so that the band-width or even the diameter of the spectrum is large). Then, roughly speaking, the norm of $\mathbf{1}_\Omega \in W(L^\infty,\ell^r)$ corresponds to $k^{1/r}$ if k is the number of intervals of unit length needed to cover W.

Theorem 16. For $r > 0$ and Q open in $\mathbb{R}^m$, with compact closure, there is some $\gamma > 0$ such that any $f \in L^2(\mathbb{R}^m)$, with spec(f) contained in at most n balls of radius r, can be completely recovered from $f\mathbf{1}_M$, if only $\inf_{x\in\mathbb{R}^m}|M\cap(x+Q)| \geq |Q| - \gamma$, i.e. if the local density of the set of missing values is not too large.

References

[BDo] M. G. Beaty, M. M. Dodson: Derivative sampling for multiband signals, Numer. Funct. Anal. and Optimiz. 10 (9&10), 1989, 875–898.

[Be] A. S. Besicovitch: Almost periodic functions. New York: Wiley, Inter-Science, 1968, Chap.2.9.

[Br] R. Bracewell: The Fourier transform and its application, 2nd Ed., Mc-Graw Hill 1986. (Chap.10: Sampling series).

[BS] R. Busby, H. A. Smith: Product-convolution operators and mixed norm spaces, Trans. Amer. Math. Soc. 263 (1981), 309–341.

[Bu] P. L. Butzer: A survey of the Whittaker-Shannon sampling theorem and some of its extensions. J. Math. Res. Expositions 3 (1983), 185–212.

[BH1] P. L. Butzer, L. Hinsen: Reconstruction of bounded signals from pseudo-periodic irregularly spaces samples. Signal Proc. 17 (1988), 1–17.

[BH2] P. L. Butzer, L. Hinsen: Two-dimensional nonuniform sampling expansions - An iterative approach. I, II. Appl. Anal. 32 (1989), 53–68 and 69–85.

[BSS] P. L. Butzer, W. Splettstößer, R. L. Stens: The sampling theorem and linear prediction in signal analysis, Jber. d. Dt. Math. Verein. 90 (1988), 1–70.

[Cl] J. J. Clark: Sampling and reconstruction of non-bandlimited signals, SPIE Visual Comm. and Image Processing, Nov. 1989, 1199–126.

[CC] D. Cochran, J. J. Clark: ICASSP 1990: D*.2, 1539–1541. On the sampling and reconstruction of time-warped band-limited signals.

[Co] A. Cordoba: La formule sommatoire de Poisson, C. R. Acad. Sci. Paris, 306 (1988), 373–376.

[DS1] D. Donoho, P. Stark: Uncertainty principles and signal recovery, SIAM J. APPL. Math 48/3 (1989), 906–931.

[DS2] D. Donoho, P. Stark: Recovery of a sparse signal when the low frequency information is missing, Techn. Rep. 179 (June 1989), Dept. Stat., UCB.

[DS] R. Duffin, A. Schaeffer: A class of nonharmonic Fourier series. Trans. Amer. Math. Soc. 72 (1952), 341–366.

[F7] H.G.Feichtinger: Wiener amalgams over Euclidean spaces and some of their applications, this volume.

[F8] H.G.Feichtinger: Discretization of convolutions and the generalized sampling principle, J. Approx. Theory 1991, to appear.

[FG5] H. G. Feichtinger, K. Gröchenig: Multidimensional irregular sampling of band-limited functions in L^p-spaces. Proc. Conf. Oberwolfach , Feb. 1989. ISNM 90 (1989), Birkhäuser, 135–142.

[FG6] H. G. Feichtinger, K. Gröchenig: Error analysis in regular and irregular sampling theory, submitted.

[Go] R. P. Gosselin: On the L^p-theory of cardinal series, Annals of Math., 78 (1963), 567–581.

[Hi] I. I. Hirschman: On multiplier transformations, Duke Math. J. 26 (1959), 221–242.

[Je] A. J. Jerri: The Shannon sampling theorem - its various extensions and applications, a tutorial review. Proc. IEEE 65 (1977), 1565–1596.

[LO] J. S. Lim, A. V. Oppenheim: Advanced Topics in Signal Processing, Prentice Hall Signal Proc. Series, 1988. Chap. 8: H.W.Schüssler and P.Steffen: Some Advanced topics in filter design.

[Ma1] F. A. Marvasti: A unified approach to zero-crossing and nonuniform sampling of single and multi-dimensional systems. Nonuniform. P.O.Box 1505, Oak Park, IL 60304, 1987.

[MC] F. A. Marvasti, L. Chuande: Parseval relationship of nonuniform samples of one- and two-dimemsional signals. Trans IEEE ASSP 36/6 (1990), 1061–1053.

[Pa1] A.Papoulis: Error analysis in sampling theory. Proc. IEEE 54/7 (1966), 947–955.

[Pa2] A. Papoulis: Signal analysis. McGraw-Hill. New York, 1977.

[Pe] J. Peetre: New thought on Besov spaces. Duke Univ. Press, Durham, 1976.

[PP] M. Plancherel, G. Polya: Fonctions entieres et integrales de Fourier mutiples, Comment. Math. Helv 9 (1937), 224–248.

[PRS1] E. I. Plotkin, L. M. Roytman, M. N. S. Swamy: Nonuniform sampling of band-limited modulated signals. Signal Proc. 4 (1982), 295–303.

[PRS2] E. I. Plotkin, L. M. Roytman, M. N. S. Swamy: Reconstruction of nonuniformly sampled band-limited signals and jitter error reduction. Signal Proc. 7 (1984), 151–160.

[Sm] K. T. Smith: The uncertainty principle on groups, IMA Preprint Series 403, Inst of Math. and Appl., Univ. of Minnesota, Minneapolis, MN.

[St] E. M. Stein: Singular integrals and differentiability properties of functions. Princeton Univ. Press, Princeton N.J., 1975.

[Tr] H. Triebel: Theory of function spaces, Birkhäuser, Basel, 1983.

[Y] R. Young: An Introduction to Nonharmonic Fourier Series. Acad. Press, New York, 1980.

This manuscript was prepared in spring 1990, while the author was holding a Max Kade research fellowship. He would like to thank the Mathematics Dept. of the University of Maryland at College Park for the hospitality.

Wiener Amalgams over Euclidean Spaces and Some of Their Applications

HANS G. FEICHTINGER, University of Vienna, Department of Mathematics, Strudlhofgasse 4, A-1090 Wien, AUSTRIA

1 INTRODUCTION

Wiener amalgams were introduced by the author in 1980 ([F1,2]). The concept was aimed at the possibility of describing local and global properties of a function or distribution independently (allowing to speak of increasing smoothness at infinity ...). For a very readable survey of 'ordinary' amalgams, built up on local L^p- and using global l^q-spaces, see the article by J. J. Fournier and J. Stewart [FSt], explaining applications of amalgams to a wide range of problems in analysis. In the present paper further applications of Wiener amalgam spaces will be explained, among them results (such as Thm.8 or several results in [F7]) which use these spaces only for the proof, but not in their statements. We also explain the properties of two families of operators, used for spline approximation and discretization of measures in the context of function spaces. Much of the material presented here was influenced by the author's experience in using Wiener amalgam spaces while proving results on atomic decompositions and during the work on the irregular sampling problem of functions (joint papers with K. Gröchenig). Besides the new results presented in this paper the reader will find a number of applications of Wiener amalgam spaces to questions arising in signal analysis, especially in regular and irregular sampling theory, in the paper [F7] in this volume.

2 A REVIEW OF WIENER AMALGAM SPACES

We start with a description of the general approach to Wiener amalgam spaces (formerly called Wiener type spaces or generalized amalgams). We mainly restrict our attention in this paper to an important subclass, i.e. to those Wiener amalgam spaces on $\mathbb{R}^m$ which are spaces of tempered distributions. In the final section we shall indicate some features of Wiener amalgams over locally compact groups and their applications.

We denote the Schwartz space of rapidly decreasing functions by $\mathcal{S}(\mathbb{R}^m)$. It is endowed with a natural family of seminorms, and the space $\mathcal{S}'(\mathbb{R}^m)$ of tempered distributions is its topological dual. $\mathcal{K}(\mathbb{R}^m)$ denotes the space of all compactly-supported continuous, complex-valued functions on $\mathbb{R}^m$, thus the usual space of test function $\mathcal{D}(\mathbb{R}^m)$ coincides with $\mathcal{S}(\mathbb{R}^m) \cap \mathcal{K}(\mathbb{R}^m)$. We write T_x for the translation operator given by $T_x f(y) := f(y-x)$. Generic constants will be denoted by $C_1, C_2, \ldots$, as they arise in the proofs; they may take different values in different proofs.

Wiener amalgam spaces arose from the idea of generating Banach spaces of functions or distributions whose elements are characterized by a certain local behavior (local membership in a certain function space) and a certain global behavior (partly expressed on terms of the local norm). If the local behavior is just L^p-summability, and the global behavior is just ℓ^q-summability these spaces where treated earlier as amalgams of L^p and ℓ^q in the literature. An informative survey has been given by J. J. Fournier and J. Stewart [FSt], cf. also [BDD]. If we want to use more general local norms (including norms of Besov- or Sobolev spaces) or if we would like a more general global notion (such as decay at a certain rate at infinity) the following concept turns out to be very useful.

Definition 1. A Banach space $(B, \|\cdot\|_B)$ is *uniformly localizable*, if

(L1) $\mathcal{S}(\mathbf{R}^m) \hookrightarrow (B, \|\cdot\|_B) \hookrightarrow \mathcal{S}'(\mathbf{R}^m)$ are continuous embeddings, and

(L2) for every $h \in \mathcal{D}(\mathbb{R}^m)$ one has $h \cdot f \in B$ for any $f \in B$ and there is some $C = C(h) > 0$ such that $\|T_x h \cdot f\|_B \leq C \cdot \|f\|_B$ for all $x \in \mathbb{R}^m$, $f \in B$.

Since all examples of interest will satisfy the following condition we shall assume for convenience that the *local component* $(B, \|\cdot\|_B)$ has property L3).

(L3) $(B, \|\cdot\|_B)$ is isometrically translation invariant, i.e. $\|T_x f\|_B = \|f\|_B$, with $L^1 * B \subseteq B$ and $\|g * f\|_B \leq \|g\|_1 \|f\|_B$ for $g \in L^1$, $f \in B$.

Remark 1. In view of (L2) the space

$$B_{loc} := \{f \in \mathcal{S}'(\mathbb{R}^m)\ ,\ h \cdot f \in B \text{ for any } h \in D(\mathbb{R}^m)\}$$

is well defined. Fixing any non-zero function $h \in \mathcal{D}(\mathbb{R}^m)$ the *control function* (f localized my means of h, and measured in the B-norm) is defined by

$$F_h(x) := \|T_x h \cdot f\|_B\ . \tag{1}$$

Following standard terminology in the theory of short time Fourier transforms h might be called a *window function*, and the most natural choice is a function h having a plateau type graph, i.e. taking values between zero and one, with $h(m) \equiv 1$ for all $m \in M$, M being some open set around the origin. However, these extra requirements are *not* part of the definition given above. Uniformly localizable spaces will be used as *local components*.

Examples. There is an abundance of uniformly localizable spaces. In addition to the ordinary L^p-spaces (used for 'ordinary amalgams', cf. [FSt]) there are the spaces $\mathcal{F}L^p$ which have been used for 'generalized amalgams' in [F3]. In fact $\mathcal{F}L^p := \{\hat{\sigma} \mid \sigma \in L^p\}$, the space of all tempered distributions which arise as Fourier transforms of L^p-functions (with the norm being taken from L^p), is uniformly localizable, since the embedding $\mathcal{S}(I\!R^m) \hookrightarrow \mathcal{F}L^1$ and the convolution relation $L^1 * L^p \subseteq L^p$ imply that the elements of (the isometrically translation invariant Banach algebra) $\mathcal{F}L^1$ define bounded multipliers on $\mathcal{F}L^p$. In order to measure local smoothness, one can use Lipschitz spaces, Besov spaces, potential spaces or, more generally, Triebel-Lizorkin spaces (see [F1] or [F4] for more examples).

For a description of the *global behavior* of the control function F_h given by (1) (which already involves the use of the local norm $\|\cdot\|_B$) we need a *global component* $(Y, \|\cdot\|_Y)$. It has to satisfy

(G1) $(Y, \|\cdot\|_Y)$ is a solid BF-space, i.e. a Banach space of locally integrable functions, continuously embedded into $L^1_{loc}(I\!R^m)$, with the extra property that for $f \in Y$ and $g \in L^1_{loc}(I\!R^m)$ the pointwise (a.e.) estimate $|g(x)| \leq |f(x)|$ implies $g \in Y$ and $\|g\|_Y \leq \|f\|_Y$.

(G2) $(Y, \|\cdot\|_Y)$ is translation invariant in the following sense:
For some $\alpha \geq 0$, $\|T_x f\|_Y \leq C \cdot w_\alpha(x)\|f\|_Y$, for all $f \in Y$, where $w_\alpha(x) := (1+|x|)^\alpha$.

(G3) $(Y, \|\cdot\|_Y)$ is a Banach convolution module with respect to the Beurling algebra $L^1_\alpha(I\!R^m) := \{f \mid fw_\alpha \in L^1(I\!R^m)\}$, i.e. $f * g \in Y$ for $f \in Y$ and $g \in L^1_\alpha(I\!R^m)$, together with the corresponding norm estimate $\|f * g\|_Y \leq \|g\|_{1,\alpha} \cdot \|f\|_Y$.

In most cases this property follows from (G2), e.g. if $\mathcal{S}(I\!R^m)$ is dense in Y.

Examples. Weighted L^q-spaces $L^q_s := \{f \mid fw_s \in L^q(I\!R^m)\}$ with natural norm $\|f\|_{q,s} := \|fw_s\|_q$, $0 < |s| \leq \alpha$ are the typical choice for Y, showing that the global space Y controls the behavior of F_h by means of decay conditions due to w_s (for $s < 0$ this may as well mean setting bounds on the rate of growth) as well as summability conditions (through the parameter q). Of course, in the case of several dimensions one might take mixed norms spaces and anisotropic weights (many other choices for Y are available then). All weights w appearing in this note (besides the weights w_s are supposed to be continuous, strictly positive, and *α-moderate* in the following sense: $w(x+y) \leq w_\alpha(y)w(x)$ for some $\alpha \geq 0$ and all $x, y \in I\!R^m$. This ensures that $L^q_s(I\!R^m) \hookrightarrow \mathcal{S}'(I\!R^m)$. The solidity of Y also implies that $Y \hookrightarrow W(L^1, Y)$ for any global component Y, hence $W(B, Y) \hookrightarrow \mathcal{S}'(I\!R^m)$.

Definition 2. Given B, Y as above we define the **Wiener Amalgam Space**

$$W(B, Y) := \{ f \in B_{loc} ,\ F_h \in Y \} \tag{2}$$

with the natural norm $\|f\,|W(B, Y)\| := \|F_h\|_Y$.

Remark 2. It has been shown in [F1] that under these circumstances (and more general ones) the space $W(B, Y)$ is a well defined Banach space. Under the assumptions made here it is continuously embedded into $\mathcal{S}'(I\!R^m)$. *Different*, non-zero window-functions h define *equivalent* norms. Since we shall make frequent use of this fact let us give a proof.

Proof. Given h_1 and h_2 let us estimate the Y-norm of F_{h_1} by means of F_{h_2}. First of all we observe that we may assume h_2 to be positive, since otherwise the local multiplier property (L2) allows us to show that for $h := |h_2|^2$ one has the estimate $\|F_h\|_Y \leq C_1 \cdot \|F_{h_2}\|_Y$. Assuming thus positivity of h_2 we find a linear combination of shifted versions of h_2, i.e. $h_3 := \sum_{n=1}^k T_{x_n} h_2$, such that $h_3 \geq \delta_0 > 0$ on $\text{supp}(h_1)$. On the other hand the translation invariance of Y allows us to estimate the norm of $F_{h_3} \leq \sum_{n=1}^k T_{-x_n} F_{h_2}$ by $C_2 \cdot \|F_{h_2}\|_Y$. Finally, we choose some $h_4 \in \mathcal{S}(I\!R^m)$ such that $h_4(t) = 1/h_3(t)$ for all $t \in \text{supp}(h_1)$. Then clearly $h_1 = h_3 \cdot (h_4 h_1)$ and therefore we end up with $\|F_{h_1}\|_Y \leq C_3 \cdot \|F_{h_2}\|_Y$. For reasons of symmetry this shows the required norm equivalence. □

Remark 3. Observe that ordinary weighted L^p-spaces are special Wiener amalgam spaces, since $L^p_w(\mathbb{R}^m) = W(L^p, L^p_w)$. We also mention that the norm on the spaces $W(B,Y)$ can be described equivalently using discrete global norms (cf. [F1]), which helps in 'guessing' what the correct inclusions between various spaces are. However, we prefer to stay with the 'continuous' norm based on (1) and (2) in this paper.

Remark 4 (*Duality*). It is also easy to verify that $\mathcal{D}(\mathbb{R}^m)$ (and with little extra arguments also $\mathcal{S}(\mathbb{R}^m)$) is embedded into $W(B,Y)$, and that each $h \in \mathcal{D}(\mathbb{R}^m)$ defines a bounded multiplier on $W(B,Y)$ by (L2). If $\mathcal{D}(\mathbb{R}^m)$ is a dense subspace of B as well as of Y, the dual space of $W(B,Y)$ can be identified in the natural way with $W(B',Y')$, and for many other operations – such as pointwise multiplication of Wiener amalgam spaces – 'coordinatwise' arguments may be used (see also [FGr], or [F2] for results on interpolation of Banach spaces). It will be convenient to use the notation $\langle \cdot, \cdot \rangle$ to describe duality between suitable pairs of Banach spaces.

Remark 5. We have the following useful result concerning convolution:

Assume that (B^1, B^2, B^3) and (Y^1, Y^2, Y^3) are Banach convolution triples, i.e. that

$$\|f * g\|_{B^3} \leq C_1 \cdot \|g\|_{B^1} \cdot \|f\|_{B^2} \quad \text{for all } g \in B^1,\ f \in B^2$$

and

$$\|F * G\|_{Y^3} \leq C_2 \cdot \|G\|_{Y^1} \cdot \|F\|_{Y^2} \quad \text{for all } G \in Y^1,\ F \in Y^2 .$$

Then the spaces $(W(B^1,Y^1), W(B^2,Y^2), W(B^3,Y^3))$ form a Banach convolution triple, i.e. for some constant $C_3 > 0$ one has

$$\|f * g \,|W(B^3,Y^3)\| \leq C_3 \cdot \|g\,|W(B^1,Y^1)\| \, \|f\,|W(B^2,Y^2)\| . \tag{3}$$

This is a generalization of Young's inequality for convolutions of L^p-functions (see [F1], [FSt], [BDD], [BS] for special cases).

Remark 6. By the convolution relations (A3) implies that $W(B,Y) * L^1_\alpha \subseteq W(B,Y)$, since we may identify L^1_α with $W(L^1, L^1_\alpha)$. It also ensures that $\mathcal{D}(\mathbb{R}^m)$ is dense in $W(B,Y)$ if $\mathcal{K}(\mathbb{R}^m)$ is dense in Y. We actually show that any $f \in W(B,Y)$ can be boundedly approximated by a net in $\mathcal{D}(\mathbb{R}^m)$.

Proof. Choosing first some $h \in \mathcal{D}(\mathbb{R}^m)$ such that $\|f - hf\,|W(B,Y)\| \leq \varepsilon/2$ we find $k \in \mathcal{D}(\mathbb{R}^m)$ with $\|hf - k * hf\,|W(B,Y)\| \leq \varepsilon/2$, thus $\|f - k * hf\,|W(B,Y)\| \leq \varepsilon$. Since $k * hf$ is the convolution product of two compactly supported distributions it is compactly supported itself. It also belongs to $\mathcal{D}(\mathbb{R}^m) * \mathcal{S}'(\mathbb{R}^m) \subseteq \mathcal{S}(\mathbb{R}^m)$, which guarantees the required smoothness. It is always possible to choose the function $h \in \mathcal{D}(\mathbb{R}^m)$ within a given a priori bound of the multiplier algebra of $W(B,Y)$ (cf. Prop.5 below). The L^1_α-boundedness of k (from some approximate unit) therefore implies that for some constant $C_0 > 0$ one has $\|k * hf\,|W(B,Y)\| \leq C_0 \|f\,|W(B,Y)\|$. □

Remark 7. There are also two version of the *Hausdorff-Young inequality* (HY in the sequel, see [BDu], [F2]). The first one, which works under the expected restrictions $1 \leq p,q \leq 2$, implies that the Fourier transform maps $W(L^p, L^q)$ into $W(L^{q'}, L^{p'})$, with $1/p' + 1/p = 1$, thus interchanging the roles of the local and the global component. The *generalized HY inequality*, which is strictly stronger, is based on the use of generalized amalgams (cf. [F2], [F3]),

$$\mathcal{F}(W(\mathcal{F}L^p, L^r)) \hookrightarrow W(\mathcal{F}L^r, L^p) \quad \text{for } 1 \leq r \leq p \leq \infty . \tag{4}$$

3 SPLINE APPROXIMATION AND DISCRETIZATION OPERATORS

In this section we assume that Ψ denotes a positive family of functions ($0 \leq \psi_i(x) \leq 1$ for all x), which is δ-*fine* and C-*bounded*, i.e. satisfies $\text{supp}(\psi_i) \subseteq B_\delta(x_i)$ for some $\delta > 0$ and all $i \in I$, and $\sum_{i\in I} \psi_i(x) \leq C < \infty$ for all $x \in I\!R^m$. In most cases we will work with δ-BUPUs (*Bounded Uniform Partitions of Unity*), i.e. families satisfying $\sum_{i\in I} \psi_i(x) \equiv 1$ a.e. on $I\!R^m$. In that case the family of 'midpoints' has to be δ-dense in $I\!R^m$, i.e. $I\!R^m = \bigcup_{i\in I} B_\delta(x_i)$. We write $|\Psi|$ for the size of Ψ which is given as $\sup_{i\in I} \text{diam}(\text{supp}(\psi_i))$, thus $|\Psi| \leq \delta$ for any δ-BUPU.

Examples. In the most simple one-dimensional case, for a given δ-dense family $(x_i)_{i\in Z}$, we could use a 'rectangular' partition of unity which consists of the indicator functions $\mathbf{1}_{(m_{i-1},m_i)}$, with $m_i := (x_i + x_{i-1})/2$ for $i \in \boldsymbol{Z}$. The analogue of this in higher dimensions is the indicator functions of the *Voronoi region* associated with x_i, defined by

$$V_i := \{\, x \mid |x_i - x| < |x - x_j| \ \forall j \neq i \,\} \ . \tag{5}$$

Another simple version of a δ-BUPU is a system of triangular functions $(\Delta_i)_{i\in I}$, which are piecewise linear on the interval $[x_i, x_{i+1}]$ and satisfy $\Delta_i(x_i) = 1$, and $\Delta_i(x_j) = 0$ for $i \neq j$.

In contrast to the original definition in [F1] (which was designed to allow a proof of the equivalence of discrete and continuous norms on Wiener amalgam spaces) we do *not* assume here that the family of points is *relatively separated* (cf. section 3 of [F7]), i.e. no restriction on the (upper) density of the family of points is made, and cases where X has accumulation points are included in the discussion. On the other hand it is clear from the assumptions, that in such cases the functions ψ_i have to be very small (some of them might be zero). Given such a family Ψ we define the following two operators.

Definition 3. The *spline quasi-interpolant* for the family Ψ is defined for any continuous function f by

$$Sp_\Psi f := \sum_{i\in I} f(x_i)\,\psi_i \ . \tag{6}$$

Along with this operator we also define a 'discretization operator', mapping locally integrable functions into discrete measures

$$D_\Psi f := \sum_{i\in I} \langle f, \psi_i \rangle \cdot \delta_{x_i} = \sum_{i\in I} \left(\int f(x)\psi_i(x)\,dx \right) \cdot \delta_{x_i}, \quad for f \in L^1_{loc}(I\!R^m) \ . \tag{7}$$

Remark 8. Actually, D_Ψ is well defined for arbitrary Radon measures, if Ψ is a family of continuous functions and if we interpret $\langle \mu, \psi_i \rangle$ as the natural duality between $\mathcal{K}(I\!R^m)$ and $\mathcal{R}(I\!R^m) = \mathcal{K}(I\!R^m)'$. In fact, for continuous Ψ the operator Sp_Ψ is continuous on $\mathcal{K}(I\!R^m)$ and D_Ψ is the dual operator, since

$$\langle D_\Psi \mu, f \rangle = \langle \mu, Sp_\Psi f \rangle \quad \text{for } f \in \mathcal{K}(I\!R^m) \text{ and } \mu \in \mathcal{R}(I\!R^m) \ . \tag{8}$$

Clearly Sp_Ψ maps $C^b(I\!R^m)$, the space of bounded continuous complex-valued functions on $I\!R^m$, endowed with the sup-norm $\|f\|_\infty := \sup_{x\in I\!R^m} |f(x)|$, into itself, with $\|Sp_\Psi f\|_\infty \leq C\|f\|_\infty$ for all $f \in C^b(I\!R^m)$ if Ψ is continuous. On the other hand both of these operators are not well defined or at least not bounded on any of the L^p-spaces for $1 \leq p < \infty$. On the other hand for 'smooth' functions in L^p one may expect that $Sp_\Psi f$ is a good approximation to f in the L^p-sense for small δ. It turns out that the spaces $W(C^0, Y)$ provides the appropriate setting for results on boundedness or approximation behavior of these operators. Band-limited function estimates of this type (based on the mean-value theorem which can be applied only for functions on Euclidean spaces) have been crucial in our approach to the irregular sampling problem (cf. [FG5]). We have the following results on these operators.

Lemma 1. For any $\delta > 0$ and $C > 0$ there exists some $C_\delta > 0$ such that for any C-bounded and δ-fine continuous family Ψ one has

$$\|Sp_\Psi f\,|W(C^0,Y)\| \leq C_\delta \cdot \|f\,|W(C^0,Y)\| \,. \tag{9}$$

Proof. Since we are free to choose our 'window' function defining the norms on both sides of the inequality, we choose any $h \in D(\mathbb{R}^m)$ for the right side, and use a $k \in \mathcal{D}(\mathbb{R}^m)$ for the left norm, with $k(t) \equiv 1$ on $\mathrm{supp}(h) + B_\delta(0)$. In this case we have the pointwise inequality $F_h(Sp_\Psi f) \leq C \cdot F_k(g)$ (a local version of the sup-norm estimate $\|Sp_\Psi f\|_\infty \leq \|f\|_\infty$), from which the required estimate follows. □

The easiest way to verify that these spline-type functions approximate f in the appropriate sense is by means of the concept of a local oscillation. For $\delta > 0$ we define the *δ-oscillation* of a continuous function by

$$\mathrm{osc}_\delta f(x) := \sup_{|y|\leq\delta} |f(x) - f(x+y)| \,. \tag{10}$$

Remark 9. The function $\mathrm{osc}_\delta f$ has many features of a modulus of continuity for the $W(C^0, L^p)$-norm , usually defined as $\omega_W(f,\delta) := sup_{|y|\leq\delta}\|f - T_y f\,|W(C^0,Y)\|$. However, despite the results described in Thm.3 below we have not been able to determine whether the two expressions are equivalent in terms of rate of convergence.

Using the concept of local oscillation we are able to formulate

Lemma 2. For any $f \in W(C^0,Y)$ one has $\mathrm{osc}_\delta f \in W(C^0,Y)$. If $\mathcal{K}(\mathbb{R}^m)$ is a dense subspace of Y then $\|\mathrm{osc}_\delta f\,|W(C^0,Y)\| \to 0$ for $\delta \to 0$, which implies $\|Sp_\Psi f - f\,|W(C^0,Y)\| \to 0$.

Proof. Observe that one has the pointwise estimate $\mathrm{osc}_\delta f \leq 2 \cdot F_h$, for any $h \in \mathcal{D}$ which satisfies $h(x) \equiv 1$ on $B_\delta(0)$. Since Y is solid and $\mathcal{K}(\mathbb{R}^m)$ is dense in Y there exists $k \in \mathcal{D}$ such that $\|F_h \cdot (1-k)\|_Y \leq \varepsilon/4$. Uniform continuity of f on compact sets implies that we have uniform convergence of $\mathrm{osc}_\delta f$ on compact subsets, therefore one has for sufficiently small $\delta > 0$.

$$\|\mathrm{osc}_\delta f \cdot k\,|W(C^o,Y)\| \;\leq\; \sup_{u\in\mathrm{supp}(k)} |\mathrm{osc}_\delta f(u)| \cdot \|k\,|W(C^0,Y)\| \;<\; \frac{\varepsilon}{2} \,.$$

Altogether this implies for sufficiently small δ

$$\|\mathrm{osc}_\delta f\,|W(C^0,Y)\| \leq \|\mathrm{osc}_\delta f \cdot (1-k)\,|W(C^0,Y)\| + \frac{\varepsilon}{2} \leq 2\,\|F_h \cdot (1-k)\,|W(C^0,Y)\| + \frac{\varepsilon}{2} < \varepsilon$$

In order to verify the second statement we use the pointwise estimate

$$|Sp_\Psi f - f| \leq \mathrm{osc}_\delta(f) \;\text{ for any } \delta\text{-BUPU } \Psi \,, \tag{11}$$

which follows easily by summation over the pointwise estimates for the terms

$$|f(x)\cdot\psi_i(x) - f(x_i)\cdot\psi_i(x)| \;\leq \mathrm{osc}_\delta f(x)\cdot\psi_i(x) \;\; \forall x \in \mathbb{R}^m \;\;.$$

□

In connection with these operators (as well as with convolution operators) and with the characterization of compact subsets in function spaces the following two notions turn out to be useful (cf. [F4]).

Definition 4. Let M be a bounded set in a Banach space $(B, \|\cdot\|_B)$ of tempered distributions.

(i) M is called *(uniformly) tight* if there exists some bounded net of multipliers $(h_\gamma)_{\gamma\in\Gamma}$ in $\mathcal{D}(\mathbb{R}^m)$, such that $\lim_\gamma h_\gamma \cdot f = f$ uniformly over M, in the B-norm sense, i.e. for every $\varepsilon > 0$ there exists $h \in \mathcal{D}(\mathbb{R}^m)$ such that

$$\|h \cdot f\|_B \leq C \cdot \|f\|_B \ \ \forall f \in B \,,$$

and

$$\|h \cdot f - f\|_B \leq \varepsilon \cdot \|f\|_B \ \ \forall f \in M \,. \tag{12}$$

(ii) M is *equicontinuous* if for every $\varepsilon > 0$ there exists $\delta > 0$ such that

$$\|T_x f - f\|_B \leq \varepsilon \ \ \forall f \in M, \ x \in \mathbb{R}^m \text{ with } |x| \leq \delta \,. \tag{13}$$

Remark 10 (*Stability of these concepts*). Note that both properties (tightness and equicontinuity) are presevered under passage to the closure, i.e. a set M is tight in B if and only if $\overline{M}$ is tight in B. Actually, the following holds. If a set can be approximated by tight (or equicontinuous) sets, then it is tight (equicontinuous) itself. More precisely, if for a set M and each $\varepsilon > 0$ there exists a tight M_ε such that $M \subseteq \bigcup_{x\in M_\varepsilon} B_\varepsilon(x)$, then M is tight itself.

Remark 11 (*Stability under convolution and multiplication*). It is an easy exercise to check the following. Let M_1, M_2 be tight subsets of B^1, B^2. Then $M_1 * M_2$ is tight in B^3, if (B^1, B^2, B^3) is a convolution triple. In a similar way equicontinuity is preserved under pointwise multiplication, if one has a pointwise multiplication triple (with appropriate norm estimates). Moreover, a convolution product is equicontinuous if only one factor is equicontinuous, and a pointwise product $M_1 \cdot M_2$ is tight if only one factor is tight (cf. [F3,4] for details).

The concept of oscillation is very useful in order to describe equicontinuity in the spaces $W(C^0, Y)$. The following result also sheds some light on the usefulness of the concept equicontinuity for the approximation behavior of spline approximations (cf. [F4] for results on equicontinuity in general spaces).

Theorem 3 (*characterization of equicontinuity in* $W(C^0, Y)$).

For a bounded subset M in $W(C^0, Y)$ the following conditions are equivalent.

(a) M is equicontinuous in $W(C^0, Y)$.

(b) For every $\varepsilon > 0$ there exists $k \in \mathcal{D}(\mathbb{R}^m)$ such that

$$\|k * f - f \,|W(C^0, Y)\| \leq \varepsilon \ \ \forall f \in M \,.$$

(c) The oscillation functions tend to zero uniformly over M, i.e. for every $\varepsilon > 0$ there exists $\delta_0 > 0$ such that $\|\mathrm{osc}_\delta f\|_Y \leq \varepsilon$ for all $f \in M$, $\delta \leq \delta_0$.

(d) The same as (c), with $\|\mathrm{osc}_\delta f \,|W(C^0, Y)\| \leq \varepsilon$.

(e) The family of spline quasi-interpolants $Sp_\Psi f$ is uniformly convergent to f in the norm of $W(C^0,Y)$, i.e. for any $\varepsilon > 0$ there exists $\delta_0 > 0$ such that $\|f - Sp_\Psi f\,|W(C^0,Y)\| \le \varepsilon$ for all $f \in M$, and all δ-BUPU's, with $\delta \le \delta_0$.

Proof.
(a) $\Rightarrow$ (b): Assume that M is equicontinuous. Choose $\delta > 0$ such that $\|T_x f - f\,|W(C^0,Y)\| \le \varepsilon$ for all $f \in M$. Then any function $k \in \mathcal{K}^+(\mathbb{R}^m)$ with $\int k(x)\,dx = 1$ and support in $B_\delta(0)$ satisfies $\|k * f - f\,|W(C^0,Y)\| \le \varepsilon$ for all $f \in M$. In fact, writing the convolution as a vector-valued integral one obtains

$$\|k * f - f \cdot \int k(y)\,dy\,|W(C^0,Y)\| \le \int \|T_y f - f\,|W(C^0,Y)\| \cdot k(y)\,dy \le \varepsilon \quad \forall f \in M \ .$$

(b) $\Rightarrow$ (d), hence (c): We start with the pointwise estimate

$$\mathrm{osc}_\delta(k * f) \le \mathrm{osc}_\delta(k) * |f| \ . \tag{14}$$

This implies $\mathrm{osc}_\delta(f) \le \mathrm{osc}_\delta(k * f) + \mathrm{osc}_\delta(f - k * f)$, hence

$$\|\mathrm{osc}_\delta(f)\,|W(C^0,L^p)\| \le C\cdot\|\mathrm{osc}_\delta k\,|W(C^0,L^1)\|\cdot\|f\,|W(L^1,L^p)\| + \|\mathrm{osc}_\delta(f - k*f)\,|W(C^0,L^p)\| \ .$$

The second term is small, since $\mathrm{osc}_\delta(f - k * f)$ is dominated by the control function of $2(f - k * f)$, which is not only small in the Y-norm but also in the $W(C^0,Y)$-norm (as a consequence of the independence of the norm on the specific window). The uniform continuity of $k \in \mathcal{K}(\mathbb{R}^m)$ allows us to choose δ in a way that the first expression on the right side is small as well.
(d) $\Rightarrow$ (a): This implication is obvious, since the definition of $\mathrm{osc}_\delta f$ implies $|T_x f - f| \le \mathrm{osc}_\delta f$ for $|x| \le \delta$, given the estimate in $W(C^0,Y)$.
(c) $\Rightarrow$ (e): Follows from (11) (cf. the proof of Lemma 1).
(e) $\Rightarrow$ (a): Actually we suppose only, that the convergence of $Sp_\Psi f$ to f is uniform for some subnet, satisfying a few extra conditions. We consider in this paragraph only δ-BUPUs which are equicontinuous (each for itself) and with a finite degree of overlap.

Explicitly we suppose that we have: Ψ is a δ-BUPU such that $(B_{3\delta}(x_i))_{i\in I}$ defines a covering of finite (maximal) height of $\mathbb{R}^m$, and such that the functions $\{\psi_i \,|\, i \in I\}$ form an equicontinuous subfamily in $C^0(\mathbb{R}^m)$. This set of conditions is of course satisfied by any family, derived by tensor products of triangular functions (with basic width δ) over a regular lattice. In this case the amount of overlap between their supports depends only on dimension of the space.

Under these circumstances we claim: For any such Ψ the set

$$\{Sp_\Psi f \,|\, f \in W(C^0,Y),\ \|f\,|W(C^0,Y)\| \le 1\} \tag{15}$$

is equicontinuous in $W(C^0,Y)$.

The proof of this is based on the following estimate. We assume that $|y| \le \delta$ Choosing first some non-negative function ϕ such that $\mathrm{supp}(\phi) \subseteq B_{3\delta}(0)$ and $\phi(t) \equiv 1$ on $B_{2\delta}(0)$ we have $\psi_i(x - y) = \psi_i(x - y) T_{x_i}\phi(x)$, since $\mathrm{supp}(T_y\psi_i) \subseteq B_{2\delta}(x_i)$. This gives the following pointwise estimate:

$$\mathrm{osc}_\delta(Sp_\Psi f)(x) = \sup_{|y|\le\delta} |Sp_\Psi f(x - y) - Sp_\Psi f(x)| \le \sup_{|y|\le\delta} \sum_{i\in I} |f(x_i)||\psi_i(x - y) - \psi_i(x)| \le$$

$$\le \sum_{i\in I} |f(x_i)| \cdot \sup_{|y|\le\delta} |\psi_i(x-y) - \psi_i(x)| \cdot T_{x_i}\phi(x) \le \sum_{i\in I} |f(x_i)| \cdot \|\mathrm{osc}_\delta \psi_i\|_\infty \cdot T_{x_i}\phi \ .$$

On the basis of the equicontinuity of Ψ we can find for any given $\eta > 0$ some $\delta_0 > 0$ such that $\|\mathrm{osc}_\delta \psi_i\|_\infty \le \eta$. Using now the fact that the finite height condition implies (and is actually equivalent to the statement that the discrete measure $\sum_{i\in I} \delta_{x_i}$ belongs to $W(M, L^\infty)$) we obtain the required norm estimate.

$$\|\mathrm{osc}_\delta(Sp_\Psi f)\,|W(C^0,Y)\| \le \sup_{i\in I} \|\mathrm{osc}_\delta \psi_i\|_\infty \cdot \|\sum_{i\in I} |f(x_i)| \cdot \delta_{x_i} * \phi\,|W(C^0,Y)\| \le$$

$$\le \eta \cdot C_1 \cdot \|\sum_{i\in I} |f(x_i)| \cdot \delta_{x_i}\,|W(M,Y)\| \cdot \|\phi\,|W(C^0,L^1)\| \le$$

$$\le \eta \cdot C_2 \cdot \|\sum_{i\in I} \delta_{x_i}\,|W(M,L^\infty)\| \cdot \|f\,|W(C^0,Y)\| \cdot \|\phi\,|W(C^0,L^1)\| \ .$$

Since M is bounded in $W(C^0, Y)$ it is possible to chose η (hence δ_0) in a way such that

$$\|\mathrm{osc}_\delta(Sp_\Psi f)\,|W(C^0,Y)\| \le \varepsilon \ \ \forall f \in M \ .$$

We have thus verified that if M satisfies (e) it can be approximated uniformly in the $W(C^0, Y)$-norm by equicontinuous set $\{Sp_\Psi f | f \in M\}$, and the proof is complete in view of Remark 10. □

Next we come to a discussion of the discretization operators D_Ψ. Without proof we state the following results (cf.[FG4]):

Lemma 4.

(i) For any $\gamma > 0$ and $C > 0$ the family of operators $\mu \mapsto D_\Psi\mu$, where Ψ runs through the system of bounded and γ-fine and C-bounded families (without necessarily giving a BUPU), is uniformly bounded on $W(M, Y)$.

(ii) As Ψ runs through the family of δ-BUPUs one has convergence of the family $D_\Psi\mu \to \mu$ for $|\Psi| \to 0$ in the vague topology, i.e. $D_\Psi\mu(k) \to \mu(k)$ for any $k \in \mathcal{K}(\mathbb{R}^m)$. In particular, $D_\Psi\mu \to \mu$ in the w*-sense, if Y is the dual of a localizable space X which contains $\mathcal{D}(\mathbb{R}^m)$ as a dense subspace (i.e. in the $\sigma(W(M,Y), W(C^0, X))$ sense, to be more precise).

Using this terminology we next describe the general compactness criterion for distribution spaces (known as the Riesz-Weil criterion in the case of L^p-spaces, cf. [F4]) for the setting of Wiener amalgams.

Proposition 5. Assume that $(B, \|\cdot\|_B)$ is a Banach space satisfying (L1) to (L3), and that $\mathcal{D}(\mathbb{R}^m)$ is dense in B as well as in Y. Then a closed bounded subset $M \subseteq W(B, Y)$ is *compact* if and only if it is uniformly tight and equicontinuous with respect to the $W(B, Y)$-norm. This holds if and only if it is possible to determine for any $\varepsilon > 0$ two functions $h, k \in \mathcal{D}(\mathbb{R}^m)$ such that

$$\|h(k * f)\,|W(B,Y)\| < \varepsilon \ \ (\text{and} \ \ \|k * (hf)\,|W(B,Y)\| < \varepsilon) \ \ \forall f \in M \ . \qquad (16)$$

Proof. We have to check a few properties as described in [F4]. First of all it is already clear that $W(B, Y)$ has continuous shift, i.e. $\|T_x f - f\|_B \to 0$ for $x \to 0$, for all $f \in W(B, Y)$.

The second requirement is (12), i.e. the existence of an approximate unit in $\mathcal{D}(\mathbb{R}^m)$ for B, bounded in the operator norm, i.e. some family $(h_\gamma)_{\gamma\in\Gamma}$ in $\mathcal{D}(\mathbb{R}^m)$ satisfying

$$\|h_\gamma \cdot f\|_B \leq C \cdot \|f\|_B \text{ and } \|h_\gamma \cdot f - f\|_B \to 0 \ \forall f \in W(B,Y) .$$

This property, however, can be derived from assumption (L2) as $\mathcal{D}(\mathbb{R}^m)$ is contained in the pointwise multiplier A of B, and the multiplier space of $W(B,Y)$ contains $W(A,L^\infty)$ (cf. [FGr]). □

For a discussion of operators preserving (or improving) the above properties we also use a related concept for (families of) operators.

Definition 5. We call a bounded family of operators $(T_\gamma)_{\gamma\in\Gamma}$ *uniformly tight* (a single operator just *tight*) if the set $T_\Gamma M := \{T_\gamma x \mid x \in M, \ \gamma \in \Gamma\}$ is uniformly tight for every tight set M. We call it *strictly tight* if there exists some compact set $K_0 \subseteq \mathbb{R}^m$ such that $\mathrm{supp}(f) \subseteq S \subseteq \mathbb{R}^m$ implies $\mathrm{supp}(T_\gamma f) \subseteq S + K_0$. Finally, we call such a family *equicontinuous* if the set $T_\Gamma M$ is equicontinuous for every bounded set M.

Remark 12 (*Stability of these concepts*). As a consequence of the stability of the notions of tightness and equicontinuity (cf. remarks 10/11) the (norm) limit of a sequence of tight (or equicontinuous) operators is tight (equicontinuous) itself. It follows that these operators form closed subspaces of the space of operators between two Banach spaces of distributions, and closed subalgebras in the operator algebra of any given space.

Remark 13. Due to the boundedness assumption on the family $(T_\gamma)_{\gamma\in\Gamma}$ the tightness is already satisfied if the tightness is satisfied for bounded sets M which have common compact support. Using the concept of the translate of an operator between translation invariant function spaces, given by $T_x(T) : f \mapsto T_{-x}(T(T_x f))$ one may describe the above notion of equicontinuity equivalently as equicontinuity in the Banach space of operators (in the operator norm sense).

By means of these concepts we can show, for example, that for any $k \in \mathcal{K}(\mathbb{R}^m)$ we have uniform convergence of $D_\Psi f * k$ to $f * k$, on the unit ball of L^p, i.e. the operators $T_\Psi : f \mapsto D_\Psi f * k$ converge to the convolution operator $f \mapsto f * k$, for $|\Psi| \to 0$.

To prove this we note that the family $\{D_\Psi f, \ |\Psi| \leq 1\}$ is a tight family in $W(M,L^p)$ for $f \in W(M,L^p)$. Since convolution by k is a tight and equicontinuous operator from $W(M,L^p)$ into $W(C^0,L^p)$ (this is an immediate consequence of the convolution relations and the fact that convolution commutes with translation) we find that $\{D_\Psi f * k, \ |\Psi| \leq 1\}$ is actually relatively compact in $W(C^0,L^p)$. Since the w*-convergence shows that we have at least pointwise convergence $D_\Psi f * k(x) \to f * k(x)$ for every $x \in \mathbb{R}^m$ the proof is complete. In fact, if we would *not* have norm convergence of this net there would be a subnet staying away from $f*k$. Applying the above argument to that subnet would produce a contradiction. What we have shown can be summarized as follows (in a typical form).

Lemma 6. For every $f \in W(M,L^p)$ and $k \in W(C^0,L^1)$ we have

$$\|(D_\Psi f - f) * k \,|W(C^0,L^p)\| \to 0 \text{ for } |\Psi| \to 0 \; ; \tag{17}$$

in particular, we have uniform as well as L^p-convergence.

Equipped with these arguments it is now easy to check that most product-convolution or convolution-product operators (for short PC or CP operators, cf. [BS]) act as compact operators on L^p-spaces (or other localizable spaces).

Lemma 7. Let $g \in W(L^{p'}, L^1)$ and $h \in W(L^p, L^\infty)$ be given. Then the CP-operator $Tf := (f * g) \cdot h$ is a bounded operator L^p. If moreover $h \in W(L^p, C^0)$, then T is a compact operator on L^p for $1 \leq p < \infty$.

Proof. Boundedness follows from the multiplier and convolution theorems

$$Tf = (f * g) \cdot h \in (W(L^p, L^p) * W(L^{p'}, L^1)) \cdot W(L^p, L^\infty) \subseteq W(C^0, L^p) \cdot W(L^p, L^\infty) \subseteq L^p \ .$$

If h belongs to $W(L^p, C^0)$ (this is the closure of $\mathcal{K}(\mathbb{R}^m)$ in $W(L^p, L^\infty)$), then pointwise multiplication by h is a tight operator (as norm limit of tight operators). Thus convolution by g maps the unit ball of L^p into an equicontinuous subset of $W(C^0, L^p)$. Pointwise multiplication by h transforms this equicontinuity (because h is also equicontinuous in the $W(L^\infty, L^p)$ sense) into equicontinuity in the L^p-sense and furthermore produces tightness. Thus altogether we obtain relative compactness of $T(M)$ in L^p for any bounded subset $M \subseteq L^p$ by the compactness criterion. □

Remark 14. The last result is a prototype of general results of this form. Many similar statements can be derived, using the convolution and multiplication theorems for weighted Wiener amalgam spaces. Results on the boundedness of $PC - CP$ operators motivated the work of Busby and Smith [BS].

Almost all of the results described so far hold for locally compact Abelian groups. The following result concerns a special problem on $\mathbb{R}^m$, where dilations play a role, i.e. a special form of spline approximations for L^p-spaces over $\mathbb{R}^m$.

The question is the following: Since the operators Sp_Ψ are not bounded on L^p-spaces it is natural to use regularization operators (convolution by decent functions) before applying the spline operators. A simple cascade argument yields the following result on approximation by splines.

Given $f \in L^p(\mathbb{R}^m)$ and $\varepsilon > 0$ there exists a $k \in \mathcal{K}(\mathbb{R}^m)$ with $\|k * f - f\|_p \leq \varepsilon/2$. Fixing k we find some $\delta > 0$ such that $\|Sp_\Psi(k * f) - k * f\|_p \leq \varepsilon/2$ for any δ-BUPU Ψ, hence $\|Sp_\Psi(k * f) - f\|_p \leq \varepsilon$, for δ small enough.

We conclude this section with a discussion of the following question, arising from this description. If we obtain the function k by suitable contraction (isometrically in L^1) of a fixed function k_0, can we choose the sampling rate (used to obtain the values $f(x_i)$ needed to built the function $Sp_\Psi f$) at the minimal expected rate, which would be given by the contraction factor. Since the answer is positive in case of L^2 (see [Au]) we can be optimistic.

The following notation for dilation operators appears to be useful:

$$St_\rho f(x) := \rho^{-m} f(x/\rho) \ . \tag{18}$$

Writing q for the dual index, given by $1/q + 1/p = 1$, it satisfies the following rules:

$$\|St_\rho f\|_p = \rho^{-m/q} \|f\|_p \quad \forall f \in L^p, \rho > 0 \ ; \tag{19}$$

$$St_\rho(f * g) = St_\rho f * St_\rho g \tag{20}$$

The following convention keeps notations short:
Writing $\rho\Psi := (\rho^m St_\rho \psi_i)_{i \in I}$ one has $|\rho\Psi| = |\rho||\Psi|$, i.e. $\rho\Psi$ is a $\rho\delta$-BUPU if Ψ is a δ-BUPU.

Our L^p-version of the spline-approximation theorem now reads as follows.

Theorem 8. For $h \in W(C^0, L^1)$ with $\int_{\mathbb{R}^m} h(x)\, dx = 1$ for any $f \in L^p(\mathbb{R}^m)$, $1 \leq p < \infty$

$$\|f - Sp_{\rho\Psi}(St_\rho h * f)\|_p \to 0 \ , \ \text{for } \rho \to 0 \ . \tag{21}$$

Proof. We are going to derive the result from the following two assertions:
(i) The operators $f \to Sp_{\rho\Psi}(St_\rho h * f)$ are uniformly bounded on $L^p(\mathbb{R}^m)$;
(ii) For $f \in \mathcal{K}(\mathbb{R}^m)$ and $h \in \mathcal{K}(\mathbb{R}^m)$ one has common compact supports of the functions $Sp_{\rho\Psi}(St_\rho h * f)$, $\rho \le 1$, and $\|f - Sp_{\rho\Psi}(St_\rho h * f)\|_\infty \to 0$ for $\rho \to 0$.
The key to our proof is the identity

$$Sp_{\rho\Psi}(St_\rho h * f) = St_\rho(Sp_\Psi(h * St_{\rho^{-1}} f)) \text{ for } \rho > 0 \ . \tag{22}$$

It implies (together with the above rules) that (i) holds for $h \in W(C^0, L^1)$:

$$\begin{aligned} \|St_\rho(Sp_\Psi(h * St_{\rho^{-1}} f))\|_p &= \rho^{-m/q}\|Sp_\Psi(h * St_{\rho^{-1}} f)\|_p \le \\ &\le C_0 \cdot \rho^{-m/q}\|Sp_\Psi(h * St_{\rho^{-1}} f)\,|W(C^0, L^p)\| = \\ &= C_0 \cdot \rho^{-m/q}\|Sp_\Psi\|\,\|h * St_{\rho^{-1}} f)\,|W(C^0, L^p)\| = \\ &= C_1 \cdot \rho^{-m/q}\|Sp_\Psi\|\,\|h\,|W(L^q, L^1)\|\,\|St_{\rho^{-1}} f)\,|W(L^p, L^q)\| = \\ &= C_2 \cdot \rho^{-m/q}\|Sp_\Psi\|\,\|h\,|W(L^q, L^1)\|\,\rho^{mq}\,\|f\|_p = C_3 \cdot \|h\,|W(L^q, L^1)\| \cdot \|f\|_p \ . \end{aligned}$$

Here $\|Sp_\Psi\|$ denotes the operator norm of Sp_Ψ. In order to verify (ii) we note that for f, $h \in \mathcal{K}(\mathbb{R}^m)$ the family $St_\rho h * f$ is not only bounded, with common compact support (for $\rho \le 1$), but also (uniformly) equicontinuous, since

$$\|T_x(St_\rho h * f) - St_\rho h * f\|_\infty = \|St_\rho h * (T_x f - f)\|_\infty \le$$

$$\le \|St_\rho h\|_1 \|T_x f - f)\|_\infty = \|h\|_1 \|T_x f - f\|_\infty \longrightarrow 0 \ ,$$

hence

$$\sup_{\rho \in (0,1)} \|Sp_\Phi(St_\rho h * f) - St_\rho h * f\|_\infty \le \|h\|_1 \sup_{|x| \le |\Phi|} \|T_x f - f\|_\infty \longrightarrow 0$$

for $|\Phi| \to 0$, and

$$\|Sp_\Psi(St_\rho h * f) - St_\rho h * f\|_\infty \ \le \ \|h\|_1 \sup_{|x| \le \rho|\Psi|} \|T_x f - f\|_\infty \longrightarrow 0 \quad \text{for } \rho \to 0 \ .$$

Given this estimate it is easy to see that it extends to $f \in L^p$ by continuity. Having verified (i) for $f \in L^p$ it is also possible to extend (i) to the case $h \in W(L^q, L^1)$. In fact, it is possible to approximate the expression $St_\rho(Sp_\Psi(h * St_{\rho^{-1}} f))$ by an equivalent expression involving only $\tilde{h} \in \mathcal{K}(\mathbb{R}^m)$, due to the estimate (resulting again from (i))

$$\|St_\rho(Sp_\Psi(h * St_{\rho^{-1}} f) - Sp_\Psi(\tilde{h} * St_{\rho^{-1}} f))\|_p \le$$

$$\le C_0 \cdot \rho^{-m/q}\|Sp_\Psi|W(C^0, L^p)\|\,\|(h - \tilde{h}) * St_{\rho^{-1}} f\,|W(C^0, L^p)\| = C_3 \cdot \|h - \tilde{h}\,|W(L^q, L^1)\|\,\|f\|_p.$$

This shows that the result holds (for given p) for any $h \in W(L^q, L^1)$. □

Remark 15. Using the last argument of the proof of Theorem 3 it is possible to verify that the stated convergence takes place uniformly on a bounded set $M \subseteq B$ if and only if M is equicontinuous in $L^p(\mathbb{R}^m)$.

4 WIENER AMALGAMS ON LC GROUPS AND WAVELET THEORY

Throughout this paper we have treated Wiener amalgam spaces over $\mathbb{R}^m$, which is of course just a typical special case of a locally compact Abelian group with respect to addition. The natural setting for Wiener amalgam spaces are, however, (at least) general lc. (= locally compact) groups, since only the availabilty of a local norm (such as a p-norm based on the existence of a left invariant Haar measure) and some global norm (again typically some L^q-space with some submultiplicative weight) are required for the definition of the spaces $W(B,Y)$. Since left and right translations don't commute in general, a number of technical problems comes up, and 'left' and 'right' Wiener amalgam spaces have to be considered, the space Y has to be assumed to be both, left- and right translation invariant.

However, amalgam spaces turn out to behave quite satisfactory, and it is not hard to verify duality and pointwise multiplier theorems. For the convolution theorem there is no problem for so-called [IN]-groups. Such groups have the property that some neighborhood Q of the identity is invariant under inner automorphisms, i.e. satisfies $y^{-1}Qy = Q$ (see [BS],[F1]). A typical example is the reduced Heisenberg group $\boldsymbol{H}_m := \mathbb{R}^m \times \mathbb{R}^m \times \boldsymbol{T}$ (cf. [FG1]). In this case the result about convolution triples is still valid (cf. [BS],[F1]), and for a while it seemed to be the natural setting. Later it turned out, that in the proof of certain results about atomic decompositions related to integrable and irreducible representations (presented in [FG1-3] and [Gr]) results of this type were required for groups such as the '$ax + b$-group' of affine transformations of the real line. Actually, the use of the oscillation concept was triggered by the successful use of this concept for questions in general wavelet theory in [Gr]. It turned out, that it is possible to prove the required convolution relations by introducing extra weights (e.g. the weight steming from the asymptotic behavior of the norm of right translation operators over Wiener amalgam spaces defined by means of left translations).

These modified convolution relations, together with the use of operators of the form Sp_Ψ and D_Ψ (as described in section 3), turned out to be extremely useful in proving general results about atomic decompositions (cf. [FG1-3]) or about recovery of a distribution from a family of coefficients, taken with respect to a coherent system of test functions (see [Gr]). In fact, these problems can be shown to be equivalent to problems concerning functions on these locally compact groups which satisfy a certain convolution relation, and are thus closely related to questions about band-limited functions over $\mathbb{R}^m$ (a summary of these connections is given in [F5]). A summary of this theory of coherent non-orthogonal series expansions is given in [CFG], a more comprehensive presentations has been given by Heil and Walnut in [HW]. In section 4 of [HW] it is also pointed out how Wiener amalgam spaces can be used within wavelet theory (see also [He] and [W]).

Amalgam spaces are also useful in many other areas of analysis, e.g. in the study of almost periodic functions [AL], [F6].

References

[AL] L. N. Argabright, J. Gil de Lamadrid: Fourier analysis of unbounded measures on locally compact abelian groups. Memoirs Amer. Math. Soc. 145 (1974).

[Au] J. P. Aubin: Applied Functional Analysis, Wiley, 1984.

[BDo] M. G. Beaty, M. M. Dodson: Derivative sampling for multiband signals, Numer. Funct. Anal. and Optimiz. 10 (9&10), 1989, 875–898.

[BDD] J. P. Bertrandias, C. Datry, C. Dupuis: Unions et intersections d'espaces L^p invariantes par translation ou convolution, Ann. Inst. Fourier, 28/2 (1978), 53–84.

[BDu] J. P. Bertrandias, C. Dupuis: Transformation de Fourier sur les espaces $\ell^p(L^{p'})$, Ann. Inst. Fourier 29/1 (1979), 189–206.

[BS] R. Busby, H. A. Smith: Product-convolution operators and mixed norm spaces, Trans. Amer. Math. Soc. 263 (1981), 309–341.

[CFG] C. Cenker, H. G. Feichtinger, K. Gröchenig: Nonorthogonal expansions of signals and some of their applications. Abstract: Proc. ECMI conference, Strobl, May 1989, Kluwer.
The full version is published in Image Acquisition and Real-Time Visualization, 14. OEAGM Treffen, ÖCG, Bd. 56, ed. G. Bernroider, A. Pinz, May 1990, 129–138.

[F1] H. G. Feichtinger: Banach convolution algebras of Wiener's type, Proc. Conf. "Functions, Series, Operators", Budapest, August 1980, Colloquia Math. Soc. J. Bolyai, North Holland Publ. Co., Amsterdam-Oxford-New York. 1983, 509–524.

[F2] H. G. Feichtinger: Banach spaces of distributions of Wiener's type and interpolation, Proc. Conf. Oberwolfach, August 1980. Functional Analysis and Approximation. Ed. P. Butzer, B. Sz. Nagy and E. Görlich. Int. Ser. Num. Math. Vol. 69, Birkhäuser-Verlag. Basel-Boston-Stuttgart, 1981, 153–165.

[F3] H. G. Feichtinger: Generalized amalgams, with applications to the Fourier transform, Can. J. Math. (1990), 395-409.

[F4] H. G. Feichtinger: Compactness in translation invariant Banach spaces of distributions and compact multipliers, J. Math. Anal. Appl. 102 (1984), 289–327.

[F5] H. G. Feichtinger: Coherent frames and irregular sampling. Proc. Conf. "Recent Advances in Fourier Analysis and Applications", NATO conference, Il Ciocco, Italy, July 1989. NATO ASI Series C, Vol. 315, ed. J. S. Byrnes and J. L. Byrnes, Kluwer, 1990, 427–440.

[F6] H. G. Feichtinger: Strong almost periodicity and Wiener-type spaces. Proc.Conf. Constr. Function Theory, Varna '81, Bulgaria. Sofia 1983, 321–327.

[F7] H.G. Feichtinger: New results on regular and irregular sampling based on Wiener amalgams, this volume.

[FGr] H. G. Feichtinger, P. Gröbner: Banach spaces of distributions defined by decomposition methods, I. Math. Nachr. 123 (1985), 97–120.

[FG1] H. G. Feichtinger, K. Gröchenig: A unified approach to atomic characterizations through integrable group representations. Proc. Conf. Lund, June 1986, Lect. Notes in Math. 1302 (1988), 52–73.

[FG2] H. G. Feichtinger, K. Gröchenig: Banach spaces related to integrable group representations and their atomic decompositions, I. J. Funct. Anal. 86 (1989), 307–340.

[FG3] H. G. Feichtinger, K. Gröchenig: Banach spaces related to integrable group representations and their atomic decompositions, II. Monatsh. f. Math. 108 (1989), 129–148.

[FG4] H.G.Feichtinger and K. Gröchenig: Reconstruction of multivariate band-limited functions from irregular sampling values. SIAM J.Math.Anal., to appear.

[FSt] J. J. Fournier, J. Stewart: Amalgams of L^p and ℓ^q, Bull. Amer. Math. Soc. 13 (1985),1–21.

[Gr] K. Gröchenig: Describing functions: atomic decompositions versus frames. preprint.

[He] C. Heil: Wiener amalgam spaces in generalized harmonic analysis and wavelet theory, Thesis, Maryland, April 1990.

[HW] C. Heil, D. Walnut: Continuous and discrete wavelet transforms. SIAM Rev. 31/4 (1989), 628–666.

[W] D. Walnut: Continuity properties of the Gabor frame operator, preprint.

This manuscript was prepared in spring 1990, while the author was holding a Max Kade research fellowship. He would like to thank the Mathematics Dept. of the University of Maryland at College Park for the hospitality.

Isometries of Musielak-Orlicz Spaces

R.J.FLEMING Department of Mathematics, Central Michigan University, Mt. Pleasant, MI 48859.

J.E.JAMISON Department of Mathematical Sciences, Memphis State University, Memphis, Tennessee 38152.

A.KAMIŃSKA Department of Mathematical Sciences, Memphis State University, Memphis, Tennessee 38152.

Introduction. In 1932, S. Banach in his monograph [2], gave a characterization of isometries in L^p, for $1 \leq p < \infty$, $p \neq 2$. A different proof and a partial generalization to Orlicz spaces of this result, was given by J.Lamperti in [9]. In 1963, G.Lumer ([10]) made an important step in studying isometries. He introduced a new method for complex spaces, based on Hermitian operators. In fact, he applied Hermitian operators to find a characterization of isometries in reflexive Orlicz spaces. His method has become a fruitful tool for further studies and generalizations. Thus, e.g. in [14], [5], [1], [4], the authors characterized isometries in vector valued function spaces or complex sequence spaces, via Hermitian operators. In [15], [16], M.G.Zaidenberg found a description of isometries in Orlicz and symmetric spaces, without assumptions of reflexivity or separability, generalizing essentially Lumer's result.

Here we will study isometries in Musielak-Orlicz spaces, generalizing the classical results of Banach to spaces which are not symmetric in general. We apply Lumer's and Zaidenberg's method.

In section 1 we characterize Hermitian operators while in section 2 isometries on L_Φ. We conclude with some examples and a description of isometries in Nakano spaces $L^{p(t)}$. Under some restrictions on Φ, which however don't exclude nonreflexive or nonseparable spaces, we will show that any Hermitian operator is of diagonal type and any surjective isometry U must be of the form $Uf = uGf$, where u is a measurable function and G a linear transformation induced by a regular set isomorphism. The function u and the operator G must satisfy some particular equation.

In what follows, let $\mathbb{C}, \mathbb{R}, \mathbb{N}$ denote the complex, real and natural numbers. Let the triple (T, Σ, μ) be a measure space, where μ is a σ- finite and atomless measure defined on Σ. The function Φ will denote a Musielak-Orlicz function i.e. a function $\Phi : \mathbb{R}_+ \times T \longrightarrow \mathbb{R}_+$ with the following two properties: (1) for all $t \in T$, $\Phi(0,t) = 0$ and $\Phi(\cdot,t) : \mathbb{R}_+ \longrightarrow \mathbb{R}_+$ is convex and strictly increasing, (2) for all $u \in \mathbb{R}_+, \Phi(u,\cdot) : T \longrightarrow T$ is Σ- measurable. The Musielak-Orlicz space L_Φ is then defined as a set of all complex valued, Σ- measurable functions $f : T \longrightarrow \mathbb{C}$ such that

$$I_\Phi(\lambda f) = \int_T \Phi(|\lambda f(t)|, t)\, d\mu < \infty$$

for some $\lambda > 0$ dependent on f. L_Φ is a Banach space equipped with the norm

$$\| f \| = inf\{\varepsilon > 0 : I_\Phi(\frac{f}{\varepsilon}) \leq 1\}.$$

If the function Φ satisfies $\Phi(u,t) = \Phi(u,r)$ for all $t, r \in T$ then L_Φ becomes an ordinary Orlicz space. The symbol Φ^* will denote a complementary function to Φ i.e. $\Phi^*(v,t) = \sup_{u \geq 0}\{uv - \Phi(u,t)\}$ and $\Phi'(u,t)$ will be a left derivative of $\Phi(u,t)$ with respect to u. Let E_Φ be a closed subspace of L_Φ defined as a family of all complex valued Σ- measurable functions f such that $I_\Phi(\lambda f) < \infty$ for all $\lambda > 0$. The subspace E_Φ has nicer properties than the whole space L_Φ, e.g. E_Φ is separable if the measure space is separable, while L_Φ may be not. It is often convenient to use E_Φ rather than L_Φ. They coincide if Φ satisfies condition Δ_2 (see eg. [8],[11]). In general the function χ_A may not belong to E_Φ or L_Φ, even if $\mu A < \infty$. But (see [6] p.64) there exists a sequence $\{T_n\}$ which is a partition of T such that

$$I_\Phi(\lambda \chi_{T_n}) < \infty \tag{0.1}$$

for all $\lambda > 0, n \in \mathbb{N}$. Simply $\chi_{T_n} \in E_\Phi$ for all $n \in \mathbb{N}$. In what follows $\{T_n\}$ will denote a sequence satisfying (0.1). Further information about the spaces of Orlicz type can be found e.g. in [7], [8], [11]. The support of $f, \{t \in T : f(t) \neq 0\}$, will be denoted by $suppf$. Finally, for $a \in \mathbb{C}, sgn\, a = \frac{\bar{a}}{|a|}$ if $a \neq 0$ and $sgn\, a = 0$ if $a = 0$.

A set map $G : \Sigma \longrightarrow \Sigma$, defined modulo null sets, is called a regular set isomorphism (see e.g.[14]) if it satisfies the following conditions

(i) $G(A^c) = G(A)^c$, where A^c is a complement of A,

(ii) $G(\bigcup_{i=1}^{\infty} A_i) = \bigcup_{i=1}^{\infty} G(A_i)$ for any sequence $\{A_n\}$ of pairwise disjoint sets,

(iii) $\mu G(A) = 0$ if and only if $\mu A = 0$.

Such a set map induces a unique linear transformation on the class of Σ - measurable functions, characterized by $\chi_A \mapsto \chi_{G(A)}$. We will also use the letter G to denote this transformation. Thus $G\chi_A = \chi_{G(A)}$.

Let us recall Zaidenberg's result, theorem 2 in [15],which has proven to be very useful in our investigations. In fact we will give here a modification of this result. But, since the proof is only slightly different than the original one, we will omit it.

A Banach space $(E, \| \ \|)$ of measurable functions is called an ideal if the condition $| f(t) | \leq | g(t) |, g \in E$, implies $f \in E$ and $\| f \| \leq \| g \|$.

Theorem 0.1. *(cp.[15]) Let E be an ideal Banach space. Suppose there exists a sequence $\{T_n\}$ of pairwise disjoint sets such that $\bigcup T_n = T$ and $\chi_{T_n} \in E$ for all $n \in \mathbb{N}$. Suppose U is a surjective isometry on E. Let G be a regular set isomorphism of Σ and u a measurable function such that*

$$U\chi_A = u\chi_{G(A)}$$

for all $A \in \Sigma$ contained in T_n. Then

$$Uf = uGf$$

for all $f \in E$.

1. Hermitian operators on L_Φ.

If X is a Banach space and X^* its dual, then a bounded linear operator H in X is said to be Hermitian if $x^*(Hx)$ is real for every pair of vectors $x \in X$ and $x^* \in X^*$ such that x^* is a support functional of x i.e. $\| x \| = \| x^* \|$ and $x^*(x) = \| x \|^2$. Equivalently the operator H is Hermitian if $e^{i\alpha H}$ is a surjective isometry on X([3]).

Before a Hermitian operator on L_Φ will be discussed, let's determine the support functional of any element of E_Φ.

Propositon 1. *For any $g \in E_\Phi$, the functional*

$$F_g(f) = C(g) \int_T f(t) sgn\, g(t) \Phi' \left(\frac{|g(t)|}{\| g \|}, t \right) d\mu$$

where $f \in L_\Phi$ and

$$C(g) = \| g \|^2 / \int_T | g(t) | \Phi' \left(\frac{|g(t)|}{\| g \|}, t \right) d\mu$$

is a support functional of g.

Proof: The proof is based on the well known inequalities for the Musielak-Orlicz functions(e.g.[8],[11]). Namely, for any $u, w \geq 0$ and any $t \in T$, $uw \leq \Phi(u,t) + \Phi^*(w,t)$,(Young's inequality), $u\Phi'(u,t) \leq \Phi(2u,t)$, and $u\Phi'(u,t) = \Phi(u,t) + \Phi^*(\Phi'(u,t),t)$. Since $g \in E_\Phi$, $C(g) < \infty$. Indeed

$$\int_T | g(t) | \Phi' \left(\frac{|g(t)|}{\| g \|}, t \right) d\mu \leq \| g \| \, I_\Phi\left(\frac{2}{\| g \|} g\right) < \infty.$$

It is obvious that $F_g(g) = \| g \|^2$. By the definition of the norm we have $I_\Phi(f) \leq 1$ for any $f \in L_\Phi$ with $\| f \| = 1$, while for $f \in E_\Phi$, $I_\Phi(f) = 1$ whenever $\| f \| = 1$. Therefore for any $f \in L_\Phi$ with $\| f \| = 1$ we have

$$| F_g(f) | \leq \| g \| \frac{\int_T \Phi(| f(t) |, t) + \int_T \Phi^*(\Phi'(\frac{|g(t)|}{\|g\|}, t), t)}{\int_T \Phi(\frac{|g(t)|}{\|g\|}, t) + \int_T \Phi^*(\Phi'(\frac{|g(t)|}{\|g\|}, t), t)} \leq \| g \| .$$

Thus, we have that $F_g(g) = \| g \|^2$ and $\| F_g \| = \| g \|$ and the proof is finished.

The main result of this section, Theorem 6, will be proceeded by several lemmas and propositions. Originally the following lemma was proved for Orlicz spaces by G.Lumer in [10]. The proof here is based exactly on the same idea.

Lemma 2. *Let H be an arbitrary Hermitian operator on L_Φ, f_1 and f_2 be elements in E_Φ with disjoint supports A_1 and A_2 respectively. Then*

$$\int_{A_1} \frac{H f_2(t) \overline{f_1(t)}}{| f_1(t) |} \Phi'\left(\frac{| f_1(t) |}{\| f_1 + f_2 \|}, t\right) d\mu = \int_{A_2} \frac{\overline{H f_1(t)} f_2(t)}{| f_2(t) |} \Phi'\left(\frac{| f_2(t) |}{\| f_1 + f_2 \|}, t\right) d\mu.$$

Proof: Let $g = f_1 + e^{i\theta} f_2$ where $\theta \in \mathbb{R}$. Then $g \in E_\Phi$ and $Hg = Hf_1 + e^{i\theta} H f_2$. Hence, by the previous proposition,

$$\frac{1}{C(g)} F_g(Hg) = \int_T H f_1(t)\, sgn\, g(t) \Phi' \left(\frac{|g(t)|}{\| g \|}, t \right) d\mu + e^{i\theta} \int_T H f_2(t)\, sgn\, g(t) \Phi' \left(\frac{|g(t)|}{\| g \|}, t \right) d\mu$$

$$= \int_{A_1} Hf_1(t)\frac{\overline{f_1(t)}}{|f_1(t)|}\Phi'\left(\frac{|f_1(t)|}{\|f_1+f_2\|},t\right)d\mu + e^{-i\theta}\int_{A_2} Hf_1(t)\frac{\overline{f_2(t)}}{|f_2(t)|}\Phi'\left(\frac{|f_2(t)|}{\|f_1+f_2\|},t\right)d\mu$$

$$+e^{i\theta}\int_{A_1} Hf_2(t)\frac{\overline{f_1(t)}}{|f_1(t)|}\Phi'\left(\frac{|f_1(t)|}{\|f_1+f_2\|},t\right)d\mu + \int_{A_2} Hf_2(t)\frac{\overline{f_2(t)}}{|f_2(t)|}\Phi'\left(\frac{|f_2(t)|}{\|f_1+f_2\|},t\right)d\mu$$

is real for all $\theta \in \mathrm{R}$. But if $a + be^{i\theta} + ce^{-i\theta}$ is real for all $\theta \in \mathrm{R}$, then $b = \bar{c}$. Thus setting

$$b = \int_{A_1} Hf_2(t)\frac{\overline{f_1(t)}}{|f_1(t)|}\Phi'\left(\frac{|f_1(t)|}{\|f_1+f_2\|},t\right)d\mu$$

$$c = \int_{A_2} Hf_1(t)\frac{\overline{f_2(t)}}{|f_2(t)|}\Phi'\left(\frac{|f_2(t)|}{\|f_1+f_2\|},t\right)d\mu$$

the proof is completed.

Immediately, we get the following

Corollary 3. *For any $A, B \in \Sigma, A \cap B = \emptyset$ such that $\chi_A, \chi_B \in E_\Phi$ and any $\alpha, \beta > 0$ with $\| \alpha\chi_A + \beta\chi_B \| = 1$ we have*

$$\frac{1}{\alpha}\int_A H(\chi_B)(t)\Phi'(\alpha,t)\,d\mu = \frac{1}{\beta}\int_B \overline{H(\chi_A)(t)}\Phi'(\beta,t)\,d\mu.$$

The next lemma appears to be somewhat technical and in fact it is a preparation for Proposition 5.

Lemma 4. *Let $A \in \Sigma$ with $\mu A > 0$. Suppose $u \mapsto \frac{\Phi'(u,t)}{u}, (u > 0)$ is nondecreasing and nonconstant for all $t \in A$. Then there exist a sequence $\{\alpha_n\}$ of positive numbers and a sequence $\{A_n\}$ of pairwise disjoint sets such that $\bigcup A_n = A$, $\chi_{A_n} \in E_\Phi, I_\Phi(\alpha_n\chi_{A_n}) < 1$ and*

$$\forall\gamma > \alpha_n\ \exists B \subset A_n,\ \mu B > 0\ \forall t \in B\quad \frac{\Phi'(\gamma,t)}{\gamma} > \frac{\Phi'(\alpha_n,t)}{\alpha_n}.$$

If we suppose that $u \mapsto \frac{\Phi'(u,t)}{u}$ is nonincreasing and nonconstant, then we get the analogous condition but with the reverse inequality i.e. $\frac{\Phi'(\gamma,t)}{\gamma} < \frac{\Phi'(\alpha_n,t)}{\alpha_n}$.

Proof: We shall show the lemma just for the nondecreasing function $u \mapsto \frac{\Phi'(u,t)}{u}$. Denote by Q the set of all positive rational numbers. For $\alpha, \beta \in Q$ define

$$C_{\alpha\beta} = \left\{t \in A : \frac{\Phi'(\alpha,t)}{\alpha} < \frac{\Phi'(\beta,t)}{\beta}\right\}.$$

Since the function $u \mapsto \frac{\Phi'(u,t)}{u}$ is nondecreasing and nonconstant, $\bigcup C_{\alpha\beta} = A$. The family $\{C_{\alpha\beta} \cap T_n\}_{n\in\mathbb{N},\alpha,\beta\in Q}$ is countable. Now make this family pairwise disjoint and denote by $\{C_n\}$. It has the following property. For all $m \in \mathbb{N}$, there exist $\alpha, \beta \in Q$ such that $\frac{\Phi'(\alpha,t)}{\alpha} < \frac{\Phi'(\beta,t)}{\beta}$ for all $t \in C_m$ and $\chi_{C_m} \in E_\Phi$ i.e. $I_\Phi(\lambda\chi_{C_m}) < \infty$ for all $\lambda > 0$. Since the set function $W \mapsto I_\Phi(\beta\chi_W), W \subset C_m$ is nonatomic, we can divide C_m into a finite number of pairwise disjoint subsets $C_m{}^i$ such that $I_\Phi(\beta\chi_{C_m{}^i}) < 1$ for any i, and the union $\bigcup_i C_m{}^i$ is the set C_m. Now define the sequence $\{A_n\}$ as a family $\{C_m{}^i\}$. This sequence is pairwise disjoint with union equal to A. Moreover for each A_n there exist $\alpha, \beta \in Q$ such that $\frac{\Phi'(\alpha,t)}{\alpha} < \frac{\Phi'(\beta,t)}{\beta}$ for each $t \in A_n$ and $I_\Phi(\beta\chi_{A_n}) < 1$. For $t \in A_n$ define

$$\alpha(t) = \sup\left\{\gamma \geq \alpha \; : \frac{\Phi'(\gamma,t)}{\gamma} = \frac{\Phi'(\alpha,t)}{\alpha}\right\}.$$

This function is measurable, because

$$\alpha(t) = \sup_{\substack{\gamma\in Q \\ \gamma\geq\alpha}} \left\{\gamma\chi_{G_\gamma}(t) + \alpha\chi_{A_n\setminus G_\gamma}(t)\right\},$$

where

$$G_\gamma = \left\{t \in A : \frac{\Phi'(\gamma,t)}{\gamma} = \frac{\Phi'(\alpha,t)}{\alpha}\right\}.$$

We have $\frac{\Phi'(\alpha(t),t)}{\alpha(t)} = \frac{\Phi'(\alpha,t)}{\alpha}$ by the left continuity of $u \mapsto \Phi'(u,t)$. Moreover $\frac{\Phi'(\gamma,t)}{\gamma} > \frac{\Phi'(\alpha(t),t)}{\alpha(t)}$ for any $\gamma > \alpha(t)$. Let

$$\alpha_n = ess\,inf_{t\in A_n}\alpha(t).$$

Then $\alpha \leq \alpha_n \leq \alpha(t) < \beta$, for $t \in A_n$. For any $\gamma > \alpha_n$ there exists $B \subset A_n$ with $\mu B > 0$ such that $\gamma > \alpha(t)$ for $t \in B$. Thus

$$\frac{\Phi'(\gamma,t)}{\gamma} > \frac{\Phi'(\alpha(t),t)}{\alpha(t)} \geq \frac{\Phi'(\alpha_n,t)}{\alpha_n},$$

for $t \in B$. Since $\alpha_n < \beta$ we have also that $I_\Phi(\alpha_n\chi_{A_n}) < 1$.

Let in the following

$$T_0 = \left\{t \in T : u \mapsto \frac{\Phi'(u,t)}{u} \text{ is constant }\right\}.$$

Our next result is a kind of a diagonalization theorem.

Proposition 5. *Suppose $u \mapsto \frac{\Phi'(u,t)}{u}$ is either nondecreasing or nonincreasing for any $t \in T$. If $H : L_\Phi \longrightarrow L_\Phi$ is a Hermitian operator, then for any $B \in \Sigma$ such that $\chi_B \in E_\Phi$, we have*

I. $suppH(\chi_B) \subset T_0 \cup B$.

II. *If* $B \cap T_0 = \emptyset$ *then* $suppH(\chi_B) \subset B$.

Proof: I. Let $\tilde{T} = T \setminus (T_0 \cup B)$ and suppose $\mu\tilde{T} > 0$. Define

$$P = \left\{ t \in \tilde{T} : u \mapsto \frac{\Phi'(u,t)}{u} \text{ is nondecreasing} \right\}$$

$$\bar{P} = \left\{ t \in \tilde{T} : u \mapsto \frac{\Phi'(u,t)}{u} \text{ is nonincreasing} \right\}.$$

The sets P and $\bar{P}$ are disjoint and $\tilde{T}$ is their union. Let

$$S = \{ t \in T : reH(\chi_B) \geq 0 \}$$

$$S' = \{ t \in T : reH(\chi_B) \leq 0 \}.$$

For all $t \in P \cap S$, $u \mapsto \frac{\Phi'(u,t)}{u}$ is nondecreasing and nonconstant. Suppose further $\mu(P \cap S) > 0$. Then, by the previous lemma, there exist a sequence $\{\alpha_n\}$ of positive numbers and a sequence $\{P_n\}$ of pairwise disjoint sets such that $\bigcup P_n = P \cap S$, $I_\Phi(\alpha_n \chi_{P_n}) < 1$, $\chi_{P_n} \in E_\Phi$ and $\frac{\Phi'(\beta,t)}{\beta} > \frac{\Phi'(\alpha_n,t)}{\alpha_n}$ for $\beta > \alpha_n$ on some subsets of P_n with positive measure. Since $\chi_B \in E_\Phi$, the function $\lambda \mapsto I_\Phi(\lambda\chi_B)$ has finite values and $I_\Phi(\lambda\chi_B) \to \infty$ if $\lambda \to \infty$. Thus there exists $\gamma > 0$ such that

$$I_\Phi(\alpha_n \chi_{P_n}) + I_\Phi(\gamma\chi_B) = 1 \tag{5.1}$$

Since $\chi_{P_n} \in E_\Phi$, there exists $\beta_n > \alpha_n$ such that $I_\Phi(\beta_n\chi_{P_n}) = 2I_\Phi(\alpha_n\chi_{P_n})$. By nonatomicity of μ, we will find two disjoint sets P_{n1}, and P_{n2} such that $P_n = P_{n1} \cup P_{n2}$, $P_{n1} \cap P_{n2} = \emptyset$ and $I_\Phi(\beta_n\chi_{P_{n1}}) = I_\Phi(\beta_n\chi_{P_{n2}}) = I_\Phi(\alpha_n\chi_{P_n})$. By (5.1) we have

$$I_\Phi(\beta_n\chi_{P_{ni}}) + I_\Phi(\gamma\chi_B) = 1 \tag{5.2}$$

for $i = 1, 2$. By (5.1) and (5.2), $\| \alpha_n\chi_{P_n} + \gamma\chi_B \| = \| \beta_n\chi_{P_{ni}} + \gamma\chi_B \| = 1$ for $i = 1, 2$. So applying Corollary 3, we get

$$\int_{P_n} H(\chi_B)(t) \frac{\Phi'(\alpha_n,t)}{\alpha_n} \, d\mu = \int_B \overline{H(\chi_{P_n})(t)} \frac{\Phi'(\gamma,t)}{\gamma} \, d\mu \tag{5.3}$$

and

$$\int_{P_{ni}} H(\chi_B)(t)\,\frac{\Phi'(\beta_n,t)}{\beta_n}\,d\mu = \int_B \overline{H(\chi_{P_{ni}})(t)}\,\frac{\Phi'(\gamma,t)}{\gamma}\,d\mu$$

for $i = 1,2$. Adding last two equations for $i = 1,2$, we get

$$\int_{P_n} H(\chi_B)(t)\,\frac{\Phi'(\beta_n,t)}{\beta_n}\,d\mu = \int_B \overline{H(\chi_{P_n})(t)}\,\frac{\Phi'(\gamma,t)}{\gamma}\,d\mu. \tag{5.4}$$

Combining (5.3) and (5.4)

$$\int_{P_n} H(\chi_B)(t)\,\frac{\Phi'(\beta_n,t)}{\beta_n}\,d\mu = \int_{P_n} \overline{H(\chi_B)(t)}\,\frac{\Phi'(\alpha_n,t)}{\alpha_n}\,d\mu.$$

Hence

$$\int_{P_n} reH(\chi_B)(t)\,\frac{\Phi'(\beta_n,t)}{\beta_n}\,d\mu = \int_{P_n} re\overline{H(\chi_B)(t)}\,\frac{\Phi'(\alpha_n,t)}{\alpha_n}\,d\mu. \tag{5.5}$$

The function $reH(\chi_B)(t)$ is nonnegative for all $t \in P_n$, because $P_n \subset P \cap S$. Moreover, by the choice of α_n and P_n, $\frac{\Phi'(\beta_n,t)}{\beta_n} > \frac{\Phi'(\alpha_n,t)}{\alpha_n}$ for t belonging to a subset with positive measure. But then, the equality (5.5) is possible only if $reH(\chi_B)(t) = 0$ for $a.a.t \in P_n$. Similarly we show that $reH(\chi_B)(t) = 0$ for $t \in P \cap S'$. Thus $reH(\chi_B)(t) = 0$ for all $t \in P$. There is no essential difference to show that $imH(\chi_B)(t) = 0$ for all $t \in P$. Thus $H(\chi_B)(t) = 0$ for all $t \in P$. To show that $H(\chi_B)(t) = 0$ for $t \in \bar{P}$, we proceed as above, applying the previous lemma for the nonincreasing function $u \mapsto \frac{\Phi'(u,t)}{u}$. Thus, we have that $H(\chi_B)(t) = 0$ for $t \in P \cup \bar{P} = \tilde{T}$, which finishes the proof of I.

II. If $\chi_B \in E_\Phi$, then by I, $supp\, H(\chi_B) \subset T_0 \cup B$. Denote

$$W = \{t \in T_0 : reH(\chi_B) \geq 0\}.$$

If $W_n = W \cap T_n$, then the sets W_n are pairwise disjoint and $\chi_{W_n} \in E_\Phi$ for all $n \in \mathbb{N}$. Without loss of generality we can suppose $\mu W_n > 0$ for all $n \in \mathbb{N}$. Since χ_B, $\chi_{W_n} \in E_\Phi$ and B and W_n are disjoint, there exist $\alpha, \beta > 0$ such that

$$I_\Phi(\alpha\chi_{W_n}) + I_\Phi(\beta\chi_B) = \| \alpha\chi_{W_n} + \beta\chi_B \| = 1.$$

By Corollary 3 we have

$$\frac{1}{\alpha}\int_{W_n} H(\chi_B)(t)\,\Phi'(\alpha,t)\,d\mu = \frac{1}{\beta}\int_B \overline{H(\chi_{W_n})(t)}\,\Phi'(\beta,t)\,d\mu$$

which implies that

$$\frac{1}{\alpha}\int_{W_n} reH(\chi_B)(t)\ \Phi'(\alpha,t)\,d\mu = \frac{1}{\beta}\int_B reH(\chi_{W_n})(t)\ \Phi'(\beta,t)\,d\mu.$$

Since $W_n \subset T_0$, $suppH(\chi_{W_n}) \subset T_0$ by part I. Thus the right side of the above equality is zero. So the left side is zero too. But $reH(\chi_B)(t) \geq 0$ and $\Phi'(\alpha,t) > 0$ for $t \in W_n$, so $reH(\chi_B)(t) = 0$ for $t \in W_n$.The set W is a union of W_n, which implies that $reH(\chi_B)(t) = 0$ for all $t \in W$. Taking the set $\{t \in T_0 : reH(\chi_B) \leq 0\}$ instead of W, we shall prove similarly that $reH(\chi_B)(t) = 0$ for all $t \in T_0$. Following the above procedure, with no essential changes, we can show that $imH(\chi_B)(t) = 0$ for all $t \in T_0$. Thus $suppH(\chi_B) \subset B$ which completes the proof.

Now with this preliminary work, we are in a position to determine Hermitian operators on L_Φ, the main result of this section.

Theorem 6. *Let $\mu T_0 = 0$. Suppose the function $u \mapsto \frac{\Phi'(u,t)}{u}$ is either nondecreasing or nonincreasing for $t \in T$. Any bounded linear operator $H : L_\Phi \longrightarrow L_\Phi$ is Hermitian if and only if*

(6.1) $$Hf = hf$$

for all $f \in L_\Phi$, where h is a real, bounded function. Then $\| H \| = \| h \|_\infty$.

Proof: In [10], Lemma 7, G.Lumer proved that any operator H of the form (6.1) defined on an Orlicz space is Hermitian. He also showed that $\| H \| = \| h \|_\infty$. In fact, his proof is so general that can be used for any ideal Banach space, so for the Musielak-Orlicz space too.

Now let H be Hermitian. By the previous proposition II, since $\chi_{T_n} \in E_\Phi$, $suppH(\chi_{T_n}) \subset T_n$. Then the function

$$h(t) = \sum_{n=1}^{\infty} H(\chi_{T_n})(t)$$

is well defined and $H(\chi_A) = h\chi_A$ for $A \in \Sigma$ contained in one of the T_n. Indeed,

$$H(\chi_{T_n})\chi_A = H(\chi_A)\chi_A + H(\chi_{T_n \setminus A})\chi_A = H(\chi_A),$$

which obviously implies that $H(\chi_A) = h\chi_A$. Thus $e^{i\alpha H}(\chi_A) = e^{i\alpha h}\chi_A$ for all real α and measurable subsets A of T_n. Since $e^{i\alpha H}$ is a surjective isometry on L_Φ, applying Theorem 0.1, we get that $e^{i\alpha H}(f) = e^{i\alpha h}f$ for any $f \in L_\Phi$. Now, from Theorem 13.35 in [13] it follows that $Hf = hf$ for every $f \in L_\Phi$.

2. Isometries on L_Φ

In this section we apply the results of Hermitian operators to discuss isometries on Musielak-Orlicz spaces L_Φ. In fact we show that any surjective isometry is a weighted-composition operator, where the set isomorphism, the weight function and the generating function Φ are related by some particular equation.

Before we formulate the main result, let's prove the following measure theoretic lemma.

Lemma 7. *Let ν be a positive measure on Σ equivalent to μ (i.e. $\mu A = 0 \leftrightarrow \nu(A) = 0$) and let c be a positive number. If $c < \nu(T) < \infty$ and f is a measurable, real function such that $\int_A f(t)\,d\mu = 0$ for any A with $\nu(A) = c$, then $f(t) = 0$ for a.a. $t \in T$.*

Proof: Since μ is atomless and ν is equivalent to μ, ν is atomless too. For the contrary suppose $\mu(supp f) > 0$. Let

$$B_1 = \{t \in T : f(t) > 0\}, \quad B_2 = \{t \in T : f(t) < 0\}.$$

Assume for a moment that $supp f = T$. Take any A with $\nu A = c$. Then $B_1 \cup B_2 = T$ and $\int_A f(t)\,d\mu = 0$. Hence

$$\int_{A\cap B_1} f(t)\,d\mu = -\int_{A\cap B_2} f(t)\,d\mu.$$

Thus $\mu(A\cap B_1)$ and $\mu(A \cap B_2)$ must be positive. Since $\nu(T \setminus A) > 0$, either $\nu(B_1 \setminus A)$ or $\nu(B_2 \setminus A)$ is positive. Let e.g. $\nu(B_1 \setminus A) > 0$. Choose $D \subset A \cap B_2$ such that $0 < \nu(D) < \nu(B_1 \setminus A)$ and choose $\bar{D} \subset B_1 \setminus A$ with $\nu(\bar{D}) = \nu(D)$. Let $\bar{A} = \bar{D} \cup (A \setminus D)$. Then $\nu(\bar{A}) = \nu(\bar{D}) + \nu(A \setminus D) = \nu(D) + \nu(A \setminus D) = c$, and

$$\begin{aligned}
\int_{\bar{A}} f(t)\,d\mu &= \int_{\bar{D}} f(t)\,d\mu + \int_{A\setminus D} f(t)\,d\mu \\
&= \int_{\bar{D}} f(t)\,d\mu + \int_{A\cap B_1} f(t)\,d\mu + \int_{(A\cap B_2)\setminus D} f(t)\,d\mu \\
&= \int_{\bar{D}} f(t)\,d\mu - \int_{A\cap B_2} f(t)\,d\mu + \int_{A\cap B_2} f(t)\,d\mu - \int_{D} f(t)\,d\mu \\
&> 0,
\end{aligned}$$

because f is positive on $\bar{D}$ and negative on D, which contradicts the assumption.

Now assume that $\mu(T \setminus suppf) > 0$. If $\nu(suppf) = \nu(B_1 \cup B_2) > c$ then it is the previous case with $suppf$ instead of T.

If $\nu(suppf) \leq c$ then there exists a set A such that $\nu(A) = c$ and $B_1 \cup B_2 \subset A$. Choose $D \subset A \cap B_2$ with $0 < \nu(D) < \nu(T \setminus A)$ and $\bar{D} \subset T \setminus A$ in such a way that $\nu(\bar{D}) = \nu(D)$. Then for $\bar{A} = \bar{D} \cup (A \setminus D)$ we have $\nu(\bar{A}) = \nu(A) = c$ and $\int_{\bar{A}} f = -\int_D f > 0$,which is a contradiction of the assumptions and the proof is finished.

If G is a regular set isomorphism of Σ, then $\mu(G^{-1}A)$ is a measure absolutely continuous with respect to μ. Let G' be its Radon-Nikodym derivative i.e. $\mu(G^{-1}A) = \int_A G'(t)\, d\mu$. By change of variables, it is obvious that

$$\int_A f\, d\mu = \int_{GA} (Gf)G'\, d\mu$$
$$\int_{GA} f\, d\mu = \int_A G^{-1}(f(G')^{-1})\, d\mu.$$

Theorem 8. *Let $\mu T_0 = 0$ and suppose the function $u \mapsto \frac{\Phi'(u,t)}{u}$ is nondecreasing or nonincreasing for $t \in T$.*

If $U : L_\Phi \longrightarrow L_\Phi$ is a surjective isometry, then there exist a measurable function $u, u(t) \neq 0$, and a set isomorphism G such that

$$Uf = uGf, \tag{8.1}$$

for all $f \in L_\Phi$ and

$$\Phi(|\, u(t)\, |\, \lambda, t) = G'(t)\, G\Phi(\lambda, t), \tag{8.2}$$

for all a.a. $t \in T$ and $\lambda \in \mathbb{R}_+$.

Conversely, if a measurable function u with $u(t) \neq 0$ a.e. and a set isomorphism G satisfy the equation (8.2), then U defined by (8.1) is a surjective isometry of L_Φ.

Proof: Suppose u and G satisfy the equation (8.2). We shall show that $I_\Phi(Uf) = I_\Phi(f)$ for all $f \in L_\Phi$. In virtue of (8.2) and the substitution formula, this is obvious for simple functions. For an arbitrary f from L_Φ, there exists a sequence $\{f_n\}$ of simple functions belonging to L_Φ satisfying $f_n(t) \longrightarrow f(t)$ a.e. and $|\, f_n\, | \leq |\, f\, |$. Hence, by the Fatou lemma $I_\Phi(f) = \underline{\lim}\, I_\Phi(f_n)$. But $I_\Phi(f_n) = I_\Phi(Uf_n)$, so $I_\Phi(f) = \lim I_\Phi(Uf_n)$. Since $Uf = uGf$ and

$Uf_n = uGf_n$, $Uf_n(t) \longrightarrow Uf(t)$ *a.e.* and $| Uf_n | \leq | Uf |$ .Thus applying the Fatou lemma again, $I_\Phi(Uf) = \lim I_\Phi(Uf_n) = I_\Phi(f)$. Hence, for all $\lambda > 0$, $I_\Phi(\lambda f) = I_\Phi(\lambda Uf)$, which implies that $\| f \| = \| Uf \|$.

Since $u(t) \neq 0$ for *a.a.* $t \in T$, it is obvious that U is onto isometry.

Now suppose U is a surjective isometry. The operator $H_A f = \chi_A f$ is Hermitian on L_Φ, so UH_AU^{-1} is Hermitian also(see [5]). By Theorem 6, there exists a real measurable function h_A such that $UH_AU^{-1}f = h_A f$. Hence $h_A{}^2 = h_A$ which implies that $h_A = 1$ or $h_A = 0$. Defining $G(A) = \{t \in T : h_A(t) = 1\}$, G is a set isomorphism on Σ. Thus $h_A = \chi_{G(A)}$. For $A \subset T_n$, we have $U(\chi_A) = UH_AU^{-1}U(\chi_{T_n}) = U(\chi_{T_n})\chi_{G(A)}$. Since the supports of $U(\chi_{T_n})$ are pairwise disjoint, the function $\sum U(\chi_{T_n})$ is well defined. Thus, setting $u = \sum U(\chi_{T_n})$ we have $U(\chi_A) = u\chi_{G(A)}$ for any $A \in \Sigma$ contained in some T_n. Now the equation (8.1) is obvious by Theorem 0.1. Since U is surjective, it is also immediate that $u(t) \neq 0$ for *a.a.* $t \in T$.

To show condition (8.2), at first define a sequence $\{D_n\}$. Let $\{B_n\}$ be an increasing sequence such that $\bigcup B_n = T$ and the function $| u |$ is bounded on each B_n. Then $\{D_n\}$ is defined as $\{G^{-1}(B_n) \cap T_m\}_{n,m \in \mathbb{N}}$. Since $\bigcup G^{-1}(B_n) = T$ and $\bigcup T_m = T$, so $\bigcup D_n = T$. Moreover for any $A \subset D_i$, A must be contained in $G^{-1}(B_n)$ and in T_m for some $n, m \in \mathbb{N}$. Hence $\chi_A \in E_\Phi$ and $G(A) \subset B_n$. The last inclusion means that $| u |$ is bounded on $G(A)$. Assume without loss of generality that $\mu D_n > 0$ for all $n \in \mathbb{N}$. Fix for a moment a set D_n. Since $\chi_{D_n} \in E_\Phi$, there exists $\lambda_n > 0$ such that

$$\int_{D_n} \Phi(\lambda_n, t) = 1.$$

Let $\lambda > \lambda_n$. For any $A \subset D_n$ with $\int_A \Phi(\lambda, t) = 1$ we have $\| \lambda\chi_A \| = 1$, which implies that $\| U\lambda\chi_A \| = 1$. But by (8.1), $U(\lambda\chi_A) = \lambda u\chi_{G(A)}$. The function $| u |$ is bounded on $G(A)$, so $\lambda u\chi_{G(A)} \in E_\Phi$. Then $I_\Phi(\lambda u\chi_{G(A)}) = 1$ which implies that

$$\int_A G^{-1}([G'(t)]^{-1}\Phi(\lambda \mid u(t) \mid, t))\, d\mu = 1.$$

Thus we have shown that

$$\int_A \{G^{-1}([G'(t)]^{-1}\Phi(\lambda \mid u(t) \mid, t)) - \Phi(\lambda, t)\}\, d\mu = 0$$

for all $A \subset D_n$ with $\int_A \Phi(\lambda, t)\, d\mu = 1$. Now apply Lemma 7 with $T = D_n$, the measure ν defined as $\nu(A) = \int_A \Phi(\lambda, t)\, d\mu$ for $A \in D_n \cap \Sigma$, $\alpha = 1$ and

$$f(t) = G^{-1}([G'(t)]^{-1}\Phi(\lambda \mid u(t) \mid, t)) - \Phi(\lambda, t).$$

So $f(t) = 0$ for *a.a.* $t \in D_n$. Hence

$$\Phi(\lambda \mid u(t) \mid, t) = G'(t) G\Phi(\lambda, t) \tag{8.3}$$

for *a.a.* $t \in D_n$ and $\lambda > \lambda_n$.

Now let $\lambda \leq \lambda_n$. Denoting $\alpha = \int_{D_n} \Phi(\lambda, t)$, we have that $0 < \alpha \leq 1$. For any $A \subset D_n$ with $\int_A \Phi(\lambda, t) = \frac{\alpha}{2}$ choose $\gamma > \lambda_n$ with $\int_{D_n \setminus A} \Phi(\gamma, t) \geq 1 - \frac{\alpha}{2}$. Then choose $B \subset D_n \setminus A$ such that $\int_B \Phi(\gamma, t) = 1 - \frac{\alpha}{2}$. Hence

$$\int_A \Phi(\lambda, t)\, d\mu + \int_B \Phi(\gamma, t)\, d\mu = 1, \tag{8.4}$$

which implies that $\| \lambda \chi_A + \gamma \chi_B \| = 1$ and $\| \lambda u \chi_{G(A)} + \gamma u \chi_{G(B)} \| = 1$. The function $\lambda u \chi_{G(A)} + \gamma u \chi_{G(B)} \in E_\Phi$, because $\mid u \mid$ is bounded on both $G(A)$ and $G(B)$. Then $I_\Phi(\lambda u \chi_{G(A)} + \gamma u \chi_{G(B)}) = 1$, which implies that

$$\int_A G^{-1}([G'(t)]^{-1} \Phi(\lambda \mid u(t) \mid, t)\, d\mu + \int_B G^{-1}([G'(t)]^{-1} \Phi(\gamma \mid u(t) \mid, t)\, d\mu = 1.$$

But since $\gamma > \lambda_n$, by the previous part we have

$$\Phi(\gamma \mid u(t) \mid, t) = G'(t)\; G\Phi(\gamma, t)$$

for *a.a.*$t \in D_n$. Hence

$$\int_A G^{-1}([G'(t)]^{-1} \Phi(\lambda \mid u(t) \mid, t))\, d\mu + \int_B \Phi(\gamma, t)\, d\mu = 1.$$

Combining the above and (8.4) we have

$$\int_A \{G^{-1}([G'(t)]^{-1} \Phi(\lambda \mid u(t) \mid, t)) - \Phi(\lambda, t)\}\, d\mu = 0$$

for any $A \subset D_n$ with $\int_A \Phi(\lambda, t)\, d\mu = \frac{\alpha}{2}$. Applying Lemma 7 similarly as before, but with $c = \frac{\alpha}{2}$, the equation (8.3) is satisfied for all $\lambda \leq \lambda_n$ and *a.a.* $t \in D_n$. Thus, since $\bigcup D_n = T$, we get the equation (8.2) which finishes the proof.

As an immediate consequence we get a characterization of isometries in Nakano spaces $L^{p(t)}$, for $p(t) \neq 2$. Recall that the Nakano space is a Musielak-Orlicz space with $\Phi(u, t) = u^{p(t)}$, where $p(t)$ is a real measurable function with its range in the interval $[1, \infty)$ (see e.g. [12]). If $p(t) \neq 2$ *a.e.*, then the function $u^{p(t)}$ satisfies the assumptions of the above theorem and we immediately have the following result.

Theorem 9. *Suppose $p(t) \neq 2$ a.e. Any surjective isometry U on the Nakano space $L^{p(t)}$ is of the form*

$$Uf = uGf, \tag{9.1}$$

$f \in L^{p(t)}$, where u is a measurable function and G is a set isomorphism satisfying

$$p(t) = Gp(t), \quad G'(t) = |\, u(t)\,|^{p(t)} \tag{9.2}$$

for a.a.$t \in T$.

Conversely, if $u(t) \neq 0$ is a measurable function and G a set isomorphism satisfying (9.2), then the operator U of the form (9.1) is a surjective isometry on $L^{p(t)}$.

We see that in the case when $p(t)$ is constant, the above result agrees with known results on isometries in L^p - spaces (cp. [2],[9]).

We conclude the paper with some remarks and examples.

(1). We will say that a linear operator $U : L_\Phi \longrightarrow L_\Phi$ is a modular isometry if $I_\Phi(Uf) = I_\Phi(f)$ for all $f \in L_\Phi$. It is obvious that each modular isometry is an isometry too. The converse is also true, under those special assumptions about Φ made in Theorem 8. Indeed, if U is an isometrty, then U must be of the form $Uf = uGf$, where u and G satisfy equation (8.2) of Theorem 8. Then $I_\Phi(Uf) = I_\Phi(f)$ as it was already shown in the proof of that theorem.

(2). If $t \in T_0$, then the function $u \mapsto \frac{\Phi'(u,t)}{u}$ is constant. So, for $t \in T_0, \Phi(u,t) = c(t)u^2$, where $c(t)$ is a positive, measurable function. Then $L_\Phi(T_0) = \{f\chi_{T_0} : f \in L_\Phi\}$ is a Hilbert space. Thus the assumption in Theorems 6 and 8, that $\mu T_0 = 0$, simply excludes Hilbert subspaces of L_Φ.

(3). If $\Phi'(u,t)$ or $\Phi(\sqrt{u},t)$ is convex or concave with respect to u, then the function $u \mapsto \frac{\Phi'(u,t)}{u}$ is nondecreasing or nonincreasing respectively. Thus such functions may serve as examples of Φ satisfying assumptions of Theorems 6 and 8.

(4). There exist Nakano spaces with nontrivial isometries.

Let $G : [0,1] \to [0,1]$ be given by $G(t) = \frac{1-t}{1+t}$. Then $G(G(t)) = t$, and G induces a regular set isomorphism on Lebesgue measurable subsets of $[0,1]$. Define

$$p(t) = \frac{1}{t + G(t)} = \frac{1+t}{1+t^2},$$

$$u(t) = \left(\frac{2}{1+t^2}\right)^{\frac{1}{p(t)}}.$$

Then $p(t) = p(G(t))$ and $p(t) \in [1, \infty)$. It is also obvious that

$$u(t)^{p(t)} = \frac{2}{1+t^2} = | G'(t) |,$$

where $| G'(t) |$ is the Radon-Nikodym derivative $\frac{dm_1}{dm}$ for the Lebesgue measure m and $m_1(A) = m(G^{-1}(A))$. Thus

$$Uf(t) = u(t)Gf(t) = \left(\frac{2}{1+t^2}\right)^{\frac{1}{p(t)}} f\left(\frac{1-t}{1+t}\right)$$

is a nontrivial isometry on $L^{p(t)}[0,1]$.

REFERENCES

1. J.Arazy, *Isometries of complex symmetric sequence spaces*, Math. Zeitschrift **188** (1985), 427–431.
2. S.Banach, "Théorie des operations lineaires," New York, Chelsea, 1955.
3. E.Berkson and H.Porta, *Hermitian operators and one-parameter groups of isometries in Hardy spaces*, Transactions of AMS **185** (1973), 331–344.
4. R.Fleming and J.Jamison, *Isometries on certain Banach spaces*, J. London Math. Soc. (2),9 (1974), 363–371.
5. J.E.Jamison and I.Loomis, *Isometries of Orlicz spaces of vector valued functions*, Math. Zeitschrift **193** (1986), 363–371.
6. A.Kamińska, *Some convexity properties of Musielak-Orlicz spaces of Bochner type*, Supp. Rendiconti del Circolo Mat. di Palermo, Serie II **10** (1985), 63–73.
7. W.Kozlowski, "Modular function spaces," Marcel Dekker Inc., New York and Basel, 1988.
8. M.A.Krasnoselskii and Ya.B.Rutickii, "Convex functions and Orlicz spaces," Groningen, 1961.
9. J.Lamperti, *On the isometries of certain function spaces*, Pacific J.Math. **8** (1958), 459–466.
10. G.Lumer, *On the isometries of reflexive Orlicz spaces*, Ann. Inst. Four (Grenoble) **13** (1963), 99–109.

11. J.Musielak, "Orlicz spaces and modular spaces," Lect. Notes in Math 1034, Springer-Verlag, 1983.

12. H.Nakano, "Topology and linear spaces," Nihonbashi, Tokyo, 1951.

13. W.Rudin, "Functional analysis," McGraw-Hill Book Company, 1973.

14. A.R.Sourour, *The isometries of* $L^p(\Omega, X)$, J.Funct. Anal. **30** (1978), 276–285.

15. M.G.Zaidenberg, *Groups of isometries of Orlicz spaces*, Sov.Mat. Dokl. **17** (**2**) (1976), 432–436.

16. M.G.Zaidenberg, *On isometric classification of symmetric spaces*, Sov. Mat. Dokl. **18** (**3**) (1977), 636–640.

Algebras of Bounded Functions on the Disc

PAMELA GORKIN, Bucknell University, Lewisburg, PA 17837

Let L^∞ denote the space of essentially bounded measurable functions on the unit circle $\mathbb{T}$. As usual, $H^\infty(\mathbb{D})$ denotes the algebra of bounded analytic functions on the open unit disc $\mathbb{D}$ and H^∞ denotes the subalgebra of L^∞ consisting of boundary values of functions in $H^\infty(\mathbb{D})$. Let C denote the algebra of continuous functions on the unit circle.

The structure of subalgebras of L^∞ is now well understood. The algebra $H^\infty + C$ was, in many ways, typical of subalgebras of L^∞. Many results indicate that an analogous structure may exist for subalgebras of $L^\infty(\mathbb{D})$ containing $H^\infty(\mathbb{D})$, while other results indicate that we must be careful when comparing algebras on the circle and algebras on the disc. We indicate some similarities and differences below, and we close with some open questions.

One of the first results on subalgebras of L^∞ is due to D. Sarason. Sarason showed that $H^\infty + C$ is a closed subalgebra of L^∞ [14]. The space $H^\infty + C$ has other important properties. For example, K. Hoffman [10 p. 193] showed that $H^\infty + C$ is the minimal closed subalgebra of

The author was supported by an NSF grant.

L^∞ containing H^∞; that is, if f is a function in L^∞ that is not in H^∞, then the closed subalgebra of L^∞ generated by H^∞ and f, denoted $H^\infty[f]$, contains $H^\infty + C$.

Interest in other subalgebras of L^∞ arose in connection with a question asked by R. G. Douglas. In what follows, let B denote a closed subalgebra of L^∞ containing H^∞ and let B_I denote the closed subalgebra of L^∞ generated by H^∞ and the complex conjugates of those inner functions that are invertible in B. Motivated by operator theoretic questions, Douglas conjectured that $B = B_I$. Algebras of the form B_I are called Douglas algebras. One example is $H^\infty[\bar{z}]$, which equals $H^\infty + C$. Note that whenever the complex conjugate of an inner function is in the algebra B, then the complex conjugate of any subfactor (which is also inner) is in B. (Let u denote the inner function and u_1 denote any subfactor. Thus $u = u_1 u_2$ for some u_2 in H^∞ and $\bar{u}_1 = \bar{u}_1 \bar{u}_2 u_2 = \bar{u}\, u_2$, which is in the algebra B.) Thus, even at first glance, the algebras B_I are quite large.

A lot has been written on this problem and good references already exist ([9] and [15]), so the discussion here will be brief. The solution to the Douglas problem depended on the theory of maximal ideal spaces, which is reviewed briefly below.

Let A denote a uniform algebra. Then $M(A)$ denotes the maximal ideal space of A. The topology on $M(A)$ is the weak-$*$ topology. With this topology $M(A)$ is a compact Hausdorff space. We will be most interested in the case in which $A = H^\infty$. Any φ in $M(H^\infty)$ can be represented by integration against a unique positive measure supported on $M(L^\infty)$. The support set of the measure is called the support of φ. In what follows, we will identify a function with its Gelfand transform.

S.Y. Chang and D.E. Marshall solved Douglas's problem. In 1975, Chang [7] showed that if A and B are two subalgebras of L^∞ containing H^∞ with $M(A) = M(B)$ and if B is a Douglas algebra, then $A = B$. A few months later, Marshall [12] showed that if B is a uniform algebra contained in L^∞ and containing H^∞, then there is a set I of inner functions such that $M(B) = M(B_I)$. In fact, Marshall was able to obtain much more information about the properties of the inner functions involved.

Let $\rho(z,w) = \left|\frac{z - w}{1 - \bar{z}w}\right|$ denote the pseudohyperbolic distance between two points z and w in $\mathbb{D}$. Recall that a Blaschke product b with zero sequence $\{z_n\}$ is interpolating if the zero sequence of b satisfies

$$\inf_n \prod_{\substack{m=1 \\ m\neq n}}^{\infty} \rho(z_m, z_n) = \inf_n\{(1 - |z_n|^2)|b'(z_n)|\} = \delta > 0;$$

that is, if the zeroes of b form an interpolating sequence for H^∞ [10]. Marshall showed that the inner functions can be taken to be interpolating Blaschke products. Marshall's results relied heavily on the theory of maximal ideal spaces. His proof depended very much on the behavior of functions on support sets of the multiplicative linear functionals in $M(H^\infty) - \mathbb{D}$.

Using the preceding information for motivation, we now consider $H^\infty(\mathbb{D})$ as a subalgebra of $L^\infty(\mathbb{D},dA)$, where dA denotes two dimensional Lebesgue area measure. In what follows, let $C(\overline{\mathbb{D}})$ denote the algebra of continuous functions on the closed unit disc $\overline{\mathbb{D}}$. We will write $H^\infty(\mathbb{D})$ when we wish to emphasize the fact that we are thinking of it as a subalgebra of $L^\infty(\mathbb{D})$. When it does not matter whether we think of H^∞ on the disc or on the circle we will simply write H^∞.

In 1975, W. Rudin [13] showed that $H^\infty(\mathbb{D}) + C(\overline{\mathbb{D}})$ is a closed subalgebra of $L^\infty(\mathbb{D})$. However, $H^\infty(\mathbb{D}) + C(\overline{\mathbb{D}})$ is not minimal. For example, if f is a function in $L^\infty(\mathbb{D})$ which is zero on an open proper subset of $\mathbb{D}$, then it is easy to see that the closed algebra of $L^\infty(\mathbb{D})$ generated by $H^\infty(\mathbb{D})$ and f, denoted $H^\infty(\mathbb{D})[f]$, does not contain $H^\infty(\mathbb{D}) + C(\overline{\mathbb{D}})$.

On the surface, then, the theory does not seem to generalize. However, in 1987, S. Axler and A. Shields [4] showed that there was an analogue of Hoffman's theorem for algebras on the disc. Axler and Shields showed that if f is a bounded function that is harmonic but not analytic on $\mathbb{D}$, then $H^\infty(\mathbb{D})[f]$ contains $C(\overline{\mathbb{D}})$. One might now also ask if there is a theorem "like" the Chang Marshall theorem for subalgebras of $L^\infty(\mathbb{D})$ containing $H^\infty(\mathbb{D})$. If there is, one difficulty will be finding out what "like" means.

First of all, the techniques needed to prove such a theorem will be quite different. For example, as mentioned above, the proof of the Chang Marshall theorem uses uniqueness of maximal ideal spaces. However, for a function f in $H^\infty(\mathbb{D})$, the maximal ideal space $M(H^\infty(\mathbb{D})[\bar{f}]) = M(H^\infty)$ [3, Lemma 4]. Support sets seem to reflect the behavior of the functions on the circle only and do not yield the information that we need here. The role that support sets played in the Chang Marshall theorem seems to be the same as the role Gleason parts play for algebras on $\mathbb{D}$.

For $\varphi_1, \varphi_2 \in M(H^\infty)$, the pseudohyperbolic distance from φ_1 to φ_2 is given by

$$\rho(\varphi_1,\varphi_2) = \sup \{|\varphi_1(f)| : f \in H^\infty(\mathbb{D}), \|f\| \le 1, f(\varphi_2) = 0\}.$$

As an example, if z and w are points in $\mathbb{D}$, then point evaluation at z and w, denoted φ_z and φ_w, are points in $M(H^\infty)$. Schwarz's Lemma shows that $\rho(\varphi_z,\varphi_w) = |\frac{z - w}{1 - \bar{w}z}|$.

Since any element φ in $M(H^\infty)$ has $\|\varphi\| \leq 1$, we see that $\rho(\varphi_1,\varphi_2) \leq 1$. If we define $\varphi_1 \sim \varphi_2$ if and only if $\rho(\varphi_1,\varphi_2) < 1$, then $\sim$ is an equivalence relation and the equivalence classes are the Gleason parts of $M(H^\infty)$. The Gleason part of a point φ in $M(H^\infty)$ will be denoted $P(\varphi)$. Thus for a point z in the open unit disc, the Gleason part of φ_z is $\mathbb{D}$. K. Hoffman [11] showed that for $\varphi \in M(H^\infty)$, there exists a map L_φ of $\mathbb{D}$ onto $P(\varphi)$ such that

(i) for f in H^∞, $f \circ L_\varphi$ is again in H^∞

and (ii) L_φ is one-one if $P(\varphi) \neq \{\varphi\}$.

We will use these results frequently in our discussion of algebras on the disc.

By using the Gelfand transform, one may think of $H^\infty(\mathbb{D})$ as contained in the algebra $C(M)$, of continuous functions on the maximal ideal space of H^∞. If u is a bounded harmonic function, u extends to a function in $C(M)$ [11], so $H^\infty(\mathbb{D})[u]$ is contained in $C(M)$. In our discussion we will use the following (slightly different) version of a lemma that appeared within a proof in [3].

Lemma 1. Let $u,f \in H^\infty(\mathbb{D})$. If $\bar{u} \in H^\infty(\mathbb{D})[\bar{f}]$, and if f is constant on a part $P(\varphi)$ for φ in $M(H^\infty)$, then u is also constant on $P(\varphi)$.

Proof. Suppose that $g_n = h_0 + h_1\bar{f} + \cdots + h_n\bar{f}^n$ for $h_0, \ldots, h_n \in H^\infty(\mathbb{D})$. Then $g_n \in C(M)$. Now f is constant on $P(\varphi)$ so that $g_n|P(\varphi) \in H^\infty$; that is, $g_n \circ L_\varphi \in H^\infty$. Since $\bar{u} \in H^\infty(\mathbb{D})[\bar{f}]$, $\bar{u}$ will be a limit of functions of the same form . Hence $\bar{u} \circ L_\varphi$ is a uniform limit of functions in H^∞. Thus

$$\overline{u \circ L_\varphi} = \bar{u} \circ L_\varphi$$

is an element of H^∞. Since $u \circ L_\varphi$ is analytic, $u \circ L_\varphi$ is constant on $\mathbb{D}$. Hence u is constant on $P(\varphi)$.

We will need to understand the behavior of functions in $C(M)$ on Gleason parts. For a function f in $H^\infty(\mathbb{D})$, one convenient way of describing the behavior of f on a Gleason part $P(\varphi)$ is to look at $(f \circ L_\varphi)'$. A standard calculation [3, p. 715] shows that if $\varphi \in M(H^\infty)-\mathbb{D}$ and $\{w_\alpha\}$ is a net of points in $\mathbb{D}$ with $w_\alpha \to \varphi$, then

$$|(f \circ L_\varphi)'(w)| = \lim_\alpha \frac{(1 - |z_\alpha|^2)|f'(z_\alpha)|}{(1 - |w|^2)}$$

where $z_\alpha = (w_\alpha + w)/(1 + \bar{w}_\alpha w)$.

Recall that the little Bloch space $\mathcal{B}_0$ is the set of analytic functions f on the disc $\mathbb{D}$ such that $(1 - |z|^2)f'(z) \to 0$ as $|z| \nearrow 1$. Thus, the formula above shows that the bounded analytic functions which are constant on Gleason parts other than the unit disc $\mathbb{D}$ is precisely the algebra $\mathcal{B}_0 \cap H^\infty(\mathbb{D})$. We shall denote this space by COPA. (This space is usually denoted COP in the literature, but we shall have reason to distinguish functions which are constant on parts and analytic from those functions which are in $C(M)$, constant on parts, but not necessarily analytic.)

Our purpose in this part of the paper is to discuss the possibility of finding a Chang Marshall theorem for algebras on the disc. Note that for $f \in \mathbb{B}_0 \cap H^\infty(\mathbb{D})$, Lemma 1 shows that any inner function with conjugate in $H^\infty(\mathbb{D})[\bar{f}]$ must be constant on parts. If b is an interpolating Blaschke product and φ is a point in the closure of the zeroes $\{z_n\}$ of b, then the formula above for the derivative of $b \circ L_\varphi$, together with the definition of interpolating shows that

$$|(b \circ L_\varphi)'(0)| = \lim (1 - |z_\alpha|^2)|b'(z_\alpha)| \geq \delta > 0.$$

In particular, $b \circ L_\varphi$ is not constant on $\mathbb{D}$. Since L_φ maps $\mathbb{D}$ onto $P(\varphi)$ we obtain the following lemma [3, Theorem 2].

Lemma 2. Let $f \in \mathbb{B}_0 \cap H^\infty(\mathbb{D})$. Then $H^\infty(\mathbb{D})[\bar{f}]$ does not contain the conjugate of any interpolating Blaschke product.

It seems as though the lack of conjugates of interpolating Blaschke products indicates that there is no Chang Marshall theorem here. However, if $f \in H^\infty(\mathbb{D}) \setminus \mathbb{B}_0$ then [3, Theorem 2] there is an interpolating Blaschke product b such that $\bar{b} \in H^\infty(\mathbb{D})[\bar{f}]$. In fact, the complex conjugate of every subproduct of the interpolating Blaschke product is also in the algebra $H^\infty(\mathbb{D})[\bar{f}]$. (Note that this is not at all clear, because the argument we used on the circle used the fact that $u\bar{u} = 1$ a.e. for an inner function on the circle. This, of course, fails on $\mathbb{D}$.) Motivated by this, our understanding of the situation on the circle and the behavior of $H^{\infty}(\mathbb{D}) + C(\overline{\mathbb{D}})$, Axler and I posed several questions.

The interpolating Blaschke product that Axler and I found in the algebra was what is called thin; that is, its zero sequence $\{z_n\}$ satisfies

$$\prod_{\substack{m=1 \\ m\neq n}}^{\infty} \rho(z_m, z_n) \to 1 \quad \text{as } n \to \infty.$$

Rather than restricting ourselves to thin interpolating Blaschke products, suppose that b is an interpolating Blaschke product. Axler and I asked whether the complex conjugate of every subfactor of b is also in $H^{\infty}(\mathbb{D})[\overline{b}]$. Unfortunately, the answer is no. K. Izuchi and I [8] have an example of an interpolating Blaschke product b such that $b = b_1 b_2$ and for some φ in $M(H^{\infty})$, b is constant on $P(\varphi)$ but b_1 and b_2 are not. Using Lemma 1, we see that neither $\overline{b}_1$ nor $\overline{b}_2$ can be in $H^{\infty}(\mathbb{D})[\overline{b}]$. Motivated again by the role support sets play in the study of subalgebras of L^{∞}, one can ask a slightly different question, and one that we have been unable to answer. Suppose that b_1 is a subfactor of an interpolating Blaschke product b and b_1 is constant on any Gleason part that b is constant on. Is $\overline{b}_1 \in H^{\infty}(\mathbb{D})[\overline{b}]$?

Given the abundance of conjugates of interpolating Blaschke products in $H^{\infty}(\mathbb{D})[\overline{f}]$ when f is in H^{∞} but not in $\mathcal{B}_0$, Axler and I asked whether or not, for $f \in H^{\infty} \sim \mathcal{B}_0$, $H^{\infty}(\mathbb{D})[\overline{f}]$ is generated by $H^{\infty}(\mathbb{D})$ and the set of complex conjugates of the interpolating Blaschke products contained in $H^{\infty}(\mathbb{D})[\overline{f}]$.

Izuchi and I [8] were able to give an example of an inner function u, not in $\mathcal{B}_0$, such that $H^\infty(\mathbb{D})[\bar{u}]$ is not generated by H^∞ and the complex conjugates of the interpolating Blaschke products which are in $H^\infty(\mathbb{D})[\bar{u}]$. However, something interesting does happen here. Let $\mathfrak{C}$ denote the set of interpolating Blaschke products whose conjugates lie in $H^\infty(\mathbb{D})[\bar{u}]$, and let $\mathfrak{D}$ denote the set of functions in $H^\infty \cap \mathcal{B}_0$ whose conjugates lie in $H^\infty(\mathbb{D})[\bar{u}]$. Then $H^\infty(\mathbb{D})[\bar{u}]$ is generated by $H^\infty(\mathbb{D})$, $\bar{\mathfrak{C}}$, and $\bar{\mathfrak{D}}$ (where the bar denotes complex conjugation). For which $f \in H^\infty$ is $H^\infty(\mathbb{D})[\bar{f}]$ generated by $H^\infty(\mathbb{D})$, the complex conjugates of functions in $H^\infty \cap \mathcal{B}_0$ which are in $H^\infty(\mathbb{D})[\bar{f}]$ and the complex conjugates of the interpolating Blaschke products which are also in the algebra? We remark that it is also not too difficult to come up with a subalgebra of $L^\infty(\mathbb{D})$ of the form $H^\infty(\mathbb{D})[\bar{f}]$ with $f \in H^\infty$ which contains the conjugate of no inner function in $\mathcal{B}_0 - C$. To create one such example, let u be any inner function which is continuous everywhere on $\mathbb{T}$ except the point $z = 1$ (where we choose u so that it is not continuous there). Suppose that there is an inner function v in $\mathcal{B}_0$ whose conjugate belongs to $H^\infty(\mathbb{D})[\bar{u}]$. We claim that the singularities of v on $\mathbb{T}$ are isolated. That will do it, because Sarason [16] has shown that this is impossible for an inner function in $\mathcal{B}_0$.

Suppose that v has a singularity at $w \neq 1$ on the unit circle. Then [9, p. 80] there exists $\{z_n\}$ with $z_n \to w$ and $v(z_n) \to 0$. Let $\varphi \in M(H^\infty) - \mathbb{D} = M(H^\infty + C)$ be a point in the closure of $\{z_n\}$ and let $g \in C$ satisfy $g(w) = 1$ and $g(1) = 0$. Then $g\bar{v}$ can be uniformly approximated by functions of the form $g(h_1 + h_2\bar{u} + h_3\bar{u}^2 + \cdots + h_n\bar{u}^n)$ where $h_1, \ldots, h_n$ are

H^∞ functions. It is not difficult to see that $g\bar{u}^j$ is continuous everywhere on $\mathbb{T}$ for every $j = 1,2,\ldots,n$. Hence $g\bar{v}$ can be uniformly approximated by functions in $H^\infty + C$. Thus

$$1 = \varphi(g) = \varphi(g(\bar{v}v)) = \varphi(g\bar{v})\varphi(v) = 0$$

where the second equality holds because $\bar{v}v = 1$ on $\mathbb{T}$, and the third equality holds because $\varphi \in M(H^\infty+C)$ and $g\bar{v} \in H^\infty + C$. Thus $\psi(v) \neq 0$ for any $\psi \in M(H^\infty + C)$ such that $\psi(z) \neq 1$. Hence the singularities of v are isolated and Sarason's result implies that $H^\infty(\mathbb{D})[\bar{u}]$ does not contain the conjugate any inner functions in $\mathcal{B}_0$. The idea in [8] is to combine these observations to produce an inner function b, not in $\mathcal{B}_0$, such that $H^\infty(\mathbb{D})[\bar{b}]$ is not generated by H^∞ and the complex conjugates of the interpolating Blaschke products which are in $H^\infty(\mathbb{D})[\bar{b}]$.

Much of what was done above was motivated by the study of the algebra $H^\infty(\mathbb{D}) + C(\bar{\mathbb{D}})$. However, results in operator theory (see [2],[3], and [19]) and duality [1] suggest that there is another algebra that might serve as motivation here.

It can be shown [18] that $H^\infty + C = H^\infty + QC$ where QC is the algebra of functions in $H^\infty + C$ whose complex conjugates are also in $H^\infty + C$. Thus one might draw one's motivation from an area version of QC rather than $C(\bar{\mathbb{D}})$. But what is the correct area version of QC?

Recall that $H^\infty + C$ functions that are real valued on a support set are constant on that set. Thus QC functions are constant on support sets. As explained above, the role support sets played in the study of algebras on the circle is now played by Gleason parts. Thus we might replace

QC by the algebra

$$\mathrm{COP} = \{f \in C(M): f \text{ is constant on parts in } M(H^{\infty}) \text{ other than } \mathbb{D}\}.$$

The problem is that so little is known about COP. Sarason [16] showed that COP contains an inner function. Recently, C. Bishop [6], C. Sundberg, and K. Stephenson [17] have been able to construct inner functions in COP. Bishop [5] has also given a characterization of the Blaschke products in COP in terms of their zeros. At this point we really only have some information about COPA $= \mathrm{COP} \cap H^{\infty}$. What is the relationship between COPA and COP?

The study of subalgebras of $L^{\infty}(\mathbb{D})$ has been motivated by the theory of subalgebras of $L^{\infty}(\mathbb{T})$. But results in operator theory and the study of $\mathcal{B}_0$ show that these algebras are interesting in their own right. We need to find the right questions. We hope that the right questions will, as in the past, inspire the development of new techniques.

REFERENCES

1. J. M. Anderson, *Bloch functions: The basic theory*, p.1-17 in Operators and Function Theory, edited S.C. Power, D. Reidel, Dordrecht, 1985.

2. Sheldon Axler, *The Bergman space, the Bloch space, and commutators of multiplication operators*, Duke Math. J. 53(1986), 305-332.

3. Sheldon Axler and Pamela Gorkin, *Algebras on the disk and doubly commuting multiplication operators*, Trans. Amer. Math. Soc. 309(1988), 711-723.

4. Sheldon Axler and Allen Shields *Algebras generated by analytic and harmonic functions*, Indiana Univ. Math. J. 36(1987) 631-638.

5. Christopher J. Bishop, *Bounded functions in the little Bloch space*, Pacific J. Math. , 142 (1990), 209-225.

6. Christopher J. Bishop, *An infinite Blaschke product in* $\mathbb{B}_o$, preprint.

7. S.Y. Chang, *A characterization of Douglas subalgebras*, Acta Math. 137 (1976) 81-89.

8. Pamela Gorkin and Keiji Izuchi, *Some counterexamples in subalgebras of* $L^\infty(\mathbb{D})$, in preparation.

9. John B. Garnett, *Bounded Analytic Functions*, Academic Press, New York, 1981.

10. Kenneth Hoffman, *Banach Spaces of Analytic Functions*, Prentice Hall, Englewood Cliffs, New Jersey, 1962.

11. Kenneth Hoffman, *Bounded analytic functions and Gleason parts* Ann. of Math. 86 (1967) 74-111.

12. D.E. Marshall, *Subalgebras of* L^∞ *containing* H^∞, Acta Math. 137 (1976), 91-98.

13. W. Rudin, *Spaces of type* $H^\infty + C$, Ann. Inst. Fourier(Grenoble) 25 (1975), 99-125.

14. Donald Sarason, *Algebras of functions on the unit circle*, Bull. Amer. Math Soc. 79 (1973) 286-299.

15. Donald Sarason, *Function Theory on the Unit Circle*, Virginia Poly. Inst.

and State Univ., Blacksburg, Virginia, 1978.

16. Donald Sarason, *Blaschke products in* $\mathbb{B}_0$, Linear and Complex Analysis Problem Book, Lecture Notes in Math., Vol. 1043, Springer-Verlag, Berlin, 1984.

17. K. Stephenson, *Construction of an inner function in the little Bloch space*, Trans. Amer. Math. Soc., to appear.

18. Thomas H. Wolff, *Two algebras of bounded functions*, Duke Math. J. 49 (1982) 321-328.

19. Dechao Zheng, *Hankel operators and Toeplitz operators on the Bergman space*, J. Funct. Anal. 83 (1989) 98-120.

On the Local Behavior of a Banach Space of Continuous Functions and Its Application to the Problem on the Range Transformations

OSAMU HATORI Department of Mathematics, Tokyo Medical College,
6 - 1 - 1 Shinjuku, Shinjuku-ku Tokyo 160, Japan

INTRODUCTION

Let X be a compact Hausdorff space and let $C(X)$ (resp. $C_R(X)$) denote the space of all complex (resp. real) valued continuous functions on X. Also let A be a complex (resp. real) Banach space continuously embedded in $C(X)$ (resp. $C_R(X)$), that is, A is a complex (resp. real) subspace of $C(X)$ (resp. $C_R(X)$) which is itself a complex (resp. real) Banach space with the norm satisfying that for some $M > 0$ the inequality $\|f\|_{\infty(X)} \leq M\|f\|_A$ holds for every f in A, where $\|\cdot\|_{\infty(X)}$ denotes the supremum norm on X. For a point x in X and a discrete topological space Λ we denote $F_x^\Lambda = \bigcap [G \times \Lambda]$, where G varies over all the compact neighborhoods of x and $[G \times \Lambda]$ denotes the closure of the direct product $G \times \Lambda$ in the Stone - Čech compactification $\beta(X \times \Lambda)$ of $X \times \Lambda$. We denote the space of all A-valued bounded (with respect to the topology induced by the norm $\|\cdot\|_A$) functions on Λ by $\tilde{A}^\Lambda$. In general the behavior of $\tilde{A}^\Lambda | F_x^\Lambda$ reflects the behavior of A near x. We show some examples. If A consists of the complex (resp. real) valued constant functions, then $\tilde{A}^\Lambda | F_x^\Lambda$ is isomorphic to the algebra $C^b(\Lambda)$ (resp. $C_R^b(\Lambda)$) of all complex (resp. real) valued bounded functions on Λ. If A coincides with $C(X)$ (resp. $C_R(X)$), then $\widetilde{A}^\Lambda | F_x^\Lambda$ coincides with $C(F_x^\Lambda)$ (resp. $C_R(F_x^\Lambda)$). Suppose that $A(B_n)$ is the ball algebra on the n-dimensional ball B_n in C^n; *i.e.*, A is the algebra of the complex valued continuous functions on the closed ball which are analytic on the interior. Then $A(B_n)^{\sim\Lambda} | F_x^\Lambda$ is isomorphic to the algebra $C^b(\Lambda)$ if x is a point in the interior. When x is a boundary point, we see by (2) of Theorem 1.1 of [6] which we will show later that

$$C^b(\Lambda) \subsetneqq A(B_n)^{\sim\Lambda} | F_x^\Lambda \subsetneqq C(F_x^\Lambda)$$

if the cardinality of Λ is at least that of an open base for the topology at x. By a theorem of Bernard - Dufresnoy [1] and (1) of the same theorem as above we also see that $(A(B_n) | \partial B_n)^{\sim\Lambda} | \widetilde{F}_x^\Lambda$ separates the points of $\widetilde{F}_x^\Lambda$ if x is a boundary point, where $\widetilde{F}_x^\Lambda = \cap [G \times \Lambda]$ of G being a compact neighborhood of x in ∂B_n. On the other hand suppose that A is a Banach algebra included in $C(X)$ and that I is an ideal (not necessary closed) of A. If x is a point off the kernel of I, then $\widetilde{A}^\Lambda | F_x^\Lambda = \widetilde{I}^\Lambda | F_x^\Lambda$. This shows that the "infinitesimally" local behavior of the Banach algebra and its ideal coincide off the kernel. In §1 we study these phenomena. First of all we show some conditions which assert $\widetilde{A}^\Lambda | F_x^\Lambda = \widetilde{B}^\Lambda | F_x^\Lambda$ for Banach spaces A and B continuously embedded in $C(X)$ (resp. $C_R(X)$). As an application of these results we show that a closed subspace of a finite codimension in an ultraseparating space is also ultraseparating.

In this paper we use the following notations and terminologies. Hereafter X denotes a compact Hausdorff space. For f in $C(X)$ and a compact subset K of X, $\|f\|_{\infty(K)}$ denotes the supremum norm of f on K. For any subsets S and T of $C(X)$, a point x in X and a compact subset K of X we denote

$$S|K = \{ f \in C(K) : \exists F \in S \ \textit{such that} \ F|K = f \},$$

where $F|K$ is the restriction of the function F to K.

$$S_x = \{ f \in S : f(x) = 0 \}$$

$$ReS = \{ u \in C_R(X) : \exists v \in C_R(X) \ \textit{such that} \ u + iv \in S \}$$

$$S + T = \{ f + g \in C(X) : f \in S, \ g \in T \}$$

$$S \cdot T = \{ fg \in C(X) : f \in S, \ g \in T \}.$$

We say that A is a complex (resp. real) Banch space included in $C(X)$ (resp. $C_R(X)$) if A is a complex (resp. real) subspace of $C(X)$ (resp. $C_R(X)$) which is a complex (resp. real) Banach space with the norm $\|\cdot\|_A$ satisfying that

$\|f\|_{\infty(X)} \leq \|f\|_A$ for every f in A. By renorming we may suppose that every Banach space continuously embedded in $C(X)$ (resp. $C_R(X)$) is equivalent to a Banach space included in $C(X)$ (resp. $C_R(X)$). Although we will state results on Banach spaces included in $C(X)$ or $C_R(X)$ those results are also valid for Banach space continuously embedded in $C(X)$ or $C_R(X)$. If V is a subset of a Banach space A included in $C(X)$ or $C_R(X)$, then we denote the closure of V in A by $cl_A V$. When A is uniformly closed we simply write $cl V$ instead of $cl_A V$. For a compact subset K of X we define the quotient norm of the restriction of A to K by

$$\|u\|_A | K = \mathbf{inf} \{ \|f\|_A : f | K = u \}$$

for $u \in A | K$. Then $A | K$ is a complex (resp. real) Banach space continuously embedded in $C(K)$ (resp. $C_R(G)$) with respect to $\| \cdot \|_A | K$. Let Λ be a discrete space. Then $\tilde{A}^\Lambda$ is a Banach space with the norm

$$\|\tilde{f}\|_{\tilde{A}^\Lambda} = \mathbf{sup} \{ \|\tilde{f}(\alpha)\|_A : \alpha \in \Lambda \}$$

for $\tilde{f}$ in $\tilde{A}^\Lambda$. When M is a subset of A, $\tilde{M}^\Lambda$ denotes the set of all bounded (with respect to $\| \cdot \|_A$) M-valued functions on Λ. If $f_\lambda \in A$ for each λ in Λ, we denote by $\langle f_\alpha \rangle$ the function in $\tilde{A}^\Lambda$ with $\langle f_\alpha \rangle (\lambda) = f_\lambda$ for every λ in Λ. In particular, for f in A, $\langle f \rangle$ is the constant function in $\tilde{A}^\Lambda$ with $\langle f \rangle (\lambda) = f$ for every λ in Λ. On the other hand, since every function f in A satisfies the inequality $\|f\|_{\infty(X)} \leq \|f\|_A$ we may suppose that every A-valued function $\tilde{f}$ in $\tilde{A}^\Lambda$ is a complex (resp. real) valued bounded (with respect to the supremum norm) continuous function on $X \times \Lambda$ which is defined by $\tilde{f}(x, \alpha) = (\tilde{f}(\alpha))(x)$ for every (x, α) in $X \times \Lambda$. So we may suppose that $\tilde{A}^\Lambda$ is a complex (resp. real) Banach space included in $C(\tilde{X}^\Lambda)$ (resp. $C_R(\tilde{X}^\Lambda)$), where $\tilde{X}^\Lambda$ is the Stone - Čech compactification of the direct product $X \times \Lambda$. Let x be a point in X. We denote $F_x^\Lambda = \bigcap [G \times \Lambda]$, where G varies over all the compact neighborhoods of x and $[\cdot]$ denotes the closure in $\tilde{X}^\Lambda$. A is said to be *ultraseparating* on X if $\tilde{A}^N$ separates the points of $\tilde{X}^N$ for the discrete space N of all positive integers. (This is equivalent to : $\tilde{A}^\Lambda$ separates the points of $\tilde{X}^\Lambda$ for an infinite Λ [6; Corollary 1].) A is said to be *ultraseparating near* a point x in X if there is a compact neighborhood G of x such that $A | G$ is ultraseparating on G. For further information about $\tilde{A}^\Lambda$ and related subjects, see [6].

1 INFINITESIMALLY LOCAL BEHAVIOR OF A BANACH SPACE OF CONTINUOUS FUNCTIONS

In this section we study the infinitesimally local behavior of a Banach space of continuous functions, that is, we show conditions on F_x^Λ for $\tilde{A}^\Lambda | F_x^\Lambda = \tilde{B}^\Lambda | F_x^\Lambda$ of a complex (resp. real) Banach space A included in $C(X)$ (resp. $C_R(X)$) and a Banach space B continuously embedded in A.

Theorem 1.1. *Let A be a complex (resp. real) Banach space included in $C(X)$ (resp. $C_R(X)$) with the norm $\|\cdot\|_A$ and B be a complex (resp. real) Banach space continuously embedded in A with the norm $\|\cdot\|_B$. Let x be a point in X and Λ be a discrete space. Suppose that for every non-zero function $\tilde{f}$ in $\tilde{A}^\Lambda | F_x^\Lambda$ there exists $\tilde{g}$ in $\tilde{B}^\Lambda | F_x^\Lambda$ such that the strict inequality*

$$\|\tilde{f} - \tilde{g}\|_{\tilde{A}^\Lambda | F_x^\Lambda} < \|\tilde{f}\|_{\tilde{A}^\Lambda | F_x^\Lambda}$$

holds, where $\|\cdot\|_{\tilde{A}^\Lambda | F_x^\Lambda}$ is the quotient norm of the restriction of $\|\cdot\|_{\tilde{A}^\Lambda}$ to F_x^Λ. Then $\tilde{A}^\Lambda | F_x^\Lambda = \tilde{B}^\Lambda | F_x^\Lambda$.

Proof. If Λ is finite, then $F_x^\Lambda = \{x\} \times \Lambda$, so the conclusion is trivial. So we consider the case that Λ is infinite. By a standard argument on Banach spaces one can prove $\tilde{A}^\Lambda | F_x^\Lambda = \tilde{B}^\Lambda | F_x^\Lambda$ by using the following: There exists a positive integer n_0 which satisfies that for every function $\tilde{f}$ in $\tilde{A}^\Lambda | F_x^\Lambda$ with $\|\tilde{f}\|_{\tilde{A}^\Lambda | F_x^\Lambda} < 1$ there is $\tilde{g}$ in $\tilde{B}^\Lambda | F_x^\Lambda$ with $\|\tilde{g}\|_{\tilde{B}^\Lambda | F_x^\Lambda} \leq n_0$ and $\|\tilde{f} - \tilde{g}\|_{\tilde{A}^\Lambda | F_x^\Lambda} \leq 1 - 1/n_0$. So we will verify this condition. Suppose that it is not true, that is, for every positive integer n there exists $\tilde{f}_n$ in $\tilde{A}^\Lambda | F_x^\Lambda$ with $\|\tilde{f}_n\|_{\tilde{A}^\Lambda | F_x^\Lambda} < 1$ such that every $\tilde{g}$ in $\tilde{B}^\Lambda | F_x^\Lambda$ satisfying

$$\|\tilde{f}_n - \tilde{g}\|_{\tilde{A}^\Lambda | F_x^\Lambda} \leq 1 - 1/n$$

must also satisfies

$$\|\tilde{g}\|_{\tilde{B}^\Lambda | F_x^\Lambda} > n.$$

By the definition of the quotient norm there is $\tilde{\varphi}_n$ in $\tilde{A}^\Lambda$ with $\tilde{\varphi}_n | F_x^\Lambda = \tilde{f}_n$ and $\|\tilde{\varphi}_n\|_{\tilde{A}^\Lambda} \leq 1$ for every n. Take a homeomorphism T of Λ onto $\Lambda \times \mathbf{N}$ with

$$T(\lambda) = (t(\lambda), s(\lambda))$$

for every λ in Λ, where $\mathbf{N}$ is the discrete space of all positive integers. If $\tilde{\varphi}$ is the function on Λ with

$$\tilde{\varphi}(\lambda) = \tilde{\varphi}_{s(\lambda)}(t(\lambda))$$

for every λ in Λ, then $\tilde{\varphi}$ is in $\tilde{A}^\Lambda$ with $\|\tilde{\varphi}\|_{\tilde{A}^\Lambda} \leq 1$. By the hypothesis of Theorem 1 there is a function $\tilde{h}$ in $\tilde{B}^{\Lambda}$ which satisfies the inequality

$$\|\tilde{\varphi}|F_x^{\Lambda} - \tilde{h}|F_x^{\Lambda}\|_{\tilde{A}^{\Lambda}|F_x^{\Lambda}} < \|\tilde{\varphi}|F_x^{\Lambda}\|_{\tilde{A}^{\Lambda}|F_x^{\Lambda}}.$$

There is a large m with

$$\|\tilde{\varphi}|F_x^{\Lambda} - \tilde{h}|F_x^{\Lambda}\|_{\tilde{A}^{\Lambda}|F_x^{\Lambda}} < 1 - 1/m.$$

Thus there is $\tilde{g}$ in $\tilde{A}^{\Lambda}$ with $\tilde{g} = 0$ on F_x^{Λ} such that

$$\|\tilde{\varphi} - \tilde{h} - \tilde{g}\|_{\tilde{A}^{\Lambda}} < 1 - 1/m,$$

so

$$\|\tilde{\varphi}(\lambda) - \tilde{h}(\lambda) - \tilde{g}(\lambda)\|_A < 1 - 1/m$$

for every λ in Λ. For every n let $\tilde{h}_n$ and $\tilde{g}_n$ be functions in $\tilde{B}^{\Lambda}$ with

$$\tilde{h}_n(\lambda) = \tilde{h}(T^{-1}(\lambda, n))$$

and

$$\tilde{g}_n(\lambda) = \tilde{g}(T^{-1}(\lambda, n))$$

for every λ in Λ. We have

$$\|\tilde{\varphi}_n(\lambda) - \tilde{h}_n(\lambda) - \tilde{g}_n(\lambda)\|_A < 1 - 1/m$$

for every λ in Λ, so

$$\|\tilde{\varphi}_n - \tilde{h}_n - \tilde{g}_n\|_{\tilde{A}^{\Lambda}} \leqq 1 - 1/m.$$

For every $\varepsilon > 0$ there is a compact neighborhoods G_ε of x with $|\tilde{g}(\tilde{x})| < \varepsilon$ for every $\tilde{x}$ in $[G_\varepsilon \times \Lambda]$ since $\tilde{g} = 0$ on F_x^{Λ} and since $F_x^{\Lambda} = \bigcap [G \times \Lambda]$, where G varies over all the compact neighborhood of x. So we have

$$|\tilde{g}(y, \lambda)| = |(\tilde{g}(\lambda))(y)| < \varepsilon$$

for every y in G_ε and λ in Λ, in particular,

$$|(\tilde{g}(T^{-1}(\lambda, n)))(y)| = |(\tilde{g}_n(\lambda))(y)| < \varepsilon$$

for each n and each y in G_ε and each λ in Λ. We have that $|\tilde{g}_n| \leqq \varepsilon$ on $[G_\varepsilon \times \Lambda]$. We see that $\tilde{g}_n = 0$ on F_x^{Λ} for every n, so the inequalities

$$\|\tilde{f}_n - \tilde{h}_n|F_x^{\Lambda}\|_{\tilde{A}^{\Lambda}|F_x^{\Lambda}}$$
$$= \|\tilde{\varphi}_n|F_x^{\Lambda} - \tilde{h}_n|F_x^{\Lambda}\|_{\tilde{A}^{\Lambda}|F_x^{\Lambda}} \leqq \|\tilde{\varphi}_n - \tilde{h}_n - \tilde{g}_n\|_{\tilde{A}^{\Lambda}} \leqq 1 - 1/m$$

hold. It follows that for every n with $n > m$ we have that

$$\|\tilde{h}_n\|_{\tilde{B}^{\Lambda}} \geqq \|\tilde{h}_n|F_x^{\Lambda}\|_{\tilde{B}^{\Lambda}|F_x^{\Lambda}} > n.$$

On the other hand

$$\|\tilde{h}_n\|_{\tilde{B}^{\Lambda}} \leqq \|\tilde{h}\|_{\tilde{B}^{\Lambda}} < \infty$$

hold for every n, which is a contradiction.

Corollary 1.2. *Let A be a complex (resp. real) Banach space included in $C(X)$ (resp. $C_R(X)$) and B be a complex (resp. real) Banach space continuously embedded in A. Let x*

be a point in X *and* Λ *be a discrete space. If* $\tilde{B}^{\Lambda}|F_x^{\Lambda}$ *is dense with respect to the topology induced by the quotient norm* $\|\cdot\|_{\tilde{A}^{\Lambda}|F_x^{\Lambda}}$ *in* $\tilde{A}^{\Lambda}|F_x^{\Lambda}$, *then* $\tilde{B}^{\Lambda}|F_x^{\Lambda} = \tilde{A}^{\Lambda}|F_x^{\Lambda}$.

Corollary 1.3. *Under the assumption of Theorem* 1.1 *suppose also that cardinality of* Λ *is at least that of an open base for the topology of* X *at* x *and that* $\tilde{A}^{\Lambda}|F_x^{\Lambda} = C(F_x^{\Lambda})$ *(resp.* $C_R(F_x^{\Lambda})$*). Then there exists a compact neighborhood* G *of* x *such that* $B|G = A|G = C(G)$ *(resp.* $C_R(G)$*).*

Proof. Without loss of generality we may suppose that B is a complex (resp. real) Banch space contained in $C(X)$ (resp. $C_R(X)$). It follows by (2) of Theorem 1 in [6] that the conclusion holds.

2 RELATION BETWEEN THE INFINITESIMALLY LOCAL BEHAVIOR AND LOCAL BEHAVIOR OF CONTINUOUS FUNCTIONS

As we have shown in (2) of Theorem 1 in [6] the coincidence of $\tilde{A}^{\Lambda}|F_x^{\Lambda}$ and $C(F_x^{\Lambda})$ for certain Λ provides the coincidence of A and $C(X)$ near x. We consider such a phenomenon in the general case, that is, we study a relationship between the behavior of functions on F_x^{Λ} and the behavior of functions near x.

Definition 2.1. *Let* A *be a complex (resp. real) Banach space included in* $C(X)$ *(resp.* $C_R(X)$*). Let* L *be a subset of* A *and* x *be a point in* X. *We say that* L *is* A-*equicontinuous at* x *if for every positive number* ε *there exists a compact neighborhood* G_ε *such that the inequality* $|f(x) - f(y)| < \varepsilon\|f\|_A$ *holds for every* f *in* L *and for every* y *in* G_ε.

Theorem 2.2. *Let* A *be a complex (resp. real) Banach space included in* $C(X)$ *(resp.* $C_R(X)$*) and* B *be a complex (resp. real) Banach space continuously embedded in* A. *Let* x *be a point in* X. *Suppose that either of the following conditions holds : there exists a function* s *in* B *such that* $s(x) \neq 0$; $t(x) = 0$ *for every function* t *in* A. *Let* $\{L_n\}$ *be a countable family of* A-*equicontinous subset of* A *at* x. *If* $A = B + \bigcup_{n=1}^{\infty} L_n$, *then* $\tilde{A}^{\Lambda}|F_x^{\Lambda} = \tilde{B}^{\Lambda}|F_x^{\Lambda}$ *holds for every discrete space* Λ.

Proof. Since B is continuously embedded in A, $\tilde{B}^{\Lambda}|F_x^{\Lambda} \subset$

$\widetilde{A}^{\Lambda}|F_x^{\Lambda}$. We need only prove the reverse inclusion. For each positive integer n put

$$E_n = \{ f \in A: f = g + h \ \textit{for some} \ g \in B \ \textit{with} \ \|g\|_B \leq n \ \textit{and} \ h \in \bigcup_{k=1}^{n} L_k \ \textit{with} \ \|h\|_A \leq n \}.$$

Since $A = B + \bigcup_{n=1}^{\infty} L_n$ we have $\bigcup_{n=1}^{\infty} E_n = A$. By the Baire category theorem we see that $\mathrm{cl}_A E_m$ contains a non-empty open set in A for some m, say

$$W = \{ f \in A: \|f - f_0\|_A \leq r \} \subset \mathrm{cl}_A E_m$$

for some f_0 in A and $r > 0$. Put $L = \bigcup_{n=1}^{m} L_n$, so L is A-equicontinuous at x. Using this fact we will show that $\widetilde{A}^{\Lambda}|F_x^{\Lambda} = \widetilde{B}^{\Lambda}|F_x^{\Lambda}$ for any discrete space Λ. Let $\widetilde{f}$ be a function in $\widetilde{A}^{\Lambda}|F_x^{\Lambda}$. Without loss of generality we may suppose that $\widetilde{f}(F_x^{\Lambda}) \neq \{0\}$ and $\|\widetilde{f}\|_{\widetilde{A}^{\Lambda}|F_x^{\Lambda}} < 1$. There is a $\widetilde{\varphi}$ in $\widetilde{A}^{\Lambda}$ with $\|\widetilde{\varphi}\|_{\widetilde{A}^{\Lambda}} \leq 1$ and $\widetilde{\varphi}|F_x^{\Lambda} = \widetilde{f}$. Since $W \subset \mathrm{cl}_A E_m$ we can choose t_λ in E_m with

$$\|f_0 + r\widetilde{\varphi}(\lambda) - t_\lambda\|_A < (r/2)\|\widetilde{f}\|_{\widetilde{A}^{\Lambda}|F_x^{\Lambda}}$$

for every λ in Λ, where $t_\lambda = g_\lambda + h_\lambda$ for $g_\lambda \in B$ with $\|g_\lambda\|_B \leq m$ and $h_\lambda \in L$ with $\|h_\lambda\|_A \leq m$ respectively. Let $\widetilde{g}$ in $\widetilde{B}^{\Lambda}$ and $\widetilde{h}$ in $\widetilde{A}^{\Lambda}$ be given by $\widetilde{g}(\lambda) = g_\lambda$ and $\widetilde{h}(\lambda) = h_\lambda$ respectively for every λ in Λ. Then

$$\|\widetilde{\varphi} - (1/r)(\widetilde{g} + \widetilde{h} - \langle f_0 \rangle)\|_{\widetilde{A}^{\Lambda}} \leq (1/2)\|\widetilde{f}\|_{\widetilde{A}^{\Lambda}|F_x^{\Lambda}} < \|\widetilde{f}\|_{\widetilde{A}^{\Lambda}|F_x^{\Lambda}},$$

so

$$\|\widetilde{f} - (1/r)(\widetilde{g}|F_x^{\Lambda} + \widetilde{h}|F_x^{\Lambda} - \langle f_0 \rangle|F_x^{\Lambda})\|_{\widetilde{A}^{\Lambda}|F_x^{\Lambda}} < \|\widetilde{f}\|_{\widetilde{A}^{\Lambda}|F_x^{\Lambda}}.$$

By Theorem 1.1 we will be done if $\widetilde{h}|F_x^{\Lambda}$ is in $\widetilde{B}^{\Lambda}|F_x^{\Lambda}$ since $\langle f_0 \rangle|F_x^{\Lambda}$ is a constant function (in particular, $\langle f_0 \rangle|F_x^{\Lambda} = 0$ if $t(x) = 0$ for every t in A [6; Lemma 4]), so we now prove this. When there is a function s in B with $s(x) \neq 0$ we may suppose that $s(x) = 1$, for every $\varepsilon > 0$ there exists a compact neighborhood G_ε of x such that the inequality

$$|h_\lambda(y) - h_\lambda(x)s(y)| < \varepsilon$$

holds for every y in G_ε and for every λ in Λ. Thus we see that

$$|\widetilde{h} - \langle h_\lambda(x)s \rangle| \leq \varepsilon$$

on $[G_\varepsilon \times \Lambda]$. It follows that $\widetilde{h} = \langle h_\lambda(x)s \rangle$ on F_x^{Λ} since $F_x^{\Lambda} = \bigcap [G \times \Lambda]$, where G varies over all the compact neighborhoods of x. Thus we have that

$$\widetilde{h}|F_x^{\Lambda} = \langle h_\lambda(x)s \rangle|F_x^{\Lambda}$$

is in $\widetilde{B}^{\Lambda}|F_x^{\Lambda}$. When $t(x) = 0$ for every t in A, we see, in the same way, that $\widetilde{h} = 0$ on F_x^{Λ}. In any case we conclude that $\widetilde{A}^{\Lambda}|F_x^{\Lambda} = \widetilde{B}^{\Lambda}|F_x^{\Lambda}$ for any discrete space Λ.

Corollary 2.3. *Let A be a complex (resp. real) Banach space included in $C(X)$ (resp. $C_R(X)$). Let x be a point in X with $A_x \neq A$. Let L be a subspace of finite dimension in $C(X)$ (resp. $C_R(X)$). Let Λ be a discrete space. Then we have $\tilde{A}^\Lambda | F_x^\Lambda = (A + L)^{\sim\Lambda} | F_x^\Lambda$, where the norm $\|\cdot\|_{A+L}$ on $A + L$ is defined by the following: Put $L_0 = (A + L) \ominus A$. For F in $A + L$ we define*

$$\|F\|_{A+L} = \|f\|_A + \|g\|_{\infty(X)},$$

where $F = f + g$, $f \in A$, $g \in L_0$. Then $\|\cdot\|_{A+L}$ is a complete norm since L_0 is finite dimensional.

In general A need not be represented by the sum $B + \overset{\infty}{\underset{n=1}{\cup}} L_n$, even if $\tilde{A}^\Lambda | F_x^\Lambda = \tilde{B}^\Lambda | F_x^\Lambda$ for some Λ. For example if Λ is finite and $A = C(X)$ for an infinite X and if B is the algebra of constant functions on X, then $\tilde{A}^\Lambda | F_x^\Lambda = \tilde{B}^\Lambda | F_x^\Lambda$ holds at a point x, while it is trivial that $A \neq B + \overset{\infty}{\underset{n=1}{\cup}} L_n$ for any A-equicontinuous family $\{L_n\}$ at non-isolated x. On the other hand if Λ has large cardinality, the reverse implication of Theorem 2.2 holds.

Proposition 2.4. *Let A be a complex (resp. real) Banach space included in $C(X)$ (resp. $C_R(X)$) and B be a complex (resp. real) Banach space continuously embedded in A. Let x be a point in X. Also let Λ be a discrete space with the cardinality at least that of A. Suppose that $\tilde{A}^\Lambda | F_x^\Lambda = \tilde{B}^\Lambda | F_x^\Lambda$ holds. Then there exists a countable family $\{L_n\}$ of A-equicontinuous subsets of A at x such that $A = B + \overset{\infty}{\underset{n=1}{\cup}} L_n$.*

Proof. Let T be a surjection of Λ onto $\{f \in A\colon \|f\|_A \leq 1\}$. Let $\tilde{f}$ in $\tilde{A}^\Lambda$ be given by $\tilde{f}(\lambda) = T(\lambda)$. There is $\tilde{g} = \langle g_\lambda \rangle$ in $\tilde{B}^\Lambda$ with $\tilde{f} | F_x^\Lambda = \tilde{g} | F_x^\Lambda$ since $\tilde{A}^\Lambda | F_x^\Lambda = \tilde{B}^\Lambda | F_x^\Lambda$. Put $h_\lambda = T(\lambda) - g_\lambda$ and $L_1 = \{h_\lambda\colon \lambda \in \Lambda\}$. Then $h_\lambda(x) = 0$ for every $\lambda \in \Lambda$ since $\{x\} \times \Lambda \subset F_x^\Lambda$. We now prove L_1 is A-equicontinuous at x. Since $\tilde{f} = \tilde{g}$ on F_x^Λ, for every positive ε there is a compact neighborhood G_ε with the property that the inequality

$$|\tilde{f}(p) - \tilde{g}(p)| < \varepsilon$$

holds for every p in $[G_\varepsilon \times \Lambda]$, so

$$|T(\lambda)(y) - g_\lambda(y)| < \varepsilon$$

holds for every y in G_ε and every λ in Λ. As $h_\lambda(x) = 0$ we see that

$$|h_\lambda(y) - h_\lambda(x)| < \varepsilon$$

holds for every $\lambda \in \Lambda$ and y in G_ε. We conclude that L_1 is A-equicontinuous at x. Define $L_n = nL_1 = \{nf: f \in L_1\}$ for every positive integer n. Then L_n is A-equicontinuous at x for every n and $A = B + \bigcup_{n=1}^{\infty} L_n$ by the definition of L_1.

3 ULTRASEPARATION PROPERTIES

In this section we study the ultraseparation property for a complex (resp. real) Banach space included in $C(X)$ (resp. $C_R(X)$) by using the results of previous sections.

Theorem 3.1. *Let A be a complex (resp. real) Banach space included in $C(X)$ (resp. $C_R(X)$) and B be a complex (resp. real) Banach space continuously embedded in A. Let $\{L_n\}$ be a countable family of subsets L_n of A with $A = B + \bigcup_{n=1}^{\infty} L_n$. Let x be a point in X. Suppose that A is ultraseparating near x and that $B_x \neq B$. Suppose also that each L_n is A-equicontinuous at x. Then B is ultraseparating near x.*

Proof. Let Λ be a discrete space with the same cardinality as that of an open base for the topology of X at x. Then by Theorem 2.2 we see that $\tilde{A}^\Lambda | F_x^\Lambda = \tilde{B}^\Lambda | F_x^\Lambda$. It follows by (1) of Theorem 1 in [6] that B is ultraseparating near x since $\tilde{A}^\Lambda | F_x^\Lambda$ separates the points of F_x^Λ.

As a corollary we show a generalization of theorems of Ellis [3; Theorem 12, 13].

Corollary 3.2. *Let A be a complex (resp. real) Banach space included in $C(X)$ (resp. $C_R(X)$) and B be a complex (resp. real) Banach space continuously embedded in A such that B separates the points in X. Suppose that A is ultraseparating on X and $B_x \neq B$ for every x in X. Suppose also that there is a subspace L of a countable dimension in A such that $B + L = A$. Then B is ultraseparating on X.*

Proof. We can write $L = \bigcup_{n=1}^{\infty} L_n$ where each L_n is finite dimensional subspace of L, so is A-equicontinuous at every point on X. So by Theorem 3.1 B is ultraseparating near each point of x. Then the conclusion follows by Lemma 6 in [6].

We also show that a hypodirichlet algebra is ultraseparating.

Theorem 3.3. *Let A be a function algebra on X and G be a compact subset of X such that $(\mathbf{cl}(\mathrm{Re}A))|G$ is a subspace of finite codimension in $C_R(G)$. Then $A|G$ with the quotient norm is ultraseparating on G.*

Proof. In the same way as in the proof of Theorem 3.1 we see that $(\mathbf{cl}(\mathrm{Re}A))^{\sim\Lambda}|F_x^\Lambda = C_R(F_x^\Lambda)$ ($^{\sim\Lambda}$ taken with respect to $\|\cdot\|_{\infty(X)}$) for every x in G, where Λ is a discrete space with the same cardinality as that of an open base for the topology of X at x. Let $\tilde{x}$ and $\tilde{y}$ be different points in F_x^Λ. Then there exists $\tilde{u} = \langle u_\alpha \rangle$ in $(\mathbf{cl}(\mathrm{Re}A))^{\sim\Lambda}$ with $u_\alpha \in \mathrm{Re}A$ for each α in Λ such that $\tilde{u}(\tilde{x}) \neq \tilde{u}(\tilde{y})$. There exist v_α in $\mathrm{Re}A$ with $u_\alpha + iv_\alpha \in A$ for each α in Λ, so $\exp(u_\alpha + iv_\alpha)$ is in A and

$$\|\exp(u_\alpha + iv_\alpha)\|_{\infty(X)} = \|\exp(u_\alpha)\|_{\infty(X)}$$

is uniformly bounded since u_α is. It follows that $\langle \exp(u_\alpha + iv_\alpha) \rangle$ is in $\tilde{A}^\Lambda$ and it separates $\tilde{x}$ and $\tilde{y}$. Thus we conclude that A is ultraseparating near x, so by Lemma 6 in [6] that $A|G$ is ultraseparating on G.

4 COMPLEMENTED SUBSPACES OF $C(X)$ AND $C_R(X)$

In this section we study a complex (resp. real) Banach space A included in $C(X)$ (resp. $C_R(X)$) such that $A + L = C(X)$ (resp. $C_R(X)$), where L is a subset of $C(X)$ (resp. $C_R(X)$) with certain conditions. Theorem 4.2 below is proven by the Riesz - Kakutani theorem and the following proposition, which can be proven using standard arguments.

Proposition 4.1. *Let B_1 be a complex (resp. real) Banach space and B_2 be a complex (resp. real) Banach space continuously embedded in B_1. Suppose that there is a linear subspace L of countable dimension in B_1 such that $B_1 = B_2 + L$. Then B_2 is a closed subspace of finite codimension in B_1.*

Theorem 4.2. *Let A be a complex (resp. real) Banach space included in $C(X)$ (resp. $C_R(X)$). Suppose that there is a complex (resp. real) subspace L of countable dimension in $C(X)$ (resp. $C_R(X)$) such that $A + L = C(X)$ (resp. $C_R(X)$). Then the following hold.*

(1) *For every compact subset K of X, $A|K$ is a closed subspace of finite codimension in $C(K)$ (resp. $C_R(K)$). In particular, A is a closed subspace of finite codimension in $C(X)$ (resp. $C_R(X)$).*

(2) *For every point x in X there exists a compact neighborhood G of x such that $A|G = C(G)$ (resp. $C_R(G)$) if $A \neq A_x$; $A|G = (C(G))_x$ (resp. $(C_R(G))_x$) if $A = A_x$.*

By Theorem 4.2 and a theorem of Glicksberg on function algebras with closed restriction [4] we see the following (*cf.* [2]).

Theorem 4.3. *Let A be a Banach function algebra on X, that is, a complex Banach algebra included in $C(X)$ wich separates the points of X and contains constant functions. Suppose that L is a complex subspace of countable dimension in $C(X)$ such that $A + L = C(X)$. Then $A = C(X)$.*

By using results in the previous sections we prove generalization of Theorem 4.2 and 4.3.

Theorem 4.4. *Let A be a complex (resp. real) Banach space included in $C(X)$ (resp. $C_R(X)$) and x be a point in X. Let $\{L_n\}$ be a countable family of subsets L_n of $C(X)$ (resp. $C_R(X)$). Suppose that $A + \bigcup_{n=1}^{\infty} L_n = C(X)$ (resp. $C_R(X)$) and that each L_n is A-equicontinuous at x. Then there exists a compact neighborhood G of x such that $A|G = C(G)$ (resp. $C_R(G)$) if $A_x \neq A$; $A|G = (C(G))_x$ (resp. $(C_R(G))_x$) if $A_x = A$.*

Proof. By Theorem 2.2 we see that $\tilde{A}^{\Lambda}|F_x^{\Lambda} = C(F_x^{\Lambda})$ (resp. $C_R(F_x^{\Lambda})$) if $A_x \neq A$, where Λ is a discrete space with the same cardinality as that of an open base for the topology of X at x. Thus by (2) of Theorem 1 in [6] the conclusion in the case of $A_x \neq A$ follows. On the other hand when $A_x = A$ we see in the same way that $(A + \mathbf{C})|G = C(G)$ (resp. $C_R(G)$) for some G, where $\mathbf{C}$ is the space of all constant functions. Thus the conclusion in the case of $A_x = A$ follows.

Theorem 4.5. *Let A be a complex (resp. real) Banach algebra included in $C(X)$ (resp. $C_R(X)$) which separates the points of X and is such that $A_x \neq A$ for each x in X. Let $\{L_n\}$ be a countable family of subsets L_n of $C(X)$ (resp. $C_R(X)$) such that each L_n is A-equicontinuous at every point of X. Suppose that $A + \bigcup_{n=1}^{\infty} L_n = C(X)$ (resp. $C_R(X)$). Then $A = C(X)$ (resp. $C_R(X)$).*

Proof. By Theorem 4.4 we see that there is a finite number of compact subsets $G_1, G_2, \cdots, G_m$ of X such that $A|G_k = C(G_k)$ (resp. $C_R(G_k)$) for $k = 1, 2, \cdots, m$ and $\bigcup_{n=1}^{m} G_k = X$.

It follows by the partition of unity we see that $A + \mathbf{C} = C(X)$ (resp. $C_R(X)$), where $\mathbf{C}$ is the space of all the constant functions on X. Since $A \cdot C(X)$ (resp. $A \cdot C_R(X)$) $= A \cdot (A + \mathbf{C}) \subset A$ we conclude that $A = C(X)$ (resp. $C_R(X)$).

5 A VARIATION OF THE STONE - WEIERSTRASS THEOREM

In this section we study approximation of continuous functions. We give a Stone - Weierstrass type theorem under the hyposesis of weak operation with a countable set of weights.

Lemma 5.1. *Suppose that h is a real valued continuous function on $\{z \in \mathbf{C}: |z| \leq K\}$ for a positive constant K such that h is not harmonic (resp. analytic) on any open neighborhood of the origin. For every positive number δ and every complex number z_2 with $0 < |z_2| < M/2\delta$ there exists a smoothing operator σ of class C^∞ supported in $\{z \in \mathbf{C}: |z| < \delta\}$ such that*

$$\Delta_1(h_\sigma(0, z_2) \quad (\text{resp. } \frac{\partial}{\partial \bar{z}_1}(h_\sigma(0, z_2))) \quad \neq \quad 0$$

where $h_\sigma(z_1, z_2) = \iint h(z_1 - z_2 w)\sigma(w)\,dxdy$ $(|z_1| < M/2)$ and Δ_1 *is the Laplacian with respect to the first variable z_1.*

Proof. We show that $\Delta_1(h_\sigma(0, z_2)) \neq 0$ for some σ. Suppose that there is a number ε with $0 < \varepsilon < M/2$ such that $h_\sigma(z_1, z_2)$ is harmonic on $\{z \in \mathbf{C}: |z| < \varepsilon\}$ for every σ supported in $\{z \in \mathbf{C}: |z| < \delta/2\}$. If σ_n is supported in $\{z \in \mathbf{C}: |z| < \delta/2n\}$, then $h_{\sigma_n}(z, z_2)$ converges to $h(z)$ uniformly on $\{z \in \mathbf{C}: |z| < \varepsilon/2\}$. So we see that h is harmonic on $\{z \in \mathbf{C}: |z| < \varepsilon\}$, which is a contradiction. We conclude that for every small positive number ε there is σ_ε supported in $\{z \in \mathbf{C}: |z| < \delta/2\}$ such that $h_{\sigma_\varepsilon}(z, z_2)$ is not harmonic on $\{z \in \mathbf{C}: |z| < \varepsilon\}$. In particular, for ε with $0 < \varepsilon < \delta|z_2|/2$, $h_\sigma(z, z_2)$ is not harmonic on $\{z \in \mathbf{C}: |z| < \varepsilon\}$ for some σ supported in $\{z \in \mathbf{C}: |z| < \delta/2\}$. Thus there is a complex number z_0 with $|z_0| < \varepsilon$ such that

$$\Delta_1(h_\sigma(z_0, z_2) \neq 0.$$

Put $\tilde{\sigma}(w') = \sigma(w' + z_0/z_2)$. Then $\tilde{\sigma}$ is a smoothing operator of class C^∞ supported in $\{w' \in \mathbf{C}: |w'| < \delta\}$ By a simple calculation we have $\Delta_1(h_{\tilde{\sigma}}(0, z_2)) = \Delta_1(h_\sigma(z_0, z_2)$, so we conclude that

$$\Delta_1(h_{\tilde{\sigma}}(0, z_2)) \neq 0$$

for $\tilde{\sigma}$. We can prove in the same way that $\frac{\partial}{\partial \bar{z}_1}(h_\sigma(0, z_2)) \neq 0$ for some σ.

Theorem 5.2. *Let* A *be a point separating closed subalgebra of* $C(X)$ *for a compact Hausdorff space* X *such that* $A_x \neq A$ *for every point* x *in* X. *Let* I *be a closed ideal of* A. *Suppose that* D *is a plane domain containing the origin. Suppose that* $\{h_n\}$ *is a countable set of complex valued continuous functions on* D *such that each* h_n *is not analytic on any neighborhood of the origin. Suppose that* $\{f_n\}$ *is a countable subset of* $C(X)$ *such that* $|f_n| > 0$ *on* X *for every* n. *Suppose that for every* f *in* I *with* $f(X) \subset D$ *there are functions* $h_l \in \{h_n\}$, f_m, $f_k \in \{f_n\}$ *such that*

$$h_l(f_m f) f_k \in A.$$

Then $I = \{f \in C(X) : f | \mathbf{ker} I = 0\}$.

Proof. There is a positive number M with $D \supset \{z \in \mathbf{C}: |z| \leq M\}$. By the same way as in the proof of Lemma 6.1 we see that there are n_0, m_0, l_0, k_0 and $\xi > 0$ and f in I_x with $\|\varphi\|_{\infty(X)} \leq M/2$ such that $\{f \in I_x: \|f - \varphi\|_{\infty(X)} < \xi\}$ is included in the closure of $\{f \in I_x: \|f\|_{\infty(X)} \leq M/2,\ h_{n_0}(f_{m_0}f) f_{l_0} \in A,\ \|h_{n_0}(f_{m_0}f) f_{l_0}\|_{\infty(X)} \leq k_0\}$. Fix δ with $0 < \delta < \xi$ and fix s with $0 < \delta < \mathbf{min}\{M, \xi\}/(4\delta(\|f_{m_0}\|_{\infty(X)} + 1))$. Put $z_2 = s f_{m_0}(x)$. Then by Lemma 5.1 there is a smoothing operator $\sigma(w)$ of class $\mathbf{C}^\infty$ such that

$$\frac{\partial}{\partial \bar{z}_1} h_{n_0,\sigma}(0, z_2) \neq 0,$$

where $h_{n_0,\sigma}(0, z_2) = \int h_{n_0}(z_1 - z_2 w)\sigma(w)\,dxdy$.

There exists a positive ε with

$$\left|\frac{\partial}{\partial \bar{z}_1} h_{n_0,\sigma}(z_1, z_2')\right| \geq \frac{1}{2}\left|\frac{\partial}{\partial \bar{z}_1} h_{n_0,\sigma}(0, z_2)\right|$$

on $\{(z_1, z_2') \in \mathbf{C}^2: |z_1| \leq, |z_2 - z_2'| \leq \varepsilon\}$. Define an open neighborhood

$$K_x = \{y \in X: |(f_{m_0}\varphi)(y)| < \varepsilon,\ |s f_{m_0}(x) - s f_{m_0}(y)| < \varepsilon\}.$$

We will show that for every p in $(K_x - \{x\}) \cap \mathbf{Ch}(I)$ ($\mathbf{Ch}(I)$ is the Choquet boundary for I = the set of all weak peak points for I.) there is an interpolating compact neighborhood for I. If we consider the same argument as the above, we have that for every Choquet boundary point with at most a finite exception there is an interpolating compact neighborhood. It follows by Lemma 1 of [7] that $I = \{f \in C(X) : f | \mathbf{ker} I = 0\}$. Let p be a point in $(K_x - \{x\}) \cap \mathbf{Ch}(I)$. There is a function f_p in I_x with $\|f_p\|_{\infty(X)} = 1$ such that

$$1 \geq f_p(p) > 1 - \frac{\varepsilon - |s f_{m_0}(x) - s f_{m_0}(p)|}{|s f_{m_0}(p)|}.$$

Let g be a function in $A + \mathbf{C} = \{ f + c: f \in A,\ c$ *is a constant function on* $X\}$. For every complex number β with sufficiently small absolute value we have

$$\varphi - \beta g f_p^2 - s f_p w \in \{f \in I_x: \|f - \varphi\|_{\infty(X)} < \xi\}$$

for every complex number w with $|w| < \delta$. It follows that

$$h_{n_0}(f_{m_0}\varphi - \beta g f_{m_0} f_p^2 - s f_{m_0} f_p w) f_{l_0} \in A,$$

so

$$h_{n_0,\sigma}(f_{m_0}\varphi - \beta(g/s^2 f_{m_0}) s^2 f_{m_0}^2 f_p^2,\ s f_{m_0} f_p) f_{l_0} \in A.$$

Thus by Lemma 6 in [5]

$$(g/s^2 f_{m_0})^- \cdot \frac{\partial}{\partial \bar{z}_1} h_{n_0,\sigma}(f_{m_0}\varphi, s f_{m_0} f_p) f_{l_0} \cdot (s^2 f_{m_0}^2 f_p^2)^- \in A,$$

so

$$(|g|^2/\bar{s}^2 \bar{f}_{m_0}) \cdot \frac{\partial}{\partial \bar{z}_1} h_{n_0,\sigma}(f_{m_0}\varphi, s f_{m_0} f_p) f_{l_0} \cdot (s^2 f_{m_0}^2 f_p^2)^- \in A$$

for every g in $A + \mathbf{C}$ since $(A + \mathbf{C}) \cdot A \subset A$. By the Stone - Weierstrass theorem we have

$$C(X) \cdot \frac{\partial}{\partial \bar{z}_1} h_{n_0,\sigma}(f_{m_0}\varphi, s f_{m_0} f_p) \bar{f}_p^2 \subset A.$$

It follows that there is an interpolating compact neighborhood G_p for A, *i.e.*, $A|G_p = C(G_p)$. So there is a compact neighborhood G_p' of p with $I|G_p' = C(G_p')$.

As a corollary of Theorem 5.2 we see a weighted almost Stone - Weierstrass theorem.

Corollary 5.3. *Let A be a function algebra on a compact Hausdorff space X. Let $\{f_n\}$ be a countable subset of $C(X)$ such that $|f_n| > 0$ on X for every f_n. Suppose that for every f in A there are natural numbers $m(f)$ and $n(f)$ with*

$$\bar{f}^{m(f)} \cdot f_{n(f)} \in A.$$

Then $A = C(X)$.

The above result is not true when the operation

$$f \longmapsto \bar{f}^{m(f)} \cdot f_{n(f)}$$

is merely defined on the dense subalgebra of A.

Example. *Let A_0 be the algebra of the restrictions of analytic polynomials on the unit circle $\Gamma = \{z \in \mathbf{C}: |z| = 1\}$. Put $f_n = z^n$ on Γ. Then for every f in A_0 we have $\bar{f} \cdot f_{\mathbf{deg} f}$ in A_0, where* **deg***f is the degree of f. However the uniform closure of A_0 is the disc algebra on Γ, which does not coincides with $C(\Gamma)$ at all.*

On the other hand an almost Stone - Weierstrass theorem on the dense subalgebra is true [8].

Theorem M. *Let A be a subalgebra of $C(X)$ for a compact Hausdorff space with the following conditions.*

(1) $A \neq A_x$ *for every* x *in* X

(2) *A separates the points of* X

(3) *For every* f *in* A *there is a natural number* $n(f)$ *with* $\bar{f}^{n(f)} \in A$.

Then the uniform closure of A *coincides with* $C(X)$.

6 APPLICATIONS FOR RANGE TRANSFORMATION TYPE RESULTS

In this section we show the same type of results for range transformations as §5 by using analysis on the infinitesimal neighborhood F_x^Λ which is studied in the previous sections.

Lemma 6.1. *Let* A *be a Banach algebra contained in* $C(X)$ *for a compact Hausdorff space* X. *Let* B *be an ultraseparating Banach space continuously embedded in* $C(X)$. *Let* L *be a subspace of finite dimension in* $C_R(X)$. *Let* x *be a point in* X *with* $A \neq A_x$. *Suppose that* Λ *is a discrete space with the cardinality at least that of an open base for the topology of* X *at* x. *Suppose that* $\widetilde{B}^\Lambda | F_x^\Lambda$ *is a Banach algebra with respect to the quotient norm* $\|\cdot\|_{\widetilde{B}^\Lambda | F_x^\Lambda}$. *Let* D *be a plane domain containing the origin. Suppose that* $\{h_n\}$ *is a countable set of real valued continuous functions on* D *which are not harmonic on any neighborhood of the origin. Suppose that* $\{\varphi_n\}$ *is a countable subset of* $C(X)$ *such that* $\varphi_n(x) \neq 0$ *for every* n. *Suppose that* $\bigcup_{n=1}^{\infty} T_n$ *is a union of subsets* T_n *of* $C_R(X)$ *such that each* T_n *is* $C_R(X)$*-equicontinuous at* x *and that* $\psi(x) \neq 0$ *for every* ψ *in* $\bigcup_{n=1}^{\infty} T_n$. *Suppose that there is a positive number* M *which satisfies that for every* f *in* B *with* $\|\cdot\|_B \leq M$ *there are functions* h_l *in* $\{h_n\}$, φ_m *in* $\{\varphi_n\}$ *and* ψ *in* $\bigcup_{n=1}^{\infty} T_n$ *with* $h_l(\varphi_m f)\psi \in \mathbf{Re}A + L$. *Then there is a compact neighborhood* G *of* x *such that* $A|G = C(G)$.

Proof. Without loss of generality we may assume that $\mathbf{Re}A \cap L = \{0\}$. So $\mathbf{Re}A + L$ is a Banach space with the norm $\|\cdot\|_{\mathbf{Re}A+L}$ defined by

$$\|u + v\|_{\mathbf{Re}A} = \|u\|_{\mathbf{Re}A} + \|v\|_{\infty(X)} \qquad u \in \mathbf{Re}A,\ v \in L.$$

For every positive integer m put

$$T_n^{(m)} = \{\psi \in T_n : 1/m < |\psi(x)| < m\}.$$

We have

$$\{f \in B_x : \|f\|_B \leq M/2\} = \bigcup \{f \in B_x : \|f\|_B \leq M/2,\ h_n(\varphi_m f)\psi \in \mathbf{Re}A + L \text{ for } h_n \in \{h_n\},\ \varphi_m \in \{\varphi_n\},\ \psi \in T_l^{(l')} \text{ with } \|h_n(\varphi_m f)\psi\|_{\mathbf{Re}A+L} \leq k\},$$

where n, m, l, l' and k vary over all positive integers. By the Baire category theorem there are positive numbers n_0, m_0, l_0, l_0', k_0 and a positive number ξ and a function f_0 in B_x with $\|f_0\|_B \leq M/2$ and a dense (with respect to the topology induced by $\|\cdot\|_B$) subset U of $\{f \in B_x: \|f_0 - f\|_B < \xi\}$ such that

$$U \subset \{f \in B_x: \|f\|_B \leq M/2,\ h_{n_0}(\varphi_{m_0} f)\psi \in \mathrm{Re}A + L$$
$$\mathit{for}\quad \psi \in T_{l_0}^{(l')}\quad \mathit{with}\ \|h_{n_0}(\varphi_{m_0} f)\psi\|_{\mathrm{Re}A+L} \leq k_0\}.$$

Thus we see that for every $\langle f_\lambda\rangle$ in $(U - f_0)^{\sim A}$ there is $\langle\psi_\lambda\rangle$ in $(T_{l_0}^{(l')})^{\sim A}$ ($^{\sim A}$ taken for the uniform norm) with

$$h_{n_0}(\langle\varphi_{m_0}\rangle(\langle f_0\rangle + \langle f_\lambda\rangle))\langle\psi_\lambda\rangle \in (\mathrm{Re}A + L)^{\sim A},$$

where $U - f_0 = \{g \in B_x: g + f_0 \in U\}$ and $\langle\varphi_{m_0}\rangle$ and $\langle f_0\rangle$ are the constant functions on A with the constant value φ_{m_0} and f_0 respectively. By Lemma 4 in [6] and Corollary 2.3 we have

$$h_{n_0}(\varphi_{m_0}(x)\langle f_\lambda\rangle)\langle\psi_\lambda(x)\rangle|F_x^A \in (\mathrm{Re}A)^{\sim A}|F_x^A$$

Since $1/l_0' < |\psi_\lambda(x)| < l_0'$ we have

$$h_{n_0}(\varphi_{m_0}(x)\langle f_\lambda\rangle)|F_x^A \in (\mathrm{Re}A)^{\sim A}|F_x^A.$$

Put $F = F_x^A$, $Q = [\{x\} \times A]$, $S = \tilde{B}^A|F_x^A$, $S_0 = (U - f_0)^{\sim A}|F_x^A$, $\delta = \xi$, $h_0(z) = h_{n_0}(\varphi_{m_0}(x)z)$. It follows that the condition of Lemma 2 in [7] is satisfied by Lemma 4 in [7]. We have that for every p in $F_x^A - [\{x\} \times A]$ there is a compact neighborhood O_p (with respect to the relative topology on F_x^A) of p such that

$$\{\tilde{f} \in (\mathrm{Re}A)^{\sim A}|F_x^A: \tilde{f}([\{x\} \times A]) = 0\}|O_p = C_R(O_p).$$

By the same way as in the proof of Theorem in [7] we see that there is a compact neighborhood G of x such that

$$A|G = C(G).$$

Theorem 6.2. *Let A be a point separating closed subalgebra of $C(X)$ for a compact Hausdorff space X and I be a closed ideal of A. Let L be a subspace of a finite dimension in $C_R(X)$. Let D be a plane domain containing the origin. Let $\{h_n\}$ be a countable set of real valued continuous functions on D which are not harmonic on any open neighborhood of the origin. Let $\{f_n\}$ (resp. $\{g_n\}$) be a countable subset of $C(X)$ (resp. $C_R(X)$) such that $|f_n|$ (resp. $|g_n|$) > 0 on X. Suppose that for every f in I with $f(X) \subset D$ there are functions h_n, f_m, g_l such that*

$$h_n(f_m f)g_l \in \mathrm{Re}A + L.$$

Then $I = \{f \in C(X): f|\mathrm{ker}I = 0\}$.

Proof. Put $A_1 = A + \mathbf{C} = \{f + c: f \in A,\ c$ *is a constant function on* $X\}$. Then I is a closed ideal of A_1 and $h_n(f_m f)g_l \in \mathrm{Re}A_1 + L$. So without loss of generality we may assume that A is a function algebra on X. As in the proof of Theorem 5.2 we will show that for every point x in X there is a neighborhood K_x of x which satisfies that for every point $p \in (K_x - \{x\}) \cap \mathbf{Ch}(I)$ there is a compact neighborhood G_p of p with $I|G_p = C(G_p)$. It will follow that $I = \{f \in C(X): f|\mathbf{ker}I = 0\}$. By Lemma 6.1 we need only show that $I|G_p'$ is ultraseparating for a compact neighborhood of each p in $(K_x - \{x\}) \cap \mathbf{Ch}(I)$ for every x in X. Let x be a point in X. By the same way as in the proof of Theorem 5.2 we have that there exist positive integers n_0, m_0, l_0, k_0 and a positive number ξ and a function φ in $I_x = \{f \in I: f(x) = 0\}$ with $\|\varphi\|_{\infty(X)} \leq M/2$ (we fix a positive number M such that $D \supset \{z \in \mathbf{C}: |z| \leq M\}$) which satisfies that

$$\{f \in I_x: \|f - \varphi\|_{\infty(X)} < \xi\} \subset \{f \in I_x: \|f\|_{\infty(X)} \leq M/2, \; h_{n_0}(f_{m_0}f)g_{l_0} \in \mathbf{Re}A + L, \; \|h_{n_0}(f_{m_0}f)g_{l_0}\|_{\mathbf{Re}A+L} \leq k_0\}^-,$$

where $^-$ denotes the uniform closure here. Fix a positive number δ with $\delta < \xi$ and a fix positive number s with

$$s < \frac{min\{M, \xi\}}{4\delta(\|f_{m_0}\|_{\infty(X)} + 1)}.$$

Put $z_2 = sf_{m_0}(x)$. By Lemma 5.1 we can choose a smoothing operator σ of class C^∞ supported in $\{z \in \mathbf{C}: |z| < \delta\}$ such that

$$\Delta_1(h_{n_0,\sigma}(0, z_2)) \neq 0,$$

where

$$h_{n_0,\sigma}(z_1, z_2) = \int h_{n_0}(z_1 - z_2 w)\sigma(w)\,dxdy.$$

Choose a positive ε with

$$|\Delta_1(h_{n_0,\sigma}(z_1, z_2')|z_2'|^4)| \geq \frac{1}{2}|\Delta_1(h_{n_0,\sigma}(0, z_2)|z_2|^4)$$

on $\{(z_1, z_2') \in \mathbf{C}^2: |z_1| \leq \varepsilon, |z_2 - z_2'| \leq \varepsilon\}$. Put

$$K_x = \{y \in X: |(f_{m_0}\varphi)(y)| < \varepsilon, |sf_{m_0}(x) - sf_{m_0}(y)| < \varepsilon\}.$$

Then K_x is a neighborhood of x. Let p be a point in $(K_x - \{x\}) \cap \mathbf{Ch}(I)$. There is a function f_p in I_x with $\|f_p\|_A = 1$ such that

$$1 \geq f_p(p) > 1 - \frac{\varepsilon - |sf_{m_0}(x) - sf_{m_0}(p)|}{|sf_{m_0}(p)|}.$$

Let g be a function in A. For every complex number β with sufficiently small absolute value we have

$$\varphi - \beta g f_p^2 - s f_p w \in \{f \in I_x: \|\varphi - f\|_{\infty(X)} < \xi\}$$

for every complex number w with $|w| < \delta$. It follows that

$$h_{n_0}(f_{m_0}\varphi - \beta g f_{m_0} f_p^2 - s f_{m_0} f_p w) \in \mathbf{cl}(\mathbf{Re}A) + L,$$

(since L is finite dimensional, $\mathbf{cl}(\mathbf{Re}A + L) = \mathbf{cl}(\mathbf{Re}A) + L$), so

$$h_{n_0,\sigma}(f_{m_0}\varphi - \beta \cdot \frac{g}{s^2 f_{m_0}} \cdot (sf_{m_0}f_p)^2, sf_{m_0}f_p) \in \mathbf{cl}(\mathbf{Re}A) + L.$$

By Lemma 5 of [5] we have

$$\frac{|g|^2}{|s^2 f_{m_0}|} \cdot \Delta_1(h_{n_{n_0},\sigma}(f_{m_0}\varphi, sf_{m_0}f_p)|sf_{m_0}f_p|^4) \in \mathbf{cl}(\mathbf{Re}A) + L.$$

It follows by the Stone - Weierstrass theorem that

$$C_R(X) \cdot \Delta_1(h_{n_0,\sigma}(f_{m_0}\varphi, sf_{m_0}f_p)|sf_{m_0}f_p|^4) \subset \mathbf{cl}(\mathbf{Re}A) + L.$$

So there is a compact neighborhood G of p with

$$C_R(G) \subset (\mathbf{cl}(\mathbf{Re}A) + L)|G.$$

We see by Theorem 3.3 that $A|G$ is ultraseparating. So $\widetilde{A}^\Lambda|F_p^\Lambda$ separates the points of F_p^Λ for a discrete space Λ with a large cardinality by (1) of Theorem 1 of [6]. Since $\widetilde{A}^\Lambda|F_p^\Lambda = \widetilde{I}^\Lambda|F_p^\Lambda$ we see that I is ultraseparating near p again by (1) of Theorem 1 of [6].

Corollary 6.3. *Let A be a point separating closed subalgebra of $C(X)$ and I be a closed ideal of A. Let L be a subspace of dimension finite in $C_R(X)$. Let J be an open interval containing the origin. Let $\{\varphi_n\}$ be a countable set of real valued continuous functions on J which are not affine on any open subinterval containg the origin. Let $\{f_n\}$ be a countable subset of $C_R(X)$ such that $|f_n| > 0$ on X for every n. Suppose that for every u in* $\mathbf{Re}I$ *with $u(X) \subset J$ there are functions φ_n, f_m, f_l such that*

$$\varphi_n(f_m u) f_l \in \mathbf{Re}A + L.$$

Then $I = \{f \in C(X) : f|\mathbf{ker}I = 0\}$.

Proof. Put $D = J \times R$ and let $h_n(z) = \varphi_n(\mathbf{Re}z)$ on D. Then each h_n is not harmonic near the origin. Then for every f in A width $f(X) \subset D$ there are h_n, f_m, f_l with

$$h_n(f_m f) f_l \in \mathbf{Re}A + L.$$

It follows that $I = \{f \in C(X) : f|\mathbf{ker}I = 0\}$ by Theorem 6.2.

REFERENCES

1. A. Bernard and A. Dufresnoy, Calcul symbolique sur la frontitère de Šilov de certaines algèbres de fonctions holomorphes, *Ann. Inst. Fourier(Grenoble)*, **25**(1975), 33–43.
2. J. Čerych, On linear codimensions of function algebras, *Complex analysis and application* '85(*Varna*, 1985), 141–146, Bulgar. Acad. Sci., Sofia, 1986.
3. A. J. Ellis, Separation and ultraseparation properties for continuous function spaces, *J. London Math. Soc.*, **29**(1984), 521–532.
4. I. Glicksberg, Function algebras with closed restrictions, *Proc. Amer. Math. Soc.*, **14**(1963), 158–161.

5. O. Hatori, Range transformations on a Banach function algebra, *Trans. Amer. Math. Soc.*, **297**(1986), 629–643.
6. ______, Range transformations on a Banach function algebra.II, *Pacific J. Math.*, **138**(1989), 89–118.
7. ______, Range transformations on a Banach function algebra. IV, *preprint*.
8. S. Minsker, Some applications of the Stone-Weierstrass theorem to planar rational approximation, *Proc. Amer. Math. Soc.*, **58**(1976), 94–96.

Inequalities of von Neumann Type for Small Matrices

John A. Holbrook
Department of Mathematics and Statistics
University of Guelph
Guelph, Ont N1G 2W1 Canada

1. Introduction

The examples created by N. Varopoulos **[V]** put an end to the hopes for a general von Neumann inequality:

$$\|f(C_1,\ldots,C_n)\| \leq 1, \tag{1.1}$$

for any commuting Hilbert space contractions $C_1,\ldots,C_n$ and any analytic map $f : D^n \to D$ (here D is the complex unit disc $\{z : |z| \leq 1\}$). Nevertheless, there has been a continuing interest in variants of this inequality and the extent of its validity. This note reports on some old results (they date, in fact, from before the Varopoulos counterexamples) that focus on the case of 2×2 contractions; this case, unexpectedly, retains some intriguing puzzles (see also the paper of K. Lewis and J. Wermer **[L-W]** in these Proceedings and that of S. Drury **[D]**). The results reported below were mentioned, in part, in the thesis of L. Allen **[A]** and in our paper **[H]** but the details have not previously been available in printed form. We believe that propositions 2 and 3 have long been known, with a variety of different proofs, to several workers in the field.

Recall that J. von Neumann established (1.1) for $n = 1$ in **[vN]**. B. Sz.-Nagy showed how to construct a (strong) unitary dilation for any contraction and noted that this yields a strikingly direct approach to von Neumann's inequality. In general, we say that commuting contractions $C_1,\ldots,C_n$ on a Hilbert space H possess a (simultaneous) unitary dilation if there are commuting unitary operators $U_1,\ldots,U_n$ on some Hilbert space K containing H as a subspace such that

$$\Big(\prod_{k=1}^{n} C_k^{j_k}\Big)h = P_H\Big(\prod_{k=1}^{n} U_k^{j_k}\Big)h \quad (h \in H, j_k = 0,1,\ldots). \tag{1.2}$$

T. Ando proved that any two commuting contractions possess commuting unitary dilations, and hence that (1.1) holds also for $n = 2$. If one assumes that the contractions **double-commute**, ie that $C_j^*C_k = C_kC_j^*$ as well as $C_jC_k = C_kC_j$ for $j \neq k$ then it is not hard to construct a simultaneous unitary dilation without restiction on n. S. Parrott showed that there are three commuting contractions without any corresponding unitary dilation, closing off the dilation approach to (1.1) in the general multivariate case. A good reference for this material on dilations is the first chapter of B. Sz.-Nagy and C. Foiaş **[N-F]**.

With a remarkable, partly probabilistic, argument. Varopoulos **[V]** proved the existence of commuting contractions violating (1.1). Soon after, Varopoulos and others (see, for example, M. J. Crabb and A. M. Davie **[C-D]**) found specific counterexamples with $n = 3$ and dimension of H as low as 5. In what follows we shall see that neither the Parrott nor the Varopoulos phenomenon can occur for 2×2 contractions, and that this case points to interesting possibilities for other low-dimensional matrices.

2. Results

We shall base several of our arguments on the following "model" for commuting 2×2 contractions.

Proposition 1: If $C_1, \ldots, C_n$ are commuting 2×2 contractions, then for some fixed contraction C and analytic $f_k : D \to D$

$$C_k = f_k(C) \quad (k = 1, \ldots, n). \tag{2.1}$$

Proof: Since they commute, we may simultaneously put the C_k in upper-triangular form: with respect to some fixed orthonormal basis,

$$C_k \sim \begin{pmatrix} \lambda_k & z_k \\ 0 & \mu_k \end{pmatrix} \quad (k = 1, \ldots, n). \tag{2.2}$$

If all $z_k = 0$ we have a special case where we may take $C \sim \begin{pmatrix} 1 & 0 \\ 0 & -1 \end{pmatrix}$ (for example). Since the eigenvalues $\lambda_k, \mu_k \in D$, it is a simple exercise in conformal mapping to find analytic $f_k : D \to D$ such that $f_k(1) = \lambda_k$ and $f_k(-1) = \mu_k$.

Thus we may suppose that some $z_k \neq 0$. The relation $C_k C_j = C_j C_k$ corresponds to the equation

$$z_k(\lambda_j - \mu_j) = z_j(\lambda_k - \mu_k). \tag{2.3}$$

Hence $C_j = \lambda_j I$ for any j such that $z_j = 0$; for such j we may take $f_j(z) = \lambda_j$ (no matter what C is to be chosen).

We may now suppose that all $z_j \neq 0$; it follows that $|\lambda_j|, |\mu_j| < 1$ and we define the Möbius map m_j by $m_j(z) = (z - \mu_j)/(1 - \overline{\mu_j} z)$; then

$$C_j' = m_j(C_j) \sim \begin{pmatrix} \lambda_j' & z_j' \\ 0 & 0 \end{pmatrix}. \tag{2.4}$$

The commutativity condition $\lambda_j' z_k' = \lambda_k' z_j'$ ensures that each C_k' is a multiple of C_j' where j is chosen to maximize $|\lambda_j'| + |\mu_j'|$. Thus we have $C_k' = w_k C_j'$ for certain scalars w_k with $|w_k| \leq 1$. Recall that the Möbius maps m_j are invertible analytic maps from D onto D; thus $f_k(z) = m_k^{-1}(w_k m_j(z))$ defines an analytic $f_k : D \to D$, and with $C = C_j$ we have $f_k(C) = m_k^{-1}(w_k C_j') = m_k^{-1}(C_k') = C_k$. ∎

Proposition 2: The multivariate von Neumann inequality holds for 2×2 contractions; ie if $C_1, \ldots, C_n$ are commuting 2×2 matrices with $\|C_k\| \leq 1$ for all k and $f : D^n \to D$ is analytic, then $\|f(C_1, \ldots, C_n)\| \leq 1$.

Proof: Let C, f_k be as in proposition 1. Then $h(z) = f(f_1(z), \ldots, f_n(z))$ defines an analytic map $h : D \to D$ such that $h(C) = f(C_1, \ldots, C_n)$. We need only apply the one-variable von Neumann inequality to h and C. ∎

Proposition 3: Any n-tuple of commuting 2×2 contractions has a simultaneous unitary dilation.

Proof: Given commuting 2×2 contractions $C_1, \ldots, C_n$, let f_k and C be as in proposition 1. On some Hilbert space K, containing the 2-dimensional H on which $C, C_1, \ldots, C_n$ are defined, we have a unitary U such that $C^j h = P_H U^j h \quad (h \in H, j = 0, 1, \ldots)$. It follows that

$$(2.5) \qquad (\prod_{k=1}^{n} C_k^{j_k})h = (\prod_{k=1}^{n} f_k^{j_k}(C))h = P_H(\prod_{k=1}^{n} f_k^{j_k}(U))h \quad (h \in H, j_k = 0, 1, \ldots).$$

But the contractions $f_k(U)$ on K are double commuting and thus possess a simultaneous unitary dilation to some yet larger Hilbert space $\tilde{K}$: commuting unitaries $U_1, \ldots, U_n$ on $\tilde{K}$ such that

$$(2.6) \qquad (\prod_{k=1}^{n} C_k^{j_k})h = P_H(\prod_{k=1}^{n} f_k^{j_k}(U))h = P_H P_{\tilde{K}}(\prod_{k=1}^{n} U_k^{j_k})h = P_H(\prod_{k=1}^{n} U_k^{j_k})h,$$

where H is regarded as a subspace of $\tilde{K}$. ∎

Proposition 4: Let V be an invertible 2×2 contraction and define, for any 2×2 matrix T, $|T| = \|V^{-1}T\|$. If $V, C_1, \ldots, C_n$ commute and $|C_k| \leq 1 \quad (k = 1, \ldots, n)$, then

$$(2.7) \qquad |g(C_1, \ldots, C_n)| \leq 1$$

for any analytic map $g : D^n \to D$ such that $g(\vec{0}) = 0$.

Remark: This generalizes proposition 2 and, as we shall see, it raises interesting questions about further generalizations. To see that proposition 2 follows from (2.7) take $V = I$; conclude that $\|g(C_1, \ldots, C_n)\| \leq 1$ for any commuting contractions $C_1, \ldots, C_n$ provided $g : D^n \to D$ is analytic and $g(\vec{0}) = 0$. In this case, the final condition $g(\vec{0}) = 0$ is really no restiction: for any analytic $f : D^n \to D$ we may take for g the composition $m \circ f$, where m is the Möbius map taking $f(\vec{0})$ to 0. Then $f(C_1, \ldots, C_n) = m^{-1}(g(C_1, \ldots, C_n))$ is also a contraction.

Proof: Consider the commuting contractions $V, S_1, \ldots, S_n$, where $S_k = V^{-1}C_k$. Since $g(\vec{0}) = 0$ we may write $g(z_1, \ldots, z_n) = \sum_1^n z_k g_k(z_1, \ldots, z_n)$ for some analytic g_k. Note that

$$(2.8) \qquad V^{-1}g(C_1, \ldots, C_n) = \sum_{1}^{n} S_k g_k(VS_1, \ldots, VS_n) = h(V, S_1, \ldots, S_n),$$

where $h(z, z_1, \ldots, z_n) = \sum_1^n z_k g_k(zz_1, \ldots, zz_n)$. Now $h : D^{n+1} \to D$ by the maximum modulus principle, since, for $|z| = 1$,

$$|h(z, z_1, \ldots, z_n)| = |zh(z, z_1, \ldots, z_n)| = |g(zz_1, \ldots, zz_n)| \leq 1. \tag{2.9}$$

Apply proposition 2 to h and $V, S_1, \ldots, S_n$ to conclude that $||V^{-1}g(C_1, \ldots, C_n)|| \leq 1$. ∎

The following proposition reveals that the phenomenon of proposition 4 is much more general when $n \leq 2$. Note that V need not commute with C_1 and C_2 here.

Proposition 5: Let V be an invertible contraction on any Hilbert space H and define, for any $T \in B(H)$, $|T| = ||V^{-1}T||$. If $C_1, C_2 \in B(H)$ commute and $|C_1|, |C_2| \leq 1$ then $|g(C_1, C_2)| \leq 1$ for any analytic map $g : D^2 \to D$ such that $g(\vec{0}) = 0$.

Proof: Let $W = ((V^{-1})^* V^{-1} - I)^{\frac{1}{2}}$, and given $T \in B(H)$ consider the operator on $H \oplus H$ defined by

$$\tilde{T} \sim \begin{pmatrix} 0 & WT \\ 0 & T \end{pmatrix}. \tag{2.10}$$

An easy computation reveals that $||\tilde{T}|| \leq 1 \Longleftrightarrow |T| \leq 1$. Thus $||\tilde{C}_k|| \leq 1$ and since

$$\tilde{C}_k^a \tilde{C}_j^b \sim \begin{pmatrix} 0 & WC_k^a C_j^b \\ 0 & C_k^a C_j^b \end{pmatrix}, \tag{2.11}$$

it's clear that $\tilde{C}_1$ and $\tilde{C}_2$ are commuting contractions in $B(H \oplus H)$ and that

$$g(\tilde{C}_1, \tilde{C}_2) \sim \begin{pmatrix} 0 & Wg(C_1, C_2) \\ 0 & g(C_1, C_2) \end{pmatrix}, \tag{2.12}$$

since $g(0,0) = 0$. Applying the 2-variable von Neumann inequality (Ando), we have $||g(\tilde{C}_1, \tilde{C}_2)|| \leq 1$, and hence $|g(C_1, C_2)| \leq 1$. ∎

3. Question

For 2×2 matrices, can we **remove** the restriction (in proposition 4) that V should commute with the C_k, not just for $n \leq 2$ but for **all** n? If yes, then we would have a substantial generalization of proposition 4, and of the von Neumann inequality. If no, then the construction used in the proof of proposition 5 would yield 4×4 counterexamples to the multivariate von Neumann inequality – and counterexamples quite different in structure from those presently known.

REFERENCES

[A] L. Allen, "Analytic Inequalities for Finite-dimensional Operators," thesis, University of Guelph, 1973.

[C-D] M. J. Crabb and A. M. Davie, *Von Neumann's inequality for Hilbert space operators*, Bull. London Math. Soc. **7** (1975), 49–50.

[D] S. W. Drury, *Remarks on von Neumann's inequality*, Springer Lecture Notes in Mathematics **995** (1983), 14–32.

[H] J. Holbrook, *Von Neumann's inequality and the Poisson radius for operators*, Bull. Acad. Polon. Sci., Série Math. Astr. Phys. **22** (1974), 121–127.

[L-W] K. Lewis and J. Wermer, *On the theorems of Pick and von Neumann*, these Proceedings.

[N-F] B. Sz.-Nagy and C. Foiaş, "Harmonic Analysis of Operators on Hilbert Space," North-Holland, 1970.

[V] N. Varopoulos, *Sur une inégalité de von Neumann*, C. R. Acad. Sci. Paris, Série A **277** (1973), A19–A22.

[vN] J. von Neumann, *Eine Spectraltheorie für allgemeine Operatoren eines unitären Raumes*, Math. Nachr. **4** (1951), 258–281.

On the Asymptotic-Norming Property of Banach Spaces

Zhibao Hu and Bor–Luh Lin
Department of Mathematics
The University of Iowa
Iowa City, IA 52242

1. The asymptotic–norming property was introduced by James and Ho [JH] to demonstrate that there is a larger class of Banach spaces that satisfy the Radon–Nikodym property than the class of Banach spaces that are isomorphic to subspaces of separable dual. Three different kinds of asymptotic–norming properties were introduced and was proved that they are all equivalent in separable Banach spaces. Recently, Ghoussoub and Maurey [GM] proved that for separable Banach spaces the asymptotic–norming property is equivalent to the Radon–Nikodym property. However, in general, it is an open question whether the two properties are equivalent. In this paper, we study the three asymptotic–norming properties from the points of view of renorming and show that the three asymptotic–norming properties are equivalent in a larger class of Banach space than the separable Banach spaces. As a consequence, these results yield some interesting renorming properties for some Banach spaces with the Radon–Nikodym property. For the class of dual Banach spaces, a stronger property than the asymptotic–norming property, called the weak* asymptotic–norming property, is introduced. The class of Banach spaces with the weak* asymptotic–norming property has nicer geometric properties and a Banach space X is given such that X^* has the Radon–Nikodym property but fail to have the weak* asymptotic–norming property.

For a Banach space X, let $S_X = \{x : x \in X, \|x\| = 1\}$ and $B_X = \{x : x \in X, \|x\| \le 1\}$. Let Φ be a subset of B_{X^*}. Φ is called a *norming set* of X if $\|x\| = \sup_{x^* \in \Phi} x^*(x)$ for all x in X. A sequence $\{x_n\}$ in S_X is said to be *asymptotically normed* by Φ if for any $\epsilon > 0$, there is $x^* \in \Phi$ and $N \in \mathbb{N}$ such that $x^*(x_n) > 1-\epsilon$ for all $n \ge N$.

For $\kappa = $ I, II, or III, a sequence $\{x_n\}$ in X is said to have the *property* κ if

(I) $\{x_n\}$ is convergent;

(II) $\{x_n\}$ has a convergent subsequence;

or (III) $\bigcap_{n=1}^{\infty} \overline{co}\{x_k : k \ge n\} \ne \phi$.

Let Φ be a norming set of X. X is said to have the *asymptotic–norming property* κ, $\kappa = I, II, III$, *with respect to* Φ *(Φ–ANP–κ)* if every sequence in S_X that is asymptotically normed by Φ has the property κ. X is said to have the *asymptotic–norming property* κ(ANP–κ), if there is an equivalent norm $\|\cdot\|$ on X such that there is a norming set Φ with respect to $(X,\|\cdot\|)$ such that X has the Φ–ANP–κ, $\kappa = $ I, II, III.

2. In this section, we consider a fixed norming set Φ of a Banach space X and obtain some important basic properties about X that has Φ–ANP–κ. For a set A in a Banach space, let $\overline{co}(A)$ denote the closed convex hull of A in X. For a net $\{x_\lambda\}$ in X, x^{**} in X^{**} and a set Φ in B_{X^*}, we write $\sigma(X^{**}, \Phi)\text{–}\lim_\lambda x_\lambda = x^{**}$ if $\lim_\lambda x^*(x_\lambda) = x^{**}(x^*)$ for all x^{**} in Φ.

<u>Lemma 2.1</u>. Let X be a Banach space which has the Φ–ANP–III for some norming set Φ in B_{X^*}. Let $\{x_\lambda\}$ be a net in S_X and let x^{**} in X^{**} such that

$\|x^{**}\| = 1 = \sup_{\varphi\in\Phi} x^{**}(\varphi)$. If there exists Φ_1, $\Phi \subset \Phi_1 \subset B_{X^*}$ such that Φ_1 norms $[X,x^{**}]$ and $\sigma(X^{**},\Phi_1)-\lim_\lambda x_\lambda = x^{**}$ then there is a sequence $\{x_n\}$ in $\{x_\lambda\}$ such that $\{x_n\}$ is asymptotically normed by Φ and $\{x^{**}\} = \bigcap_{n=1}^{\infty} \overline{co}\{x_k : k \geq n\}$ and so $x^{**} \in X$.

<u>Proof</u>. Choose a sequence $\{x_n^*\}$ in Φ such that $x^{**}(x_n^*) > 1 - \frac{1}{n}$, $n \in \mathbb{N}$. We claim that, by induction, there exists a sequence $\{x_n\}$ in $\{x_\lambda\}$ and a sequence $\{F_n\}$ of finite subsets in Φ_1 satisfying

(i) $F_n \subset F_{n+1}$, $x_n^* \in F_n$, $n \in \mathbb{N}$;

(ii) F_n $\left(1-\frac{1}{n}\right)$–norms $[x^{**}, x_1,\ldots,x_{n-1}]$, $n \geq 2$;

(iii) $|x^{**}(x_n) - x^{**}(x^*)| < \frac{1}{n}$ for all x^* in F_n, $n \in \mathbb{N}$.

Indeed, let $F_1 = \{x_1^*\}$. Since $\sigma(X^{**}, \Phi_1) - \lim_\lambda x_\lambda = x^{**}$, choose x_1 in $\{x_\lambda\}$ such that $|x^{**}(x_1) - x^{**}(x_1^*)| < 1$. Let $n \in \mathbb{N}$ and assume that $\{x_i\}_{i\leq n}$ and $\{F_i\}_{i\leq n}$ have been chosen with the claimed properties. Since Φ_1 norms $[X, x^{**}]$ and $[x^{**}, x_1,\ldots, x_n]$ is finite dimensional, there is a finite subset F in Φ_1 such that F $(1-\frac{1}{n+1})$ – norms $[x^{**}, x_1,\ldots, x_n]$. Let $F_{n+1} = F_n \cup F \cup \{x_{n+1}^*\}$. Then F_{n+1} is finite. Since $\sigma(X^{**}, \Phi_1) - \lim_\lambda x_\lambda = x^{**}$, there is x_{n+1} in $\{x_\lambda\}$ such that $|x^{**}(x_{n+1}) - x^{**}(x^*)| < \frac{1}{n+1}$ for all x^* in F_{n+1}. Continuous by induction, we have proved the claim.

For any $m \geq n$, since $x_n^* \in F_n \subset F_m$, we have $|x_n^*(x_m) - x^{**}(x_n^*)| < \frac{1}{m} \leq \frac{1}{n}$. Hence $x_n^*(x_m) > x^{**}(x_m^*) - \frac{1}{n} > 1 - \frac{2}{n}$ for all $m \geq n$. Thus $\{x_n\}$ is asymptotically normed by Φ. Since X has the Φ–ANP–III, it follows that $\bigcap_{n=1}^{\infty} \overline{co}\{x_k : k \geq n\} \neq \phi$.

Let $\Phi_0 = \bigcup_{n=1}^{\infty} F_n$. By (ii), Φ_0 norms $[x^{**}, x_n, n \in \mathbb{N}]$ and by (iii),

$\lim_n x^*(n) = x^{**}(x^*)$ for all $x^* \in \Phi_0$. Thus for any y in $\bigcap_{n=1}^{\infty} \overline{\text{co}}\{x_k : k \geq n\}$, $x^*(y) = x^{**}(x^*)$ for all x^* in Φ_0. Hence $x^{**} = y$ and this completes the proof of Lemma 2.1. □

__Lemma 2.2__. Let $\{x_n^*\}$ be a sequence in S_{X^*} and let Φ be a subset of B_X. If Φ is a norming set of $[x_n^* : n \in \mathbb{N}]$ and $\{x_n^*\}$ is asymptotically normed by Φ, then $\|x\|^* = \sup_{\varphi \in \Phi} x^*(\varphi) = 1$ for all x^* in the weak*–closure of $\{x_n^*\}$.

__Proof__. Let x^* be an element in the weak*–closure of $\{x_n^*\}$. We may assume that $x^* \neq x_n^*$ for all $n \in \mathbb{N}$. Given $\epsilon > 0$, since $\{x_n^*\}$ is asymptotically normed by Φ, there is $\varphi \in \Phi$ such that $x_n^*(\varphi) > 1-\epsilon$ for all sufficiently large n. Hence $x^*(\varphi) \geq 1-\epsilon$. On the other hand, it is clear that $x^*(\varphi) \leq 1$ for all φ in Φ. Hence $\|x^*\| = 1 = \sup_{\varphi \in \Phi} x^*(\varphi)$. □

__Theorem 2.3__. *Let Φ be a norming set for a Banach space X. Then the following are equivalent.*

(1)' X *has the* Φ–ANP–III;

(2) $X^{**} \setminus X = \{x^{**} : x^{**} \in X^{**}, \|x^{**}\| > \sup_{\varphi \in \Phi} x^{**}(\varphi)\}$;

(3) Every sequence in S_X that is asymptotically normed by Φ has a weakly convergent subsequence.

__Proof__. (1) $\Rightarrow$ (2). Let $x^{**} \in X^{**}$ such that $\|x^{**}\| = \sup_{\varphi \in \Phi} x^{**}(\varphi) = 1$. Let $\{x_\lambda\}$ be a net in B_X such that $w^* - \lim x_\lambda = x^{**}$. Then $\lim_\lambda \|x_\lambda\| = \|x^{**}\| = 1$. Hence we may assume that $\|x_\lambda\| = 1$ for all λ. Since X has the Φ–ANP–III, by taking $\Phi_1 = B_{X^*}$ in Lemma 2.1, it follows that $x^{**} \in X$. On the other hand, Φ is a norming set of X, hence $X^{**}\backslash X = \{x^{**} : x^{**} \in X^{**}, \|x^{**}\| > \sup_{\varphi\in\Phi} x^{**}(\varphi)\}$.

(2) $\Rightarrow$ (3). Let $\{x_n\}$ be a sequence in S_X that is asymptotically normed by Φ. By Lemma 2.2 and (2), it follows that $\{x_n\}$ is relatively weakly compact in X. Hence $\{x_n\}$ has a weakly convergent subsequence.

(3) $\Rightarrow$ (1). Obvious. □

A Banach space X is said to have the *Kadec property (K)* if $(S_X, w) = (S_X, \|\cdot\|)$. X is said to have the *Kadec–Klee property (KK)* if for any sequence $\{x_n\}$ and $\{x\}$ in B_X such that $\lim_n \|x_n\| = \|x\| = 1$ and $w - \lim_n x_n = x$ then $\lim_n \|x_n - x\| = 0$.

Theorem 2.4. *Let Φ be a norming set of a Banach space X. Then the following are equivalent.*

(1) X *has the* Φ–*ANP–II;*

(2) X *has the* Φ–*ANP III and* X *has the Kadec property;*

(3) X *has the* Φ–*ANP III and* X *has the Kadec–Klee property.*

Proof. (1) $\Rightarrow$ (2). Suppose X fails to have the property (K). Let $\{x_\lambda\}$ be a net in S_X and $\epsilon > 0$ such that $w\text{–}\lim x_\lambda = x$ for some x in S_X and $\|x_\lambda - x\| > \epsilon$ for all λ. By Lemma 2.1, there is a sequence $\{x_n\}$ in $\{x_\lambda\}$ such that $\{x_n\}$ is asymptotically normed by Φ and $\{x\} = \bigcap_{n=1}^{\infty} \overline{\text{co}}\{x_k : k \geq n\}$. However, X has the Φ–ANP–II, $\{x_n\}$ has a convergent subsequence and its limit must be x. This is impossible.

(2) $\Rightarrow$ (3). Obvious.

(3) $\Rightarrow$ (1). Let $\{x_n\}$ be a sequence in S_X that is asymptotically normed by Φ. By Theorem 2.3, $\{x_n\}$ has a weakly convergent subsequence, say $\{x_{n_k}\}$. Let $x = w - \lim x_{n_k}$. Then $\|x\| = 1$ by Lemma 2.2. Since X has the property (KK), it follows that $\lim_n \|x_{n_k} - x\| = 0$. □

A Banach space X is said to have the property (G) if every point of S_X is a denting point of B_X.

<u>Theorem 2.5</u>. *Let Φ be a norming set of a Banach space* X. *Then the following are equivalent.*

(1) X *has the* Φ*–ANP–I;*

(2) X *has the* Φ*–ANP–II and* X *is strictly convex;*

(3) X *has the* Φ*–ANP–III and* X *has the property* (G).

<u>Proof</u>. (1) $\Rightarrow$ (2). Suppose $x = \frac{1}{2}(y+z)$ for some x,y,z in S_X. For each $n \in \mathbb{N}$, let $x_n^* \in \Phi$, $x_n^*(x) > 1 - \frac{1}{n}$. Then $x_n^*(y) > 1 - \frac{2}{n}$. Hence the sequence $\{x_n\}$ with $x_{2n-1} = x$ and $x_{2n} = y$, $n \in \mathbb{N}$, is asymptotically normed by Φ. Since X is Φ–ANP–I, hence $x = y$.

(2) $\Rightarrow$ (3). By Theorem 2.4, X has the property (K). Since X is also strictly convex, it follows from [LLT], X has the property (G).

(3) $\Longrightarrow$ (1). Let $\{x_n\}$ be a sequence in S_X that is asymptotically normed by Φ. Then the set $D = \bigcap_{n=1}^{\infty} \overline{co}\{x_k : k \geq n\} \neq \phi$. Since X has the property (G), X is strictly convex and X has the property (K). By Theorem 2.4, X has the Φ–ANP–II. It follows that $\{x_n\}$ has cluster points and all cluster points must be in D. But $D \subset S_X$ and X is strictly convex, so D consists of one point that is the limit of the sequence $\{x_n\}$. □

Lemma 2.6. Let $\|\cdot\|_i$, $i = 1,2$ be equivalent norms on a Banach space X and let Φ_i be a norming set of $(X,\|\cdot\|_i)$, $i = 1,2$. For all x in X, define

$$|||x||| = \left(\|x\|_1^2 + \|x\|_2^2\right)^{1/2}$$

and

$$\Phi = \{\lambda_1 x_1^* + \lambda_2 x_2^* : x_i^* \in \Phi_i,\ \lambda_i \geq 0,\ \ i = 1,2,\ \ \lambda_1^2 + \lambda_2^2 = 1\}.$$

Then $|||\cdot|||$ is an equivalent norm on X and Φ is a norming set of $(X,|||\cdot|||)$ with the following properties.

(1) If one of $(X,\|\cdot\|_i)$, $i = 1,2$ is strictly convex, so is $(X,|||\cdot|||)$;

(2) If one of $(X,\|\cdot\|_i)$, $i = 1,2$ has the Kadec–Klee proeprty, so is $(X,|||\cdot|||)$;

(3) If one of $(X,\|\cdot\|_i)$, $i = 1,2$ has the Φ_i–ANP–III then $(X,|||\cdot|||)$ has the Φ–ANP–III.

Furthermore, if X is a dual space and both $\|\cdot\|_i$, $i = 1,2$ are dual norms then $|||\cdot|||$ is also a dual norm on X.

Proof. Except (3), the rest of the claims in Lemma 2.6 are easy to prove. We use Theorem 2.3 to prove (3). Suppose $(X,\|\cdot\|_1)$ has the Φ_1–ANP–III. Let $x^{**} \in X^{**}$ such that

$|||x^{**}||| = \sup_{\varphi \in \Phi} x^{**}(\varphi)$. Let $a_i = \sup_{\varphi \in \Phi_i} x^{**}(\varphi)$, $i = 1,2$. Then

$\left(||x^{**}||_1^2 + ||x^{**}||_2^2\right)^{1/2} = |||x^{**}||| = \sup_{\varphi \in \Phi} x^{**}(\varphi) \le \left(a_1^2 + a_2^2\right)^{1/2}$ and $a_i \le ||x^{**}||_i$, $i = 1,2$.

Hence $a_i = ||x^{**}||_i$, $i = 1,2$. Since $(X,||\cdot||_1)$ has the Φ_1–ANP–III, by Theorem 2.3, $x^{**} \in X$. Hence $(X,|||\cdot|||)$ has the Φ–ANP–III. □

Theorem 2.7. *Let* X *be a Banach space. Then*

(1) *The following are equivalent.*

(i) X *has the ANP–I;*

(ii) X *has the ANP–II and* X *is strictly convexifiable;*

(iii) X *has the ANP–III and there is an equivalent norm* $||\cdot||$ *on* X *such that* $(X,||\cdot||)$ *has the property* (G).

(2) X *has the ANP–II if and only if* X *has the ANP–III and there is an equivalent norm* $||\cdot||$ *on* X *such that* $(X,||\cdot||)$ *has the Kadec–Klee property.*

Corollary 2.8. If X is a Banach space with ANP–I, then X admits an equivalent locally uniformly convex norm $|||\cdot|||$ and there is a norming set Φ in $B_{(X^*,|||\cdot|||)}$ such that $(X,|||\cdot|||)$ has the Φ–ANP–I.

Proof. Suppose $(X,||\cdot||_1)$ has the ANP–I. By taking equivalent norm, we may assume that $(X,||\cdot||_1)$ has the property (G). By [Tr], X has an equivalent locally uniformly convex norm $||\cdot||_2$. Let $||\cdot||$ be the norm defined in Lemma 2.6. It is easy to see that $(X,||\cdot||)$ is locally uniformly convex, hence has the proeprty (G). By Theorem 2.5, there is a norming set Φ of $(X,||\cdot||)$ such that $(X,||\cdot||)$ has the Φ–ANP–I. □

Remark. It is well–known that every separable Banach space X has an equivalent locally uniformly convex norm. Thus Theorem 2.7 is a generalization of Theorem 1.2 in [JH].

If X is a Banach space with the Radon–Nikodym property but fails to have the ANP–I. Then X does not have an equivalent norm that has the property (G). This leads to the following problem.

Problem. Does every Banach space with the Radon–Nikodym property have an equivalent norm so that it has the property (G)?

3. Let X^* be a dual Banach space. We say that X^* has the *weak* asymptotic norming property* κ (w^*–ANP–κ), if there is an equivalent norm $\|\cdot\|$ on X and a norming set Φ in $B_{(X,\|\cdot\|)}$ such that $(X^*,\|\cdot\|)$ has the Φ–ANP–κ, $\kappa = $ I, II or III. Let $x^* \in S_{X^*}$. x^* is called a weak* denting point of B_{X^*} if for every $\varepsilon > 0$ there are an element x in X and $\delta > 0$ such that diameter of $S(x^*,x,\delta) = \{y^*: y^* \in B_{X^*}, y^*(x) > x^*(x) - \delta\}$ is less than ε

Theorem 3.1. *Let* X^* *be a dual Banach space and let* Φ *be a norming set of* X^* *in* B_X. *Then*

(1) X^* *has the* Φ*–ANP–I if and only if every point of* S_{X^*} *is a weak* denting point of* B_{X^*}.

(2) X^* *has the* Φ*–ANP–II if and only if the weak* topology and the norm topology coincide on* S_{X^*};

(3) X^* *has the* Φ*–ANP–III if and only if the weak* and weak topology coincide on* S_{X^*}.

Proof. As in section 2, we start on ANP–III and use properties on ANP–III to obtain properties of ANP–II, then ANP–I.

(3). Suppose there is a net $\{x_\lambda^*\}$ in S_{X^*} such that $w^* - \lim x_\lambda^* = x^*$ for some $x^* \in S_{X^*}$ but $\{x_\lambda^*\}$ fails to converge to x^* weakly. Without loss of generality, we may assume that there are $x^{**} \in X^{**}$ and $\epsilon > 0$ such that $|x^{**}(x_\lambda^*) - x^{**}(x^*)| > \epsilon$ for all λ. By Lemma 2.1, there is a sequence $\{x_n^*\}$ in $\{x_\lambda^*\}$ that is asymptotically normed by Φ and $\{x^*\} = \bigcap_{n=1}^{\infty} \overline{co}\{x_k^* : k \geq n\}$. By Theorem 2.3, $\{x_n^*\}$ has a weakly convergent subsequence, say $\{x_{n_k}^*\}$. However, the weak limit of $\{x_{n_k}^*\}$ must be x^* which is impossible.

Conversely, suppose $(S_{X^*}, w^*) = (S_{X^*}, w)$. Let $\{x_n^*\}$ be any sequence in S_{X^*} that is asymptotically normed by Φ. By Lemma 2.2, $\{x_n^*\}$ is relatively weak* compact in S_{X^*}, hence is relatively weak compact. It follows that $\bigcap_{n=1}^{\infty} \overline{co}\{x_k^* : k \geq n\} \neq \phi$ and so X^* has the Φ–ANP–III.

(2) Suppose X^* has the Φ–ANP–II. By Theorem 2.4, X^* has the property (K). Hence, by (3), we have $(S_{X^*}, w^*) = (S_{X^*}, w) = (S_{X^*}, \|\cdot\|)$.

Conversely, if $(S_{X^*}, w^*) = (S_X, \|\cdot\|)$, then $(S_{X^*}, w^*) = (S_{X^*}, w) = (S_{X^*}, \|\cdot\|)$. Hence, by (3), X^* has the Φ–ANP–III and X^* also has the property (K). It follows from Theorem 2.4, X^* has the Φ–ANP–II.

(1) Suppose X^* has the Φ–ANP–I. By Theorem 2.5, X^* is strictly convex and by (2), we have $(S_{X^*}, w^*) = (S_{X^*}, \|\cdot\|)$. By following the same argument as in [LLT], it follows that every point in S_{X^*} is a weak* denting point of B_{X^*}.

Conversely, if every point of S_{X^*} is a w^*–denting point of B_{X^*}. Then X^* is strictly convex and by (2), X has the Φ–ANP–II. Hence X^* has the Φ–ANP–I by Theorem 2.5. □

Since the properties in Theorem 3.1 are independent of the choice of Φ, we have the following corollary.

Corollary 3.2. Let Φ be a norming set of X^* in B_X. Then

(1) For $\kappa = I, II$, or III, X^* has Φ–ANP–κ if and only X^* has the B_X–ANP–κ, $\kappa = I, II$ or III;

(2) X^* has the Φ–ANP–III if and only if $X^{***}\backslash X^* = \{x^{***} \in X^{***} : \|x^{***}\| > \|P(x^{***})\|\}$ where P is the projection from X^{***} to X^* defined by $P(x^{***})$ = the restriction of x^{***} on X.

Proof. (2) follows from (1) and $\|P(x^{***})\| = \sup_{\|x\|=1} x^{***}(x)$ for all $x^{***} \in X^{***}$. □

Following [JH], a dual Banach space X^* is said to have the *convex–w^*–Kadec Klee property (cw^*–KK)* if whenever $w^*\text{–}\lim_n x_n^* = x^*$ and $\lim_n \|x_n^*\| = \|x^*\|$ in X^* then $x^* \in \bigcap_{n=1}^{\infty} \overline{co}\{x_k^* : k \geq n\}$. X^* has the *weak*–Kadec–Klee property (w^*–KK)* if for any sequence $\{x_n^*\}$ and x^* in X^*, $w^* - \lim_n x_n^* = x^*$, $\lim_n \|x_n^*\| = \|x^*\|$ then $\lim_n \|x_n^* - x^*\| = 0$. A set A in X^* is said to be *convex weak* sequentially compact* if for any sequence $\{x_n^*\}$ in A, there exist $y_n^* \in co\{x_n^* : k \geq n\}$, $n \in \mathbb{N}$, such that the $w^* - \lim y_n^*$ exists in A.

Theorem 3.3. *Let X be a Banach space. Then*

(1) X^* *has the* w^**–ANP–III if and only if* B_{X^*} *is convex weak** *sequentially compact and* X^* *admits an equivalent dual norm with the* cw^**–KK property.*

(2) X^* *has the* w^**–ANP–II if and only if* B_{X^*} *is weak** *sequentially compact and* X^* *admits an equivalent dual norm with the* w^**–KK property.*

Proof. (1) Suppose X^* has the w^*–ANP–III. Then X^* has the RNP [JH]. Hence B_{X^*} is weak*–sequentially compact (see e.g. Prop. 3.12 in [B]) and so B_{X^*} is convex weak* sequentially compact. Furthermore, by Theorem 3.1, X^* admits an equivalent dual norm $\|\cdot\|$ such that $(S_{(X^*,\|\cdot\|)},w^*) = (S_{(X^*,\|\cdot\|)},w)$. It is clear then that $(X^*,\|\cdot\|)$ has the cw^*–KK property.

Conversely, let $|||\cdot|||$ be an equivalent dual norm on X^* such that $(X^*,|||\cdot|||)$ has the cw^*–KK property. Without loss of generality, we may assume that $\Phi = B_{(X,|||\cdot|||)}$ and $B_{(X^*,|||\cdot|||)} \subset B_{(X^*,\|\cdot\|)}$. Let $\{x_n^*\}$ be a sequence in $S_{(X^*,|||\cdot|||)}$ that is asymptotically normed by Φ. Since B_{X^*} is convex weak* sequentially compact, there is a sequence $\{y_n^*\}$, $y_n^* \in \text{co}\{x_k^* : k \geq n\}$, $n \in \mathbb{N}$, such that $w^*-\lim_n y_n^* = y^*$ for some y^* in B_{X^*}. Then $|||y^*||| \leq \underline{\lim}_n |||y_n^*||| \leq \overline{\lim}_n |||y_n^*||| \leq 1$. Since $\{x_n^*\}$ is asymptotically normed by Φ, for any $\epsilon > 0$, there is x in Φ and $N \in \mathbb{N}$ such that $x_n^*(x) > 1-\epsilon$ for all $n \geq N$. Hence $y_n^*(x) > 1-\epsilon$ for all $n \geq N$ and so $y^*(x) \geq 1-\epsilon$. Therefore $1 = |||y^*||| = \lim_n |||y_n^{**}|||$. By the fact that $(X^*,|||\cdot|||)$ has the cw^*–KK property, we have that $y^* \in \bigcap_{n=1}^{\infty} \overline{\text{co}}\{y_n^* : \kappa \geq n\} \subset \bigcap_{n=1}^{\infty} \overline{\text{co}}\{x_k^* : k \geq n\}$. This completes the proof that X^* has the w^*–ANP–III.

(2) If X^* has the w^*–ANP–II, then X^* has the RNP. As in (1), it follows that B_{X^*} is weak* sequentially compact. By Theorem 3.1, there is an equivalent dual norm $|||\cdot|||$ on X^* such that $\left(S_{(X^*,|||\cdot|||)}, w^*\right) = \left(S_{(X^*,|||\cdot|||)}, |||\cdot|||\right)$. Thus $(X^*,|||\cdot|||)$ has the w^*–KK property.

Conversely, let $|||\cdot|||$ be an equivalent dual norm on X^* such that $(X^*,|||\cdot|||)$ has the w^*–KK property. We may assume that $\Phi = B_{(X,|||\cdot|||)}$ and $B_{(X^*,|||\cdot|||)} \subset B_{(X^*,\|\cdot\|)}$. Let $\{x_n^*\}$ be a sequence in $S_{(X^*,|||\cdot|||)}$ that is asymptotically normed by Φ. Since B_{X^*} is weak* sequentially compact, there is a subsequence $\{y_n^*\}$ of $\{x_n^*\}$ such that $w^*-\lim_n y_n^* = y^*$ for some y^* in B_{X^*}. As in the proof of (1), we have that $\lim_n |||y_n^*||| = |||y^*||| = 1$. Since $(X^*,|||\cdot|||)$ has the w^*–KK property, it follows that $\lim_n |||y_n^* - y^*||| = 0$. Thus $(X^*,|||\cdot|||)$ has the w^*–ANP–II.

<u>Corollary 3.4</u>. Let X be a Banach space. Then the following are equivalent.

(1) X^* has the w^*–ANP–III;

(2) X^* has the RNP and X^* admits an equivalent dual norm with the cw^*–KK property;

(3) X^* admits an equivalent dual norm with the cw^*–KK property and ℓ_1 is not isomorphic to a subspace of X.

<u>Proof</u>. (1) $\Rightarrow$ (2). This follows from Theorem 3.3 and the fact that ANP implies RNP.

(2) $\Rightarrow$ (3). Since X^* has RNP if and only if X is an Asplund space. Thus X contains no copy of ℓ_1.

(3) $\Rightarrow$ (1). Suppose X contains no copy of ℓ_1, then B_{X^*} is convex weak* sequentially

compact (see e.g. Prop. 3.11 in [B]). By Theorem 3.3, it follows that X^* has the w^*–ANP–III. □

Corollary 3.5. Let X be a Banach space. Then the following are equivalent.

(1) X^* has the w^*–ANP–II;

(2) X^* has the RNP and X^* admits an equivalent dual norm with the w^*–KK property.

(3) X^* admits an equivalent norm with the w^*–KK property and ℓ_1 is not isomorphic to a subspace of X.

Let X^* be a dual Banach space. X^* is said to have the property (G^*) if every point of S_{X^*} is a weak* denting point of B_{X^*}.

Corollary 3.6. Let X be a Banach space. Then X^* has the property (G^*) if and only if X^* is strictly convex and X^* has the w^*–KK property.

Proof. It is clear that if X^* has the property (G^*) then X^* is strictly convex and X^* has the w^*–KK property.

Conversely, if X^* is strictly convex, then B_{X^*} is weak* sequentially compact [HS]. If X^* also has the w^*–KK property, by part (2) of Theorem 3.3, we see that X^* has the Φ–ANP–II where $\Phi = B_X$. Hence X^* has the Φ–ANP–I by Theorem 2.3. It follows from Theorem 3.1, X^* has the property (G^*).

Remarks (1). It is interesting to compare Corollary 3.6 with the fact that a Banach space X has the property (G) if and only if X is strictly convex and X has the property (K).

Furthermore, if in addition that ℓ_1 is not isomorphic to a subspace of X, then X has the property (K) if and only if X has the property (KK). [LLT]

(2). In Corollary 2.6, we show that if X has the ANP–I, then X has an equivalent locally uniformly convex norm. However, we don't know if X^* has the w^*–ANP–I, does there exist an equivalent *dual* norm on X^* which is locally uniformly convex?

We, now, give an example of a Banach space X, such that X^* has the RNP but X^* fails to have w^*–ANP–III.

Let Ω be the first uncountable ordinal and let $X = C[0,\Omega]$. Then X^* has the RNP [T]. To show that X^* fails to have w^*–ANP–III. Let $|||\cdot|||$ be any equivalent dual norm on X^* and let $||\cdot||$ be the usual norm on X^*. Let $M > 0$ such that $M||x^*|| \geq |||x^*||| \geq \frac{1}{M}||x^*||$ for all $x^* \in X^*$. For each $\alpha \in [0,\Omega]$, let δ_α be the dirac measure, that is, $\delta_\alpha(x) = x(\alpha)$ for all $x \in X$. Let $r = \lim_{\alpha<\Omega} \inf_{\alpha\leq\beta<\Omega} |||\delta_\beta|||$. Then $M \geq r \geq \frac{1}{M}$. For each $n \in \mathbb{N}$, choose $\alpha_n \in [0,\Omega)$, such that $r - \frac{1}{n} < \inf_{\alpha_n\leq\beta<\Omega} |||\delta_\beta||| \leq |||\delta_{\alpha_n}||| < r + \frac{1}{n}$ and $\alpha_n < \alpha_{n+1}$. It follows that $r = \lim_n |||\delta_{\alpha_n}|||$. Let $\alpha = \sup_n \alpha_n$. Then $\alpha \in [0,\Omega)$ and $|||\delta_\alpha||| \geq \inf_{\alpha_n\leq\beta<\Omega} |||\delta_\beta|||$ for all n. Hence $|||\delta_\alpha||| \geq r$.

Since $\{\alpha_n\}$ converges to α in the order topology, for every x in X, $\lim_n \delta_{\alpha_n}(x) = \delta_\alpha(x)$, i.e. $w^*\text{–}\lim \delta_{\alpha_n} = \delta_\alpha$. Hence $|||\delta_\alpha||| \leq \lim_n |||\delta_{\alpha_n}||| = r$. It follows that $|||\delta_\alpha||| = \lim_n |||\delta_{\alpha_n}|||$. However, for any $\lambda_n \geq 0$, $\sum_n \lambda_n = 1$, we have

$$||\delta_\alpha - \sum_{n=1}^{\infty} \lambda_n \delta_{\alpha_n}|| = 2.$$ Hence $\delta_\alpha \notin \overline{co}\{\alpha_n : n \in \mathbb{N}\}$. This shows that $(X^*, |||\cdot|||)$ does not have the cw^*–KK property and so X^* fails to have the w^*–ANP–III.

References

[B] J. Bourgain, La propriéte de Radon–Nikodým, Publication de l'University Pierre et Marie Curie, No. 36 (1979).

[D] J. Diestel, Geometry of Banach spaces, selected topics, Lect. Notes in Math., Springer–Verlag, 485 (1975).

[GM] N. Ghoussoub an d B. Maurey, The asymptotic–norming and the Radon–Nikodym properties are equivalent in separable Banach spaces, Proc. Amer. Math. Soc, 94 (1985), 665–671.

[HS] J. Hagler and F. Sullivan, Smoothness and weak* sequential compactness, Proc. Amer. Math. Soc. 78 (1980), 497–503.

[JH] R.C. James and A. Ho, The asymptotic–norming and the Radon–Nikodym properties for Banach spaces, Ark. Mat. 19 (1981), 53–70.

[LL] Bor–Luh Lin and Pei–Kee Lin, Property (H) in Lebesgue–Bochner function spaces, Proc. Amer. Math. Soc. 95 (1985), 581–584.

[LLT] Bor–Luh Lin, Pei–Kee Lin and S.L. Troyanski, Characterizations of denting points, Proc. Amer. Math. Soc. 102 (1988), 526–528.

[S] M.A. Smith, Some examples concerning rotundity in Banach space theory, Math. Annalen 233 (1978), 155–161.

[T] M. Talagrand, Renormages de quelques C(K), Israel J. Math. 54 (1986), 327–334.

[Tr] S.L. Troyanski, On a property of the norm which is close to local uniform rotundity, Math. Annalen, 271 (1985), 305–313.

Characterizations of C(X, τ) Among Its Subalgebras

SUNWOOK HWANG Department of Mathematics, Soongsil University, Seoul, Korea

1. Introduction

Let X be a compact Hausdorff space, and let $C(X)$ (resp. $C_{\mathbf{R}}(X)$) be the complex (resp. real) Banach algebra of all complex-valued (resp. real-valued) continuous functions on X with the pointwise operations and the supremum norm $\| \|_X$. Let $\tau : X \to X$ be a homeomorphism such that $\tau \circ \tau = \tau^2 = id_X =$ the identity map on X. We will call τ an **involution on** X. Define

$$C(X,\tau) = \{f \in C(X) : f(\tau(x)) = \overline{f(x)} \text{ for all } x \in X\},$$

where $\overline{f(x)}$ is the complex conjugate of $f(x)$ for $x \in X$. Then $C(X,\tau)$ is a *real* commutative Banach algebra with identity 1 which separates the points of X.

A **real function algebra** A **on** (X,τ) is a uniformly closed real subalgebra of $C(X,\tau)$ which separates the points of X and contains 1. This definition was first introduced by Kulkarni and Limaye in [8].

For complex function algebras B on X, the question of when $B = C(X)$ is closely related to the nature of ReB, the space of real parts of functions in B, and has been studied by many authors, in particular, Hoffman and Wermer [6], Wermer [12], Sidney [11], and Hatori [5]. For a real function algebra A on (X,τ), it is natural to ask whether the nature of ReA determines when $A = C(X,\tau)$ in a similar way.

In 1988, Kulkarni and Srinivasan [9] proved an analogue of the Hoffman-Wermer theorem for real function algebras, which states the following:

Proposition 1.1 ([9, Theorem 3.1]). Let A be a real function algebra on (X,τ). If ReA is uniformly closed in $C_R(X,\tau)$, then $A = C(X,\tau)$.

In this paper, using Proposition 1.1, we obtain analogues of Wermer's theorem and Sidney's theorem for real function algebras (Corollary 3.5 and Theorem 4.2, resp.).

2. Basic Results

Throughout this paper, let X be a compact Hausdorff space, τ an involution on X. Define

$$C_R(X,\tau) = \{u \in C_{\mathbf{R}}(X) : u = u \circ \tau\} \quad \text{and} \quad C_S(X,\tau) = \{v \in C_{\mathbf{R}}(X) : v = -v \circ \tau\}.$$

Then, it is easy to see that

$$C(X,\tau) = C_R(X,\tau) + iC_S(X,\tau).$$

Let E be a subset of $C(X,\tau)$. We say that E **separates weakly the points of** X if E separates the points of X except possibly for some distinct pairs x and $\tau(x)$ for x in X.

For complex subalgebras B of $C(X)$, the fact that B separates the points of X is equivalent to the fact that ReB separates the points of X. For real subalgebras of $C(X,\tau)$, we have the following fact.

Proposition 2.1. Let A be a real subalgebra of $C(X,\tau)$. If A separates the points of X, then ReA separates weakly the points of X.

Proof. Suppose ReA does not separate weakly the points of X, that is, there are points $x, y \in X$ with $x \neq y, \tau(y)$ such that

$$\mathrm{Re}f(x) = \mathrm{Re}f(y) \quad \text{for all} \quad f \in A. \tag{2.1}$$

Since $f^2 \in A$ and $\mathrm{Re}(f^2) = (\mathrm{Re}f)^2 - (\mathrm{Im}f)^2$,

$$(\mathrm{Re}f(x))^2 - (\mathrm{Im}f(x))^2 = \mathrm{Re}(f^2)(x) = \mathrm{Re}(f^2)(y) = (\mathrm{Re}f(y))^2 - (\mathrm{Im}f(y))^2,$$

so $(\mathrm{Im}f(x))^2 = (\mathrm{Im}f(y))^2$ by (2.1), whence $\mathrm{Im}f(x) = \pm\,\mathrm{Im}f(y)$. Thus we have, for given $f \in A$

$$\text{either} \quad \mathrm{Im}f(x) = \mathrm{Im}f(y) \quad \text{or} \quad \mathrm{Im}f(x) = -\mathrm{Im}f(y).$$

This implies by (2.1) that for given $f \in A$, either

$$f(x) = \mathrm{Re}f(x) + i\mathrm{Im}f(x) = \mathrm{Re}f(y) + i\mathrm{Im}f(y) = f(y)$$

or

$$f(x) = \mathrm{Re}f(x) + i\mathrm{Im}f(x) = \mathrm{Re}f(y) - i\mathrm{Im}f(y) = \mathrm{Re}f(\tau(y)) + i\mathrm{Im}f(\tau(y)) = f(\tau(y)).$$

Hence we have

$$\text{for each } f \in A, \quad \text{either} \;\; f(x) = f(y) \;\; \text{or} \; f(x) = f(\tau(y)). \tag{2.2}$$

Now, since A separates the points of X, we can choose $f_1, f_2 \in A$ such that

$$f_1(x) \neq f_1(\tau(y)) \;\; \text{and} \;\; f_2(x) \neq f_2(y). \tag{2.3}$$

Then by (2.2), we have

$$f_1(x) = f_1(y) \;\; \text{and} \;\; f_2(x) = f_2(\tau(y)). \tag{2.4}$$

Take $g = f_1 + f_2 \in A$. Then from (2.3) and (2.4),

$$g(x) = f_1(x) + f_2(x) = f_1(y) + f_2(x) \neq f_1(y) + f_2(y) = g(y)$$

and

$$g(x) = f_1(x) + f_2(x) = f_1(x) + f_2(\tau(y)) \neq f_1(\tau(y)) + f_2(\tau(y)) = g(\tau(y)).$$

So, (2.2) fails for $f = g$. This contradiction shows that (2.1) is false and that proves the proposition.

Corollary 2.2. Let A be a real function algebra on (X, τ). Then $\mathrm{Re}A$ separates weakly the points of X.

The converse of Proposition 2.1 is not true in general, even if we assume that A is uniformly closed and contains the real constants as in the following example.

Example 2.3. Let $X = [0,1]$ and let $\tau(x) = 1 - x$ for $x \in [0,1]$. Then

$$C(X,\tau) = \{f \in C([0,1]) : f(1-x) = \overline{f(x)} \;\; \text{for all } x \in X\}.$$

Take $A = \{f \in C(X,\tau) : f(0) \in \mathbf{R}\}$. Then A is a uniformly closed real subalgebra of $C(X,\tau)$ which contains the real constants. Define

$$f(x) = \begin{cases} x & \text{if} \;\; 0 \leq x \leq 1/2, \\ 1 - x & \text{if} \;\; 1/2 \leq x \leq 1. \end{cases}$$

Then $f \in C_R(X,\tau) \subset C(X,\tau)$ and $f(0) = 0 \in \mathbf{R}$, so $f \in A$. ReA separates weakly the points of X since $f = \overline{\text{Re}f} \in \overline{\text{Re}A}$ does it. But A fails to separate the points 0 and 1. For if $g \in A$, then $g(1) = \overline{g(\tau(1))} = \overline{g(0)} = g(0)$ since $g(0) \in \mathbf{R}$.

Remark. For a real subalgebra A of $C(X,\tau)$ which separates weakly the points of X, in order for A to separate the points of X, A must satisfy the following condition:

$$\text{for each } x \in X \text{ with } x \neq \tau(x), \text{ there is } f \in A \text{ such that } \text{Im}f(x) \neq 0.$$

We can see that the algebra A of Example 2.3 violates this condition.

Two points x and y in X are said to be **equivalent** (denoted by $x \sim y$) if $x = y$ or $x = \tau(y)$. It is clear that $\sim$ is an equivalence relation on X. Denote the equivalence class of $x \in X$ under τ by $< x >$, and denote by Y the set of equivalence classes of X under τ.

Give Y the quotient topology induced by the quotient map $x \rightarrow < x >$ for $x \in X$. Then Y is a compact Hausdorff space, and each $u \in C_R(X,\tau)$ induces a continuous function $u_1 : Y \rightarrow \mathbf{R}$ such that

$$u_1(< x >) = u(x) \text{ for } x \in X. \tag{2.5}$$

Then, we have

$$C_{\mathbf{R}}(Y) = \{u_1 : u \in C_R(X,\tau)\}.$$

And, it is easy to check that $\|u_1\|_Y = \|u\|_X$ for $u \in C_R(X,\tau)$.

Proposition 2.4 (Stone-Weierstrass Theorem). Let E be a subring of $C_R(X,\tau)$ which separates weakly the points of X and contains the real constants. Then E is uniformly dense in $C_R(X,\tau)$.

Proof. Let Y be the set of equivalence classes of X under τ. Put

$$E_1 = \{u_1 : u \in E\},$$

where u_1 is defined as in (2.5). It is clear that for $u, v \in E$

$$u_1 + v_1 = (u+v)_1 \text{ and } u_1 v_1 = (uv)_1.$$

Thus E_1 is a subring of $C_{\mathbf{R}}(Y)$ which contains the real constants. Moreover, E_1 separates the points of Y. Indeed, if $< x > \neq < y >$, then $x \neq y, \tau(y)$, hence there exists $u \in E$ such that $u(x) \neq u(y)$ since E separates weakly the points of X. Then $u_1 \in E_1$ and $u_1(< x >) \neq u_1(< y >)$. Thus by the real Stone-Weierstrass theorem, E_1 is uniformly dense in $C_{\mathbf{R}}(Y)$.

Now, we claim that E is uniformly dense in $C_R(X,\tau)$. To see this, let $v \in C_R(X,\tau)$ and $\varepsilon > 0$ be given. Then $v_1 \in C_{\mathbf{R}}(Y)$. Since E_1 is dense in $C_{\mathbf{R}}(Y)$, there exists $u \in E$ such that $\|u_1 - v_1\|_Y < \varepsilon$. Then

$$\|u - v\|_X = \|(u - v)_1\|_Y = \|u_1 - v_1\|_Y < \varepsilon.$$

Therefore, E is dense in $C_R(X, \tau)$.

Corollary 2.5. Let A be a real subalgebra of $C(X, \tau)$ with $1 \in A$. If ReA separates weakly the points of X and if A is conjugate closed, then ReA is uniformly dense in $C(X, \tau)$.

Proof. Since A is conjugate closed, for each $f \in A$ we have $\bar{f} \in A$. So Re$f = (f+\bar{f})/2 \in A$, and hence Re$f \in A \cap C_R(X, \tau)$. Thus, we have

$$\text{Re}A = A \cap C_R(X, \tau).$$

Then ReA is a real subalgebra of $C_R(X, \tau)$ which separates weakly the points of X and contains the real constants, so ReA is uniformly dense in $C_R(X, \tau)$ by Proposition 2.4.

3. Wermer's Theorem for Real Function Algebras

Materials in this section are based on the work of Mehta and Vasavada [10], where they obtained similar results for real Banach algebras lying in $C(X)$ with identity 1.

Let E be a (real or complex) normed linear space with norm $\| \ \|_E$. Denote by $\tilde{E} = l_\infty(\mathbf{N}, E)$ the space of all bounded functions from the set $\mathbf{N} = \{1, 2, 3, \cdots\}$ to E normed as follows: $\mathrm{N}(\{f_n\}_{n=1}^\infty) \equiv \sup\{\|f_n\|_E : n \in \mathbf{N}\} < \infty$ for a sequence $\{f_n\}_{n=1}^\infty$ in $\tilde{E}$.

The following lemma, which was originated by A. Bernard, is crucial in the proofs of Theorem 3.4 and Theorem 4.2.

Lemma (Bernard [1]). Let E and F be (real or complex) normed linear spaces such that $E \subset F$ and the inclusion is continuous. Then $\tilde{E} \subset \tilde{F}$.

If, in addition, E is complete and $\tilde{E}$ is dense in $\tilde{F}$, then $E = F$.

Denote by $\tilde{X} = \beta(\mathbf{N} \times X)$ the Stone-Cech compactification of the product space $\mathbf{N} \times X$. Then we have a natural embedding $1 \times \tau : \mathbf{N} \times X \to \tilde{X}$ defined by

$$(1 \times \tau)(n, x) = (n, \tau(x)) \text{ for } (n, x) \in \mathbf{N} \times X,$$

which has a continuous extension $\tilde{\tau} : \tilde{X} \to \tilde{X}$ (cf. [4, Section 6.5]). It is easy to check that $\tilde{\tau}$ is an involution on $\tilde{X}$.

For a real function algebra A on (X,τ), the space $\tilde{A} = l_\infty(\mathbf{N}, A)$ is the algebra of all *bounded* sequences of functions of A (*bounded* in the sense that $\{f_n\}_{n=1}^\infty \in \tilde{A}$ if and only if $\mathrm{N}(\{f_n\}_{n=1}^\infty) \equiv \sup\{\|f_n\|_X : n \in \mathbf{N}\} < \infty$). It is easy to check that $C(X,\tau)^\sim = C(\tilde{X},\tilde{\tau})$ and that $\tilde{A}$ is a closed real subalgebra of $C(\tilde{X},\tilde{\tau})$ which contains 1. We will use $\tilde{f}$ to denote an element of $C(\tilde{X},\tilde{\tau})$.

Lemma 3.1. Let A be a real function algebra on (X,τ). Define, for $f \in A$,

$$\|\mathrm{Re} f\|_q \equiv \inf\{\|g\|_X : g \in A,\ \mathrm{Re} f = \mathrm{Re} g\}.$$

Then $(\mathrm{Re}A, \|\ \|_q)^\sim = \mathrm{Re}\,((A, \|\ \|_X)^\sim)$, where $\|\ \|_q$ is a (complete) quotient norm on $\mathrm{Re}A$.

Proof. Let $\{u_n\}_{n=1}^\infty \in (\mathrm{Re}A, \|\ \|_q)^\sim$. Then there is $M > 0$ such that $\|u_n\|_q \le M$ for all $n = 1, 2, \cdots$. By the definition of the quotient norm, for each $n \in \mathbf{N}$, there is $f_n \in A$ such that $\mathrm{Re} f_n = u_n$ and $\|f_n\|_X \le \|u_n\|_q + 1$. Then

$$\mathrm{N}(\{f_n\}_{n=1}^\infty) = \sup\{\|f_n\|_X : n \in \mathbf{N}\} \le \sup\{\|u_n\|_q : n \in \mathbf{N}\} + 1 \le M + 1 < \infty,$$

so $\{f_n\}_{n=1}^\infty \in \tilde{A}$ and

$$\{u_n\}_{n=1}^\infty = \{\mathrm{Re} f_n\}_{n=1}^\infty = \mathrm{Re}(\{f_n\}_{n=1}^\infty) \in \mathrm{Re}((A, \|\ \|_X)^\sim).$$

On the other hand, let $\{u_n\}_{n=1}^\infty = \mathrm{Re}(\{f_n\}_{n=1}^\infty) \in \mathrm{Re}((A, \|\ \|_X)^\sim)$. Then there is $K > 0$ such that $\|f_n\|_X \le K$ for all $n \in \mathbf{N}$. So, for each $n \in \mathbf{N}$,

$$\|\mathrm{Re} f_n\|_q = \inf\{\|g\|_X : g \in A,\ \mathrm{Re} g = \mathrm{Re} f_n\} \le \|f_n\|_X \le K,$$

and hence $\{u_n\}_{n=1}^\infty \in (\mathrm{Re}A, \|\ \|_q)^\sim$. This proves the lemma.

It is easy to check from this lemma that $C_R(X,\tau)^\sim = C_R(\tilde{X},\tilde{\tau})$ since the quotient norm of $C_R(X,\tau)$ is same as the supremum norm on X.

Now, we will consider the separation property of $\tilde{A}$ when A is a real function algebra on (X,τ). For complex function algebras B on X, it is well-known that if $\mathrm{Re}B$ is uniformly dense in $C_{\mathbf{R}}(X)$, then $\tilde{B}$ (and hence $\mathrm{Re}\tilde{B}$) separates the points of $\tilde{X}$. For real function algebras on (X,τ), we have the following weaker result. Its proof is a slight modification of that given in [2, Lemma 4.11].

Proposition 3.2. Let A be a real function algebra on (X,τ). If $\mathrm{Re}A$ is uniformly dense in $C_R(X,\tau)$, then $\mathrm{Re}\tilde{A}$ separates weakly the points of $\tilde{X}$.

Proof. First, note that if $f \in C_R(X,\tau)$ and $f > 0$ on X, then $\log \circ f \in C_R(X,\tau)$. Since $\mathrm{Re}A$ is uniformly dense in $C_R(X,\tau)$, for given $\varepsilon > 0$, there exists $g \in A$ such that $\|\log \circ f - \mathrm{Re} g\|_X < \varepsilon$, and hence we may assume that

$$\|f - |\,e^g\,|\,\|_X \le \varepsilon. \tag{3.1}$$

Now, let $p, q \in \tilde{X}$ be so that $p \neq q, \tilde{\tau}(q)$. Then by Urysohn's lemma, there is $\tilde{\phi} \in C_R(\tilde{X}, \tilde{\tau})$ with $\tilde{\phi} > 0$ such that $\tilde{\phi}(p) = 1$ and $\tilde{\phi}(q) = 4$. Regard $\tilde{\phi}$ as a sequence $\{\phi_n\}_{n=1}^{\infty}$ in $C_R(X, \tau)$. Then each $\phi_n > 0$ and hence from (3.1) we have a sequence $\{g_n\}_{n=1}^{\infty}$ in A such that $\|\phi_n - | h_n | \|_X \leq 1$, where $h_n = e^{g_n} \in A$. Since $\|\tilde{\phi}\|_{\tilde{X}} \equiv \sup \{\|\phi_n\|_X : n \in \mathbf{N}\} < \infty$,

$$\|h_n\|_X \leq \|\phi_n\|_X + 1 \leq \|\tilde{\phi}\|_{\tilde{X}} + 1$$

for all $n = 1, 2, \cdots$, so $\{h_n\}_{n=1}^{\infty} \in \tilde{A}$. Put $\tilde{h} = \{h_n\}_{n=1}^{\infty}$. Then $\|\tilde{\phi} - | \tilde{h} | \|_{\tilde{X}} \leq 1$, and so,

$$| \tilde{\phi}(p) - | \tilde{h}(p) | | \leq 1 \text{ and } \tilde{\phi}(p) = 1 \text{ imply } | \tilde{h}(p) | \leq 2,$$

$$| \tilde{\phi}(q) - | \tilde{h}(q) | | \leq 1 \text{ and } \tilde{\phi}(q) = 4 \text{ imply } | \tilde{h}(q) | \geq 3.$$

Thus, $| \tilde{h}(p) | \neq | \tilde{h}(q) |$. In other words, $| \tilde{A} |$ separates weakly the points of $\tilde{X}$.

Now, suppose Re$\tilde{A}$ fails to separate p and q with $p \neq \tilde{\tau}(q)$. If $g = u + iv \in \tilde{A}$, then $g^2 = u^2 v^2 + 2iuv$, so $u^2 - v^2 \in \text{Re}\tilde{A}$. Hence

$$(u^2 - v^2)(p) = (u^2 - v^2)(q). \tag{3.2}$$

Since $u = \text{Re}g \in \text{Re}\tilde{A}$, $u(p) = u(q)$, so $u^2(p) = u^2(q)$, and hence from (3.2) we have $v^2(p) = v^2(q)$. But then, since $| g |^2 = u^2 + v^2$,

$$| g |^2 (p) = u^2(p) + v^2(p) = u^2(q) + v^2(q) = | g |^2 (q),$$

and hence $| g | (p) = | g | (q)$, which contradicts the fact that $| \tilde{A} |$ separates weakly the points of $\tilde{X}$. Therefore, Re$\tilde{A}$ separates weakly the points of $\tilde{X}$.

Question. Under the hypothesis of Proposition 3.2, does $\tilde{A}$ separate the points of $\tilde{X}$?

The proof of the following lemma is basically the same as that of [2, Lemma 3.2] by noticing that the norm of A dominates the supremum norm (see Proposition 2.5 and Theorem 4.4 of [7]). So, we will omit it.

Lemma 3.3. Let A be a real Banach algebra lying in $C(X, \tau)$. If ReA is a ring, then there exists a constant $K > 0$ such that

$$\|uv\|_q \leq K \|u\|_q \|v\|_q \text{ for all } u, v \in \text{Re}A.$$

Here, $\| \ \|_q$ is the quotient norm defined as in Lemma 3.1.

Theorem 3.4. Let A be a real Banach algebra lying in $C(X, \tau)$ with $1 \in A$ such that ReA is a ring and Re$\tilde{A}$ separates weakly the points of $\tilde{X}$. Then Re$A = C_R(X, \tau)$.

Proof. Let $\tilde{u} = \{u_n\}_{n=1}^{\infty}, \tilde{v} = \{v_n\}_{n=1}^{\infty} \in \text{Re}\tilde{A}$. Since Re$A$ is a ring, for each $n \in \mathbf{N}$ we have $u_n v_n \in \text{Re}A$ and

$$\|u_n v_n\|_q \leq K\|u_n\|_q\|v_n\|_q \leq K\mathrm{N}(\tilde{u})\mathrm{N}(\tilde{v})$$

by Lemma 3.3. (Here, N is the norm on Re$\tilde{A}$ = (ReA)$^\sim$ as in Lemma 3.1.) So, we have

$$\mathrm{N}(\tilde{u}\tilde{v}) = \sup\{\|u_n v_n\|_q : n \in \mathbf{N}\} \leq K\mathrm{N}(\tilde{u})\mathrm{N}(\tilde{v}) < \infty,$$

and therefore, $\tilde{u}\tilde{v} = \{u_n v_n\}_{n=1}^{\infty} \in$ Re$\tilde{A}$. This shows that (ReA)$^\sim$=Re$\tilde{A}$ is a subring of $C_R(\tilde{X}, \tilde{\tau}) = C_R(X, \tau)^\sim$. Since Re$\tilde{A}$ separates weakly the points of $\tilde{X}$ and contains the real constants, Re$\tilde{A}$ is unifomly dense in $C_R(\tilde{X}, \tilde{\tau})$ by Proposition 2.4. Then by Bernard's lemma, we have Re$A = C_R(X, \tau)$.

Corollary 3.5 (Wermer's Theorem). Let A be a real function algebra on (X, τ). If ReA is a ring, then $A = C(X, \tau)$.

Proof. By Corollary 2.2, ReA separates weakly the points of X. Since ReA is a subring of $C_R(X, \tau)$ which contains the real constants, ReA is uniformly dense in $C_R(X, \tau)$ by Proposition 2.4. Then by Proposition 3.2, Re$\tilde{A}$ separates weakly the points of $\tilde{X}$, and thus Re$A = C_R(X, \tau)$ by Theorem 3.4. Finally, Proposition 1.1 implies that $A = C(X, \tau)$.

Remark. A referee informed the author that S. Kulkarni and N. Srinivasan also presented the material in this section at the fifth annual Conference of the Ramanujan Mathematical Society held at University of Mysore, India on June 7–9, 1990.

4. Sidney's Theorem for Real Function Algebras

Let E be a subset of $C_{\mathbf{R}}(X)$, and let h be a function from a set $S \subset \mathbf{R}$ into $\mathbf{R}$. We say that h **operates in** E if $h \circ f \in E$ whenever $f \in E$ and $f(X) \subset S$.

The following lemma is an analogue of the deLeeuw-Katznelson theorem [3].

Lemma 4.1. Let E be a uniformly closed real vector subspace of $C_R(X, \tau)$ which separates weakly the points of X and contains the real constants. If any continuous non-affine function on $\mathbf{R}$ operates in E, then $E = C_R(X, \tau)$.

Proof. First, recall from Section 2 that Y is the set of equivalence classes of X under τ, which is a compact Hausdorff space with the quotient topology. Put $E_1 = \{u_1 : u \in E\}$, where u_1 is defined as in (2.5). Then clearly E_1 is a uniformly closed real vector subspace of $C_{\mathbf{R}}(Y)$ which separates the points of Y (for this, see the proof of Proposition 2.4) and contains the real constants.

Note that any continuous non-affine function on $\mathbf{R}$ which operates in E also operates in E_1. Thus by the deLeeuw-Katznelson theorem [3], we have $E_1 = C_{\mathbf{R}}(Y)$, and hence $E = C_R(X,\tau)$. This proves the lemma.

Theorem 4.2 (Sidney's Theorem). Let A be a real function algebra on (X,τ). Suppose that h is a continuous function from an interval $I \subset \mathbf{R}$ into $\mathbf{R}$ and that I contains a non-degenerate subinterval J such that h is not affine on any non-degenerate subinterval of J. If h operates in ReA, then $A = C(X,\tau)$.

Proof. Choose $a < b$ so that $[a,b] \subset J$. Let

$$D = \{u \in \mathrm{Re}A : a \le u \le b\},$$

and for each $n \in \mathbf{N}$ let

$$D_n = \{u \in D : \|h \circ u\|_q < n\}.$$

Here, $\|v\|_q \equiv \inf\{\|f\|_X : f \in A,\ \mathrm{Re}f = v\}$ for $v \in \mathrm{Re}A$ is the quotient norm on ReA. Since D is closed in ReA, D is complete. Since $D = \bigcup_{n=1}^{\infty} D_n$, the closure in Re$A$ of some D_n has a non-empty interior in D by the Baire Category theorem. Thus there are $u^0 \in D, \eta > 0$, and $r \in \mathbf{N}$ such that $D \cap D_r$ is dense in $U \cap D$, where $U = \{u \in \mathrm{Re}A : \|u - u^0\|_q < 3\eta\}$. If necessary, we may replace u^0 by $su^0 + t$ for appropriate numbers $s \in (0,1)$ and t, and shrink η somewhat, to arrange that $U \subset D$ and that

$$U \cap D_r \text{ is dense in } U. \tag{4.1}$$

Now, let V denote the uniform closure in $C_R(\tilde{X},\tilde{\tau})$ of $(\mathrm{Re}A)^{\sim} = \mathrm{Re}\tilde{A}$ (the equality from Lemma 3.1), and let

$$\begin{aligned} B &= \{\tilde{u} \in C_R(\tilde{X},\tilde{\tau}) : \tilde{u}\tilde{v} \in V \text{ for all } \tilde{v} \in \mathrm{Re}\tilde{A}\} \\ &= \{\tilde{u} \in C_R(\tilde{X},\tilde{\tau}) : \tilde{u}\tilde{v} \in V \text{ for all } \tilde{v} \in V\}. \end{aligned} \tag{4.2}$$

Then B is a closed subalgebra of $C_R(\tilde{X},\tilde{\tau})$ which contains the real constants and is contained in V. Let

$$W_\varepsilon = \{\tilde{u} \in \mathrm{Re}\tilde{A} : \mathrm{N}(\tilde{u}\{u^0\}_{n=1}^{\infty}) < \varepsilon\}, \tag{4.3}$$

where $\{u^0\}_{n=1}^{\infty}$ denotes the constant sequence $\{u^0, u^0, \cdots\}$ in Re$\tilde{A}$. Here,

$$\mathrm{N}(\{u_n\}_{n=1}^{\infty}) \equiv \sup\{\|u_n\|_q : n \in \mathbf{N}\}$$

for $\{u_n\}_{n=1}^{\infty} \in \mathrm{Re}\tilde{A}$ as defined in Section 3.

<u>Claim 1.</u> $h \circ \tilde{u} \in V$ for every $\tilde{u} \in W_{3\eta}$:
To show this, let $\tilde{v} = \{v_n\}_{n=1}^{\infty} \in \mathrm{Re}\tilde{A}$ be so that $\mathrm{N}(\tilde{v}) < 3\eta$. Since $u^0 + v_n \in U$, from (4.1) for

each $(n,k) \in \mathbf{N} \times \mathbf{N}$, we can choose $u_{nk} \in U \cap D_r$ such that $\|u_{nk} - (u^0 + v_n)\|_q < 1/k$. For each $k \in \mathbf{N}$, let $\tilde{u}_k = \{u_{nk}\}_{n=1}^{\infty} \in \mathrm{Re}\tilde{A}$. Since $u_{nk} \in D_r$, $\|h \circ u_{nk}\|_q < r$, so for each $k \in \mathbf{N}$,

$$h \circ \tilde{u}_k = \{h \circ u_{nk}\}_{n=1}^{\infty} \in \mathrm{Re}\tilde{A}.$$

As $k \to \infty$, $h \circ \tilde{u}_k$ converges uniformly on $\tilde{X}$ to $h \circ (\{u^0\}_{n=1}^{\infty} + \tilde{v})$, hence $h \circ (\{u^0\}_{n=1}^{\infty} + \tilde{v}) \in V$. Thus for $\tilde{u} \in W_{3\eta}$, by taking $\tilde{v} = \tilde{u} - \{u^0\}_{n=1}^{\infty}$ we have $h \circ \tilde{u} \in V$.

For $0 < \delta < \eta$, let λ_δ be a non-negative continuously differentiable function on $\mathbf{R}$ supported in (δ, δ) and satisfying $\int_\delta^\delta \lambda_\delta(t)dt = 1$. Let ϕ_δ denote the convolution of h and λ_δ:

$$\phi_\delta(x) = \int_{-\delta}^{\delta} h(xt)\lambda_\delta(t)dt = \int_a^b h(t)\lambda_\delta(xt)dt.$$

Then ϕ_δ is continuously differentiable on a neighborhood of $[a+\eta, b-\eta]$, and as $\delta \to 0$, ϕ_δ converges to h uniformly on $[a+\eta, b-\eta]$. (For these facts, refer to [2, Lemmas 4.19 and 4.20].)

Claim 2. $\phi_\delta{}' \circ \tilde{u} \in B$ whenever $\tilde{u} \in W_{2\eta}$:
To see this, let $\tilde{u} \in W_{2\eta}$. If $t \in [-\delta, \delta]$, then $\tilde{u} - t \in W_{3\eta}$ by (4.3) and so $h \circ (\tilde{u} - t) \in V$ by Claim 1. Hence $\phi_\delta \circ \tilde{u} \in V$ since $\phi_\delta \circ \tilde{u} = \int (h \circ (\tilde{u} - t))\lambda_\delta(t)dt$ and V is uniformly closed. Now if $\tilde{v} \in \mathrm{Re}\tilde{A}$, then for small non-zero t, we have $\tilde{u} + t\tilde{v} \in W_{2\eta}$, hence $\frac{1}{t}\{\phi_\delta \circ (\tilde{u} + t\tilde{v}) - \phi_\delta \circ \tilde{u}\} \in V$ and, letting $t \to 0$, $(\phi_\delta{}' \circ \tilde{u})\tilde{v} \in V$. Therefore, $\phi_\delta{}' \circ \tilde{u} \in B$ by (4.2). This proves Claim 2.

To prove that $A = C(X, \tau)$, it suffices to show that B separates weakly the points of $\tilde{X}$. If so, then $B = C_R(\tilde{X}, \tilde{\tau})$ by Proposition 2.4, and so defining $^-$ to be the uniform closure, we have

$$C_R(\tilde{X}, \tilde{\tau}) = B \subset V = \overline{(\mathrm{Re}A)^\sim} = \overline{\mathrm{Re}\tilde{A}} \subset C_R(\tilde{X}, \tilde{\tau}) = C_R(X, \tau)^\sim.$$

Hence $(\mathrm{Re}A)^\sim$ is dense in $C_R(X,\tau)^\sim$. By Bernard's lemma, $\mathrm{Re}A = C_R(X,\tau)$ and hence $A = C(X,\tau)$ by Proposition 1.1, and this proves the theorem.

The proof that B separates weakly the points of $\tilde{X}$ follows from the following two claims.

Claim 3. $\mathrm{Re}\tilde{A}$ separates weakly the points of $\tilde{X}$:
By Corollary 2.2, $\mathrm{Re}A$ separates weakly the points of X since A separates the points of X. And by the hypothesis, h operates in $\mathrm{Re}A$, hence $\mathrm{Re}A$ is uniformly dense in $C_R(X,\tau)$ by Lemma 4.1. Thus by Proposition 3.2, $\mathrm{Re}\tilde{A}$ separates weakly the points of $\tilde{X}$.

Claim 4. B separates weakly the points of $\tilde{X}$:
To show this, let $p, q \in \tilde{X}$ be so that $p \neq q, \tilde{\tau}(q)$. By Claim 3, $\mathrm{Re}\tilde{A}$ separates weakly the points of $\tilde{X}$, so we can choose $\tilde{v} \in W_\eta$ such that $\tilde{v}(p) \neq \tilde{v}(q)$. Choose $\varepsilon > 0$ so that $\varepsilon < \eta \mid \tilde{v}(p) - \tilde{v}(q) \mid /(2\mathrm{N}(\tilde{v})) \leq \eta$. And then choose numbers t_1, t_2, and t_3 in $(\tilde{v}(p) - \varepsilon, \tilde{v}(p) + \varepsilon)$ so that three points $(t_j, h(t_j))$ are not collinear. (This is possible because that interval is contained in J.) Then choose a positive $\delta < \eta$ small enough so that the points $(t_j, \phi_\delta(t_j))$ are close enough to the points $(t_j, h(t_j))$ to prevent them from being collinear. Thus $\phi_\delta{}'$ is not constant on $(\tilde{v}(p) - \varepsilon, \tilde{v}(p) + \varepsilon)$.

If $\phi_\delta'(\tilde{v}(p)) \neq \phi_\delta'(\tilde{v}(q))$, let $\tilde{u} = \tilde{v}$. If $\phi_\delta'(\tilde{v}(p)) = \phi_\delta'(\tilde{v}(q))$, choose $s \in (\tilde{v}(p) - \varepsilon, \tilde{v}(p) + \varepsilon)$ such that $\phi_\delta'(s) \neq \phi_\delta'(\tilde{v}(q))$, and let

$$\tilde{u} = \tilde{v} + (\tilde{v} - \tilde{v}(q))(s - \tilde{v}(p))/(\tilde{v}(p) - \tilde{v}(q)).$$

Then we have $\mathrm{N}(\tilde{u} - \tilde{v}) < \eta, \tilde{u}(q) = \tilde{v}(q)$ and $\tilde{u}(p) = s$. In either case, $\tilde{u} \in W_{2\eta}$, and $\phi_\delta'(\tilde{u}(p)) \neq \phi_\delta'(\tilde{u}(q))$, that is, $\phi_\delta' \circ \tilde{u} \in B$ (by Claim 2) separates p and q. This completes the proof.

Remark. The proof of the above theorem is a modification of the original proof of the complex case obtained by Sidney [11, Theorem 2]. The only difference occurs in Claim 3 of the above proof.

Question. Is an analogue of Hatori's Theorem [5] for real function algebras, given as follows, true ?

Let A be a real function algebra on (X, τ), and suppose that h is a non-affine continuous function on an interval I. If h operates in ReA, then $A = C(X, \tau)$.

Acknowledgement. Results of this paper are contained in the author's doctoral dissertation (The University of Connecticut, 1990) written under the direction of Professor Stuart J. Sidney. The author would like to appreciate his guidance.

References

1. A. Bernard, "Une caractérisation de $C(X)$ parmi les algèbres de Banach," *C.R. Acad. Sc., Paris Sér. A,* 267(1968), 634–635.

2. R. Burckel, *Characterizations of $C(X)$ among Its Subalgebras,* Marcel Dekker Inc., New York, 1972.

3. K. deLeeuw and Y. Katznelson, "Functions That Operates on Non-self-adjoint Algebras," *Jour. d'Anal. Math.,* 11(1963), 207–219.

4. L. Gillman and M. Jerison, *Rings of Continuous Functions,* Van Nostrand Company, 1960.

5. O. Hatori, "Functions Which Operate on the Real part of a Function Algebra," *Proc. Amer. Math. Soc.,* 83(1981), 565–568.

6. K. Hoffman and J. Wermer, "A Characterization of $C(X)$," *Pacific J. Math.,* 12(1962), 941–944.

7. L. Ingelstam, "Real Banach Algebra," Ark. Mat., 5(1964), 239–270.

8. S. Kulkarni and B. Limaye, "Gleason Parts of Real Function Algebras," *Canad. J. Math.,* 33(1981), 181–200.

9. S. Kulkarni and N. Srinivasan, "An Analogue of Hoffman-Wermer Theorem for a Real Function Algebra," *Indian J. Pure Appl. Math.,* 19(1988), 154–166.

10. R. Mehta and M. Vasavada, "Wermer's Type Result for a Real Banach Function Algebra," *Math. Today,* 4(1986), 43–46.

11. S. Sidney, "Functions Which Operate on the Real Part of a Uniform Algebra," *Pacific J. Math.,* 80(1979), 265–272.

12. J. Wermer, "The Space of Real Parts of a Function Algebra," *Pacific J. Math.,* 13(1963), 1423–1426.

Small Perturbations of Algebras of Analytic Functions on Polydiscs

KRZYSZTOF JAROSZ, Department of Mathematics and Statistics, Southern Illinois University at Edwardsville, Edwardsville, Illinois

§0. Introduction.

Let $A = A(D^n)$ be a polydisc algebra, that is, the algebra of all continuous functions on $\overline{D}^n \subset \mathbb{C}^n$, which are analytic on D^n. We study A from the point of view of the deformation theory of uniform algebras [KJ4]. We show that if the Banach-Mazur distance between A and a Banach algebra B is sufficiently small then B inherits a lot of properties of A; in particular, the spectrum of B has a structure of n dimensional complex analytic manifold.

We begin by recalling some results from [KJ4] and [RR7]. Let A be a uniform algebra. A deformation of A is a new normed algebra obtained by putting on the vector space A a new associative multiplication $\times$, which for some small positive ε satisfies

(1) $\quad \| f \times g - fg \| \leq \varepsilon \| f \| \, \| g \| \quad$ for all f, g in A.

If $\varepsilon < 1$ and this new algebra is renormed with its spectral norm, we then obtain a new uniform algebra, $A_\times$ [KJ4]. If Id is the identity map from A into $A_\times$ then

$$\|Id\| \ \|Id^{-1}\| \leq \frac{1+\varepsilon}{1-\varepsilon} \to 1 \text{ as } \varepsilon \to 0,$$

so the Banach-Mazur distance between A and $A_\times$ tends to 1 with $\varepsilon \to 0$.

By theorem 3 of [KJ4] the converse is also true. If the Banach-Mazur distance between A and a Banach algebra B is smaller than $1+\varepsilon'$, then $B=A_\times$ for some new multiplication $\times$ on A, which satisfies (1) with $\varepsilon \to 0$ as $\varepsilon' \to 0$.

Small deformations of uniform algebras were studied in [BJ1-2, KJ1-5, RR1-7] and other papers. The main question concerns stability of various properties of uniform algebras. We say that a property $\mathcal{P}$ is stable if for any uniform algebra A having the property $\mathcal{P}$ there is an $\varepsilon > 0$ such that any ε-deformation $A_\times$ of A has also this property. The following properties are stable.

1. A is Dirichlet [KJ4].
2. Choquet boundary of A is compact [KJ4].
3. $A = C(S)$, S a compact Hausdorff space [KJ4].
4. $A = A(D)$ [RR6].
5. $A = H^\infty(D)$, [KJ5].
6. A is an algebra of analytic functions of a finite bordered, possibly singular, Riemann surface [RR7].

The stability of the last property and related questions were studied by R. Rochberg in a series of papers [RR1—7] ([RR7] gives the most comprehensive exposition).

Our knowledge about deformations of algebras of analytic functions of one variable is still incomplete (especially for non-separable algebras), but already quite extensive. By contrast we know almost nothing about deformations of algebras of analytic functions of many variables. In this note we prove that a small deformation of a polydisc algebra $A(D^n)$ produces a uniform algebra, whose structure is quite similar to that of $A(D^n)$. The result should be seen only as a very small first step toward a comprehensive description of deformations of algebras of analytic functions of many variables.

§1. Notation.

We use standard Banach space terminology. For a compact space X, $C(X)$ denotes Banach algebra of all continuous, complex valued functions on X, with sup norm. A uniform algebra is a unital subalgebra of $C(X)$, which separates points of X. Further, $\mathfrak{M}(A)$, ∂A and ChA denote the maximal ideal space (= spectrum), Shilov boundary, and the Choquet boundary of A, respectively. We frequently identify A, via the Gelfand transform, with a subalgebra of $C\big(\mathfrak{M}(A)\big)$ or of $C(\partial A)$. Hence for $F \in \mathfrak{M}(A)$ and $f \in A$ we may write $F(f)$ as well as $f(F)$. A^{-1} is the set of all invertible elements of A.

For a compact subset K of X we put $A_{|K} = \{f_{|K}\colon f \in A \subset C(X)\}$. $A_{|K}$ is a subalgebra of $C(K)$. By a linear extension from $A_{|K}$ into A we mean a linear, continuous map $\Psi\colon A_{|K} \to A$ such that

$$\Psi(g)_{|K} = g \quad \text{for all } g \in A_{|K}, \quad \text{and} \quad \Psi(1) = 1.$$

K is called a peak set for A if there is a sequence of functions f_n in A such that

$$f_n = 1 \text{ on } K, \quad \| f_n \| = 1 \quad \text{for all } n \in \mathbb{N},$$

and $f_n \to 0$ uniformly on any compact subset of $X \backslash K$.

If K consists of a single point, this point is called a peak point. It is well-known that, for a separable uniform algebra A, ChA is equal to the set of all peak points for A. It is also well-known that the algebra $A_{|K}$ is complete if K is a peak set for A.

For $f_0 \in C(X)$, $f_0 \cdot A = \{f_0 \cdot f\colon\ f \in A \subset C(X)\}$. If Ω is an open and bounded subset of $\mathbb{C}^n$ then $A(\Omega) = \{ f \in C(\overline{\Omega})\colon\ f_{|\Omega}$ is holomorphic$\}$.

The unit disc in $\mathbb{C}$ we denote by D. If $f\colon U \to \mathbb{C}$, where $U \subseteq \mathbb{C}$, then ${}^1f\colon U \times \mathbb{C} \to \mathbb{C}$ $\left({}^2f\colon \mathbb{C} \times U \to \mathbb{C}\right)$ is defined by ${}^1f(z,w) = f(z)$ $\left({}^2f(z,w) = f(w)\right.$, respectively$\left.\right)$.

For $z \in D$, L_z is the corresponding Blaschke factor, that is

$$L_z(w) = \frac{z - w}{1 - \bar{z}w}, \qquad w \in \overline{D}.$$

Hence ${}^1L_z$ is the function defined on $\overline{D} \times \mathbb{C}$ by

$${}^1L_z(w_1, w_2) = \frac{z - w_1}{1 - \bar{z}w_1}, \qquad w_1 \in \overline{D}, \quad w_2 \in \mathbb{C}.$$

The Banach-Mazur distance $d_{B\text{-}M}$ between Banach spaces A and B is defined by

$$d_{B\text{-}M}(A,B) = \inf \{ \| T \| \ \| T^{-1} \| \colon \quad T\colon A \to B \ \text{ is an isomorphism}\}.$$

§2. The Results.

Theorem 1. *Let B be a complex function algebra such that $d_{B\text{-}M}\left(A(D^n),B\right) < 1+\varepsilon$ with $\varepsilon < \varepsilon_0$. Then $\mathfrak{M}(B) \cong \overline{D}^n$, so B can be seen as a subalgebra of $C(\overline{D}^n)$. There is a linear isomorphism $T\colon A(D^n) \to B$ such that*

$$(1) \quad \| Tf - f \| \ \le \ \varepsilon' \, \| f \| \qquad \text{for} \quad f \in A(D^n).$$

Moreover, D^n can be given a structure τ of an n-dimensional complex manifold such that all functions from B are τ-holomorphic.

Here $\varepsilon_0 > 0$ is an absolute constant and $\varepsilon' \to 0$ as $\varepsilon \to 0$.

Before we prove the theorem we discuss some general results concerning small perturbations of functions algebras.

Proposition 1. *Let A be a complex function algebra and B a complex Banach algebra. If there is a linear isomorphism $T\colon A \to B$ with $\|T\|\,\|T^{-1}\| < 1+\varepsilon$ and $T(1) = 1$ then B is a uniform algebra and*

$$\|g\| \geq \sigma_B(g) \geq (1-\varepsilon')\,\|g\| \qquad g \in B,$$

where $\varepsilon' \to 0$ as $\varepsilon \to 0$.

Proof. We define a second multiplication $\times$ on the Banach space A by

$$f \times g = T^{-1}(Tf \cdot Tg) \qquad f, g \in A.$$

We have

$$\|f \times g\| \leq (1+\varepsilon)^3 \|f\|\,\|g\| \qquad f, g \in A.$$

Both multiplications on A have the same unit. Hence by Theorem 3.1 $\big((v)\Leftrightarrow(iii)\big)$ of [KJ4] there is a function algebra B_1 and a linear isomorphism $T_1\colon A \to B_1$ with $\|T_1\|\,\|T_1^{-1}\| \leq 1+\varepsilon_1$ $(\varepsilon_1 \to 0$ as $\varepsilon \to 0)$ and such that

$$T^{-1}(Tf \cdot Tg) = f \times g = T_1^{-1}(Tf \cdot Tg) \qquad f, g \in A.$$

So $T_1 \circ T^{-1}\colon B \to B_1$ is an algebra isomorphism and

$$\sigma_B(g) = \sigma_{B_1}\big(T_1 \circ T^{-1}(g)\big) = \|T_1 \circ T^{-1}(g)\| \geq (1+\varepsilon)^{-1}(1+\varepsilon_1)^{-1}\|g\|$$

for any $g \in B$.

Proposition 2. *Let A be a function algebra, let $f_0 \in A \backslash A^{-1}$ be such that $|f_0| \equiv 1$ on ∂A and such that $f_0 A = \{f \in A\colon f_{|K} \equiv 0\}$, where $K = \{x \in \mathfrak{M}(A)\colon f_0(x) = 0\}$. Assume also that there is a linear, norm one extension $\Psi\colon A_{|K} \to A$. If T is a linear isomorphism from A onto a function algebra $B \subseteq C(\partial A)$ such that*

$$(2) \quad |Tf(x) - f(x)| \leq \varepsilon \|f\| \qquad f \in A,\ x \in \partial A,$$

with $\varepsilon \le \varepsilon_0$ *(absolute constant), then*

(*i*) $g_0 = Tf_0 \in B \backslash B^{-1}$,

(*ii*) $g_0 B = \{ g \in B\colon\ g_{|L} \equiv 0 \}$, *where* $L = \{ x \in \mathfrak{M}(B)\colon\ g_0(x) = 0 \}$,

and

(*iii*) $\Phi\colon A_{|K} \to B_{|L}$ *defined by* $\Phi(f) = T\big(\Psi(f)\big)_{|L}$ *is a surjective linear isomorphism with* $\|\Phi\|\ \|\Phi^{-1}\| \le 1+\varepsilon'$, *where* $\varepsilon' \to 0$ *as* $\varepsilon \to 0$.

Proof. The part (*i*) follows from Proposition 15.3 of [KJ4], we show (*ii*) and (*iii*).

Let $f_1, f_2 \in A$ be such that $f_0 f_1 = f_0 f_2$. Since $f_0 \neq 0$ on ∂A, it follows that $f_1 - f_2 \equiv 0$ on ∂A, so $f_1 = f_2$. Hence we can define a linear, surjective map $S\colon f_0 A \to g_0 B$ by $S(f_0 \circ f) = g_0 \circ Tf$. By (2) we have

(3) $$| Sf(x) - f(x) | \le 2\varepsilon \| f \| \qquad f \in f_0 A.$$

Hence $g_0 B$ is a closed ideal of B contained in the closed ideal $\{g \in B\colon g_{|L} \equiv 0\}$.

If these two ideals were not equal then $B/_{g_0 B}$ would not be a uniform algebra, since the spectral radius of any element of $\{g \in B\colon g_{|L} \equiv 0\}/_{g_0 B}$ would be equal to zero (note that any linear-multiplicative functional that anihilates $g_0 B$ anihilates also $\{g \in B\colon\ g_{|L} \equiv 0\}$). We will show that $B/_{g_0 B}$ is a uniform algebra which will prove (*ii*).

We define a map $\widetilde{S}\colon A \to B$ by

$$\widetilde{S}(f) = T\left(\Psi(f_{|K})\right) + S\left(f - \Psi(f_{|K})\right), \qquad f \in A.$$

By (2) and (3), for any $f \in A$ we have

$$\|\widetilde{S}(f) - f\|_{\partial A} = \|\widetilde{S}(f) - \Psi(f_{|K}) - \left(f - \Psi(f_{|K})\right)\|_{\partial A}$$

$$\leq \varepsilon \|\Psi(f_{|K})\| + 2\varepsilon \|f - \Psi(f_{|K})\| \leq 5\varepsilon \|f\|.$$

Hence, and by (2), $\widetilde{S}$ is a linear surjective isomorphism of A onto B with $\|\widetilde{S}\| \, \|\widetilde{S}^{-1}\| \leq (1+5\varepsilon)/(1-5\varepsilon) = 1+\varepsilon_1$. On the subspace $f_0A = \{f \in A\colon f_{|K} \equiv 0\}$ of A, $\widetilde{S}$ coincides with S and maps f_0A onto g_0B. Hence the map $\Phi\colon A/_{f_0A} \to B/_{g_0B}$ defined by $\Phi(f+f_0A) = \widetilde{S}(f) + g_0B$ is a linear surjective isomorphism of the Banach algebra $A/_{f_0A}$ onto $B/_{g_0B}$ such that $\|\Phi\| \, \|\Phi^{-1}\| \leq 1+\varepsilon_1$ and $\Phi(\mathit{1}) = \mathit{1}$. The map on $A_{|K}$ defined by $f \mapsto \Psi(f) + f_0A \in A/_{f_0A}$ is an isometry and algebra isomorphism of a function algebra $A_{|K}$ onto $A/_{f_0A}$. Hence, by Proposition 1, $B/_{g_0B}$ is a uniform algebra and the spectral radius and the norm almost coincide in $B/_{g_0B}$. On the other hand the maximal ideal space of $B/_{g_0B}$ can be identified with L and the spectral norm is given by the supremum over the maximal ideal space so $B/_{g_0B}$ and $B_{|L}$ are almost isometric. This gives (iii).

Proposition 3. *Let B be a complex function algebra such that $d_{B\text{-}M}\left(A(D^n), B\right) < 1+\varepsilon$ with $\varepsilon < \varepsilon_0$. If $K \subseteq \partial A(D^n) = \partial B$ is a peak set for the algebra $A(D^n)$ then K is also a peak set for the algebra B and*

$$d_{B\text{-}M}(A_{|K}, B_{|K}) \leq 1+\varepsilon_1 \longrightarrow 1 \quad \text{as} \quad \varepsilon \to 0.$$

The above proposition follows immediately from Theorem 16.7 of [KJ4].

§3. Proof of Theorem 1.

To simplify the notation we prove the result for $n = 2$. It will be quite transparent how to write the general case. Put $X = \partial D \times \partial D = \partial A(D^2)$. By Theorem 3.1 of [KJ4] $\partial B \cong X$ so B can be seen as a subalgebra of $C(X)$. There is also a linear isomorphism $T\colon A(D^2) \to B$ such that

(4) $$| Tf(z,w) - f(z,w) | \le \varepsilon' \| f \| \qquad f \in A(D^2),\ (z,w) \in X.$$

To prove the result we have to show that $\mathfrak{M}(B) \cong \overline{D}^2$, that (4) holds for all $(z,w) \in \overline{D}^2$ and that D^2 can be given a structure τ of 2-dimensional complex manifold such that all functions from B are τ-analytic.

We construct a homeomorphism φ from $\overline{D}^2$ onto $\mathfrak{M}(B)$ in several steps. $\overline{D}^2$ is a union of three disjoint sets: X, $\partial D \times D \cup D \times \partial D = \partial(D^2) \setminus X$, and D^2.

Step 1. *Definition of* φ *on* X.

We put $\varphi(x) = x$ *for* $x \in X$.

Step 2. *Definition of* φ *on* $\partial(D^2) \setminus X$.

We need the following lemma which is a combination of lemmas 3.1, 3.2 and 3.3 of [RR6].

Lemma 1. *Let* S *be a linear isomorphism from* $A(D)$ *onto a uniform algebra* $B_1 \subseteq C(\partial D)$ *such that*

$$|Sf(w) - f(w)| \leq \varepsilon \|f\| \qquad f \in A(D),\ w \in \partial D.$$

For any $z \in D$ *we put* $\psi(z) = \{g \cdot S(L_z) \in B_1 : g \in B_1\}$ *and for* $z \in \partial D$ *we put* $\psi(z) = z \in \mathfrak{M}(B_1)$. *Then for any* $z \in \overline{D}$ *we have*

(i) $\psi(z) \in \mathfrak{M}(B_1)$,

(ii) ψ *is a homeomorphism of* $\overline{D}$ *onto* $\mathfrak{M}(B_1)$, *and*

(iii) $|Sf(\psi(z)) - f(z)| \leq 2\varepsilon \|f\|, \qquad f \in A(D),\ z \in \overline{D}.$

To define φ on $\overline{D} \times \partial D$ let $w_0 \in D$ and put $K_{w_0} = \overline{D} \times \{w_0\}$. K_{w_0} is a peak set for $A(D^2)$ and $A(D^2)_{|K_{w_0}}$ is isometrically isomorphic with the disc algebra. The maximal ideal space of $A(D^2)_{|K_{w_0}}$ can be identified with K_{w_0}. By Proposition 3, Theorem 3.1 of [KJ4], and Lemma 1 we get a homeomorphism $\varphi(\cdot, w_0)$ from K_{w_0} onto $\mathfrak{M}(B_{|K_{w_0}}) \subseteq \mathfrak{M}(B)$. For $z \in D$, $\varphi(z, w_0)$ is the maximal ideal of B consisting of all functions $g \in B \subseteq C\big(\mathfrak{M}(B)\big)$ such that $g_{|\varphi(K_{w_0})}$ is divisible by $T({}^1L_z)_{|\varphi(K_{w_0})}$. Recall that according with the notation described in the previous section ${}^1L_z \colon \overline{D}^2 \to \mathbb{C}$ is defined by ${}^1L_z(\alpha,\beta) = L_z(\alpha)$, where L_z is the Blaschke factor.

In the same way we define φ on $\partial D \times \overline{D}$.

Step 3. *Definition of* φ *on* D^2.

Let $(z_0, w_0) \in D^2$. We put

$$\varphi(z_0,w_0) = \left\{T({}^1L_{z_0})g_1 + T({}^2L_{w_0})g_2 : \quad g_1, g_2 \in B\right\}.$$

We have to show that $\varphi(z_0,w_0)$ is a proper maximal ideal of B. We first show that $\varphi(z_0,w_0)$ is a proper ideal. To this end assume that there are $g_1, g_2 \in B$ such that

$$T({}^1L_{z_0})g_1 + T({}^2L_{w_0})g_2 \equiv 1. \tag{5}$$

Put $L = \left\{x \in \mathfrak{M}(B) : \quad T({}^1L_{z_0})(x) = 0\right\}$. By (5) we have

$$T({}^2L_{w_0}) \cdot g_2 \equiv 1 \qquad \text{on} \quad L. \tag{6}$$

Put $f_0 = {}^1L_{z_0}$ and $K = \left\{x \in \overline{D}^2 : \ {}^1L_{z_0}(x) = 0\right\} = \{z_0\} \times \overline{D}$. By Proposition 2, there is an almost isometry Φ from $A(D) \cong A(D^2)_{|K}$ onto $B_{|L}$ given by $\Phi(f) = T({}^2f)_{|L}$. By [RR6] $\Phi(L_{w_0}) = T({}^2L_{w_0})_{|L}$ is a noninvertible element of $B_{|L}$, which contradicts (6) and proves that $\varphi(z_0,w_0)$ is a proper ideal.

We show that $\varphi(z_0,w_0)$ is a maximal ideal. Set $S_1: A(D^2) \to A(D)$, $S_1(f) = f(\cdot, w_0)$; $S_2: A(D^2) \to A(D^2)$, $S_2(f) = (f - S_1(f))/{}^2L_{w_0}$. Any $f \in A$ can be decomposed as follows

$$f = f(z_0,w_0) + {}^1L_{z_0} \cdot \frac{S_1(f) - f(z_0,w_0)}{L_{z_0}} + {}^2L_{w_0} \cdot S_2(f).$$

So we can define a linear map $\Psi_{z_0,w_0}: A(D^2) \to B$ by

$$\Psi_{z_0,w_0}(f) = f(z_0,w_0) + T({}^1L_{z_0}) \cdot T\left(\frac{S_1(f) - f(z_0,w_0)}{L_{z_0}}\right) + T({}^2L_{w_0}) \cdot T(S_2(f)).$$

By (4) Ψ is close to the identity map so it is a surjective isomorphism. Hence

$B_{z_0,w_0} = \Psi_{z_0,w_0}\big(\{f \in A(D^2)\colon f(0,0)=0\}\big)$ is a closed subspace of B of dimension one. We also have $B_{z_0,w_0} \subseteq \varphi(z_0,w_0) \not\subseteq B$. Hence $B_{z_0,w_0} = \varphi(z_0,w_0)$ is a proper maximal ideal of B.

Step 4. *φ is continuous on D^2.*

A maximal ideal of a function algebra B can be identified with a linear-multiplicative functional on B or with an element from the domain of a function $g \in B$. From the definitions of φ and Ψ_{z_0,w_0} we get

$$g(\varphi(z_0,w_0)) = (\Psi_{z_0,w_0})^{-1}(g)(z_0,w_0). \tag{7}$$

Hence, for any $g \in B$, with $\|g\| \leq 1$ we have

$$|\, g(\varphi(z_0,w_0)) - g(\varphi(z_1,w_1))\,| \leq d_H\big((z_0,w_0),\,(z_1,w_1)\big)\,\|(\Psi_{z_0,w_0})^{-1} - (\Psi_{z_1,w_1})^{-1}\|$$

where $d_H(\cdot,\cdot)$ is the hyperbolic distance on D^2. Hence $\varphi_{|D^2}\colon D^2 \to \mathfrak{M}(B)$ is a continuous map if $\mathfrak{M}(B)$ is equipped with the norm topology, it is more so continuous if we take $\mathfrak{M}(B)$ with its original weak $*$ topology.

Step 5. *φ is continuous.*

The following is an obvious topological observation.

Let X be a compact metric space, Y a topological space, G a dense subspace of X and ϕ a function from X into Y. If $\phi_{|G}$ is continuous and if for any $x_0 \in X\backslash G$ and any sequence x_n in G convergent to x_0 and such that $\phi(x_n)$ is convergent we have $\lim_{n\to\infty} \phi(x_n) = \phi(x_0)$, then ϕ is *continuous.*

To end the proof of the continuity of φ let $(z_0,w_0) \in \overline{D}^2 \setminus D^2$ and let (z_n,w_n) be a sequence in D^2 convergent to the point (z_0,w_0) and such that $\varphi(z_n,w_n)$ is convergent to $x_0 \in \mathfrak{M}(B)$. From (7), (4) and the definition of Ψ_{z_0,w_0} by a direct computation we get

(8) $$| Tf(\varphi(z_0,w_0)) - f(z_0,w_0) | \leq 10\varepsilon' \| f \|, \qquad f \in A(D^2), \ (z_0,w_0) \in D^2.$$

Hence

(9) $$| Tf(x_0) - f(z_0,w_0) | \leq 10\varepsilon' \| f \| \qquad f \in A(D^2).$$

Assume first that $(z_0,w_0) \in X = \partial A(D^2)$. From (10) and (4) we have

(10) $$| Tf(x_0) - Tf(z_0,w_0) | \leq \varepsilon' \| f \|, \qquad f \in A(D^2).$$

Since (z_0,w_0) is a peak point for $A(D^2)$ it is also a peak point for B ([KJ], Theorem 16.7) so the distance, in the norm topology, between the functional $(z_0,w_0) \in \mathfrak{M}(B)$ and any other functional x_0 from $\mathfrak{M}(B)$ is equal to two. Hence, by (10) we get $(z_0,w_0) = x_0$. Assume now that $(z_0,w_0) \in \partial D^2 \setminus X$, say $z_0 \in D$ and $w_0 \in \partial D$. Put as before $K_{w_0} = \overline{D} \times \{w_0\}$. We know that K_{w_0} is a peak set for $A(D^2)$ and that $\varphi(K_{w_0}) \subseteq \mathfrak{M}(B)$ is a peak set for B. Assume that $x_0 \in \varphi(K_{w_0})$. Let μ be a probabilistic measure on $X = \partial B$ which represents the functional x_0, this is such that $\int_X g d\mu = g(x_0)$ for $g \in B$. Since $\varphi(K_{w_0})$ is a peak set not containing x_0, μ is concentrated outside of $K_{w_0} \cap X = \partial D \times \{w_0\}$, so $\mu(K_{w_0}) = 0$. Let f_n be a sequence of norm one elements of $A(D^2)$ such that $f_n \equiv 1$ on K_{w_0} and $f_n \to 0$ uniformly on any open subset of $\overline{D}^2 \setminus K_{w_0}$. By (4) we get

$$| Tf_n(x_0) | = | \int_X f_n d\mu | \to \varepsilon' \qquad \text{as} \qquad n \to \infty.$$

On the other hand, from (10), for any $n \in \mathbb{N}$ we get

$$|\, Tf_n(x_n) - 1\,| \;=\; |\; Tf_n\,(z_0,w_0)\;| \;\leq\; 10\varepsilon'.$$

Hence $x_0 \in \varphi(K_{w_0})$, $x_0 = \varphi(\alpha,w_0)$. We need to show that $\alpha = z_0$. By the definitions of $\Psi_{z,w}$ and φ on D^2, for any $n \in \mathbb{N}$ we have

$$T\,({}^1L_{z_n})\,(\,\varphi(z_n,w_n)) = 0.$$

The sequence $T({}^1L_{z_n})$ is convergent, in norm, to ${}^1L_{z_0}$ and $\varphi(z_n,w_n)$ is weak $*$ convergent to $\varphi(\alpha, w_0)$, hence

$$T({}^1L_{z_0})(\,\varphi(\alpha, w_0)) = 0.$$

By the definition of φ on $K_{w_0} = \overline{D} \times \{w_0\}$, $\varphi(z_0,w_0)$ is the only point of $\varphi(K_{w_0})$ where $T({}^1L_{z_0})$ is equal to zero. Hence $\varphi(z_0,w_0) = \varphi(\alpha, w_0)$. The map φ is a homeomorphism of K_{w_0} onto $\varphi(K_{w_0})$ so $z_0 = \alpha$.

Step 6. *φ is one to one.*

By the definition of φ and Theorem 16.7 of [KJ4] φ is one to one on ∂D^2 and $\varphi(\partial D^2) \cap \varphi(D^2) = \emptyset$. We have to show that φ is one to one on D^2. Assume $(z_i,\ w_i) \in D^2$, $i = 1,2$ are such that $\varphi(z_1,w_1) = \varphi(z_2,\ w_2)$. Expending ${}^1L_{z_2}$ in powers of ${}^1L_{z_1}$ we get

$${}^1L_{z_2} \;=\; L_{z_2}(z_1) \;+\; (1 - |\,L_{z_2}(z_1)\,|^2) \cdot {}^1L_{z_1} \;+ \overline{L_{z_2}\,(z_1)} \cdot ({}^1L_{z_1})^2 \cdot f\,,$$

where f $\in A(D^2)$, $\|\ f\ \| \leq 2$.

Applying the operator T to the above equation, then evaluating at the point

$\varphi(z_1,w_1) = \varphi(z_2,w_2) \in \mathfrak{M}(B)$ yields

$$0 \;=\; L_{z_2}(z_1) \;+\; \overline{L_{z_2}(z_1)} \;\cdot\; T\Big(({}^1L_{z_1})^2 \cdot f\Big)\Big(\varphi(z_1,w_1)\Big).$$

By (9) we have $|\ T\Big(({}^1L_{z_1})^2 \cdot f\Big)\ | \;\leq\; 10\varepsilon' \|\ f\ \|$ so finally we get

$$|\ L_{z_2}(z_1)\ | \;\;\leq\;\; |\ L_{z_2}(z_1)\ |\ 10\varepsilon'.$$

If $10\varepsilon' < 1$, this proves that $z_1 = z_2$; the same argument shows that $w_1 =$

w_2.

Step 7. *For any* $(z,w) \in D^2$ *there is a neighborhood* U of $\varphi(z_0,w_0)$ in $\mathfrak{M}(B)$ *and a homeomorphism* τ of U onto D^2 *such that* $g \circ \tau^{-1}$ *is analytic for any* $g \in B$.

Note that after we prove that φ is surjective this will give us the desired analytic structure on $D^2 \cong \varphi(D^2)$.

It follows from the definition of $\varphi(z_0,w_0)$, for $(z_0,w_0) \in D^2$ that for any $g \in B$ with $g(\varphi(z_0,w_0)) = 0$ there are $g_1, g_2 \in B$ such that $\| g_i \| \leq 3 \| g \|$, $i = 1,2$ and

$$g = T({}^1L_{z_0}) \cdot g_1 + T({}^2L_{w_0}) \cdot g_2 . \tag{11}$$

By the Gleason Embedding Theorem ([TG], p.154) there is a neighborhood U of $\varphi(z_0,w_0)$ in $\mathfrak{M}(B)$ and a homeomorphism τ of U onto an analytic variety $V \subseteq \mathbb{C}^2$. Since φ defines a continuous embedding of D^2 into $\mathfrak{M}(B)$, the topological dimension of U, and so of V is at least 4, hence V is an open subset of $\mathbb{C}^2$. The homeomorphism τ given by the Gleason Theorem is such that $g \circ \tau^{-1}$ is analytic for any $g \in B$.

Step 8. *Let* $F \in \mathfrak{M}(B)$ *and* $(z_0,w_0) \in D^2$. *Assume that* $\| F - \varphi(z_0,w_0) \| < \frac{1}{27}$, *where* F and $\varphi(z_0,w_0)$ *are considered to be functionals on* B. *Then* $F \in \varphi(D^2)$.

The above statement will follow from Step 7 and from the proof of the Gleason Embedding Theorem ([TG], p.154-155). Since $\| g_i \| \leq 3 \| g \|$, i = 1,2 in (11), we get

$$U = \left\{ x \in \mathfrak{M}(B): \ | T({}^1L_{z_0})(x) | < \tfrac{1}{24}, \quad | T({}^2L_{w_0})(x) | < \tfrac{1}{24} \right\},$$

$$V = D^2(\tfrac{1}{24}) \qquad \text{and}$$

$$\tau: U \to V \qquad \text{is given by} \qquad \tau(x) = \left(T({}^1L_{z_0})(x),\ T({}^2L_{w_0})(x) \right). \tag{12}$$

Put $G = \left\{(z,w) \in D^2\colon\ max\ \{|\ L_{z_0}(z)\ |,\ |\ L_{w_0}(w)\ |\} < \frac{1}{25}\right\}$. G is an open subset of $\mathbb{C}^2$, diffeomorphic with D^2. By (8), if ε' is small enough, $\varphi(G) \subseteq U$. Hence $\tau \circ \varphi\colon G \to V$ is a continuous, injective map from G into D^2. We define $\kappa\colon D^2\left(\frac{1}{25}\right) \to \mathbb{C}^2$ by

$$\kappa(z,w) = \tau \circ \varphi\left(L_{z_0}^{-1}(z),\ L_{w_0}^{-1}(w)\right).$$

The map κ is continuous, one to one and by (12) and (9) we have

$$d_E(\ \kappa(z,w) - (z,w)\ ) \ \leq\ 10\sqrt{2}\,\varepsilon', \qquad (z,w) \in V,$$

where $d_E(\cdot\,,\ \cdot)$ is the Euclidian metric on $\mathbb{C}^2$. Hence we get

$$D^2(\tfrac{1}{26}) \subseteq \kappa\left(D^2(\tfrac{1}{25})\right) \subseteq D^2(\tfrac{1}{24}).$$

This shows that

$$\varphi(D^2) \supseteq \left\{x \in \mathfrak{M}(B)\colon\ |\ T(^1L_{z_0})(x)\ | < \tfrac{1}{26},\ |\ T(^2L_{w_0})(x)\ | < \tfrac{1}{26}\right\},$$

and since the norms of $T(^1L_{z_0})$ and $T(^2L_{w_0})$ are not smaller than $1\text{-}\varepsilon'$ this ends the proof of this Step.

Step 9. *φ maps $\overline{D}^2$ onto $\mathfrak{M}(B)$.*

We need the following Theorem of B. Johnson [BJ2].

Theorem. *Let F be linear functional on $A(D^2)$ such that*

$$|\ F(f \cdot g) - F(f) \cdot F(g)\ | \ \leq\ \varepsilon\,\|f\|\,\|g\| \qquad f, g \in A(D^2),$$

where $\varepsilon < \varepsilon_0$ (positive absolute constant). Then there is $(z,w) \in \overline{D}^2$ such that

$$| F(f) - f(z,w) | \leq \varepsilon' \| f \|, \qquad f \in A(D^2),$$

where $\varepsilon' \to 0$ *as* $\varepsilon \to 0$.

Let $F \in \mathfrak{M}(B)$. Put $\widetilde{F} = F \circ T$. By (4) $\widetilde{F}$ is almost multiplicative on $A(D^2)$ so by Theorem of Johnson there is $(z_0, w_0) \in \overline{D}^2$ such that

$$| \widetilde{F}(f) - f(z_0, w_0) | \leq \varepsilon_1 \| f \|, \qquad f \in A(D^2).$$

By (8) we have

$$(13) \quad | Tf(\varphi(z_0, w_0)) - g(\varphi(z_0, w_0)) | \leq \varepsilon_2 \| g \|, \qquad g \in B.$$

If $(z_0, w_0) \in D^2$ then $F \in \varphi(D^2)$ by the previous step so assume $(z_0, w_0) \in \partial(D^2)$. Without loss of generality we may assume that $w_0 \in \partial D$. As before $K_{w_0} = \overline{D} \times \{w_0\}$ is a peak set for $A(D^2)$ and $A(D^2)_{| K_{w_0}}$ can be identified with the disc algebra. By the definition of φ on K_{w_0}, and since $A(D)$ is stable [RR6], $\varphi(K_{w_0})$ is the maximal ideal space of $B_{| \varphi(K_w)}$. By Theorem 16.1 of [KJ4] $\varphi(K_{w_0})$ is a peak set for B so

$$\| F_1 - \varphi(z,w) \| = 2$$

for any $(z,w) \in K_{w_0}$ and functional $F_1 \in \mathfrak{M}(B) \backslash \varphi(K_{w_0})$. By (13) this gives $F \in \varphi(K_{w_0})$.

§4. Open Problems.

Problem 1. Let B be as in Theorem 1. Is (D^n, τ) holomorphically equivalent to an open subset Ω of $\mathbb{C}^n$? Or, equivalently, is B isometric with $A(\Omega)$?

Problem 2. Let B be as in Theorem 1. Is there a continuous (real analytic) family of deformations from $A(D^n)$ to B. That is, does it exist a continuous (real analytic) map $\alpha \mapsto T_\alpha$ from the unit segment into $\mathcal{L}(A(D^n), C(\overline{D}^n))$, such that $Im(T_\alpha)$ is a uniform algebra for $\alpha \in [0,1]$?

Problem 3. Extend the result of Theorem 1 to other domains in $\mathbb{C}^n$.

Problem 4. Can Theorem 1 be extended to cover non-separable algebra $H^\infty(\Omega)$ of analytic, bounded functions on Ω?

References

[A&B] E. Artin and H. Braun, *Introduction to Algebraic Topology*, Charles E. Merrill, Inc., 1969.

[BJ1] B. Johnson, "Perturbations of Banach algebras", Proc. London Math. Soc. **35** (1977), 439-458.

[BJ2] B. E. Johnson, "Approximately multiplicative functionals", J. London Math. Soc. **35** (1986), 489-510.

[KJ1] K. Jarosz, "Perturbations of uniform algebras", Bull. London Math. Soc., **15** (1983), 133-138.

[KJ2] K. Jarosz, "Metric and algebraic perturbations of function algebras", Edinburgh Math. Soc. **26** (1983), 383-391.

[KJ3] K. Jarosz, "Perturbations of uniform algebras, II", J. London Math. Soc. **31** (1985), 555-560.

[KJ4] K. Jarosz, *Perturbations of Banach Algebras*, Springer-Verlag, Lecture Notes in Mathematics **1120**, 1985.

[KJ5] K. Jarosz, "$H^\infty(D)$ is stable", J. London Math. Soc. **37** (1988), 490-498.

[LA] L. Ahlfors, *Lectures on Quasiconformal Mappings*, Van Nostrand, New York, 1966.

[RR1] R. Rochberg, "Almost isometries of Banach Spaces and moduli of planar domains", Pacific J. Math., **49** (1973), 445-466.

[RR2] R. Rochberg, "Almost isometries of Banach spaces and moduli of Riemann surfaces", Duke Math . J., **40** (1973), 41-52.

[RR3] R. Rochberg, "Almost isometries of Banach spaces and moduli, II", Duke Math. J., **42** (1975), 167-182.

[RR4] R. Rochberg, "The Banach-Mazur Distance Between Function Algebras on Degenerating Riemann Surfaces," Lecture Notes in Math. **604** (1977), 82-94.

[RR5] R. Rochberg, "Deformation of uniform algebras", Proc. Amer. Math. Soc. **39** (1979), 93-118.

[RR6] R. Rochberg, "The disc algebra is rigid", Proc. Amer. Math. Soc. **39** (1979), 119-129.

[RR7] R. Rochberg, "Deformation of uniform algebras on Riemann surfaces", Pacific J. Math. **121** (1986), 135-181.

[TG] T. Gamelin, *Uniform Algebras*, Prentice-Hall, Inc., 1969.

Isometries and Small Bound Isomorphisms of Function Spaces

Krzysztof Jarosz, Department of Mathematics and Statistics, Southern Illinois University, Edwardsville, Il 62026, USA;

Vijay D. Pathak, Department of Applied Mathematics, M.S. University of Baroda, Baroda 390001, India.

1. Introduction.

The aim of this paper is to present some results and certain open problems concerning isometries and small-bound isomorphisms of function spaces. Within the frame of this note we can not give a comprehensive survey of this subject, we rather focus on topics of our own, personal interests. We discuss mainly the following two questions.

1. Assuming that two function spaces are linearly isometric can we say that the underlying domains over which the functions are defined are topologically homeomorphic and that the isometry can be described in terms of this homeomorphism?

2. A similar question can be asked regarding underlying domains when the function spaces are merely isomorphic under an isomorphism with a small bound.

The main source of all the results discussed in this article is the classical Banach-Stone theorem. It was proved in 1932 by S. Banach for isometries of spaces of real valued continuous functions and since then extended in various directions. The development of the theory accelerated during the last two decades due to the pioneering works of M. Cambern, W. Holsztyński and Amir in 1965 and 1966.

Almost at the same time isometries of other classical Banach spaces, including L^p, l_p, H^p, Orlicz spaces, etc., began to be investigated. Again, the starting point was a result concerning isometries of L^p spaces due to S. Banach. We did not include these results in our discussion. However, interested readers can be referred to [B1], [B2], [B11], [C6], [C7], [C12], [C23], [C24], [C25], [C28], [D2], [E2], [F1], [F2], [G4], [G5], [G7], [G8], [J1], [J9], [K3], [L1], [L4], [L23], [R3], [S2], [T1].

The authors would like to express their sincere thanks to Professor Vasavada for a fruitful discussion during preparation of this article.

2. Definitions and Notations.

We denote the unit circle and the disk by $\Gamma = \{z \in \mathbb{C} : |z| = 1\}$ and $\mathbb{D} = \{z \in \mathbb{C}: |z| \leq 1\}$, respectively. We consider real or complex Banach spaces E, F, ... We let $B(E)$ to be the closed unit ball of the Banach space E, and $ext(E)$ the set of all extreme points of $B(E)$.

By an isometry from E onto F we mean a norm preserving linear isomorphism from E onto F. We denote by $I(E, F)$ the set of all isometries from E onto F; when $E = F$ we write it shortly as $I(E)$. We define the *Banach-Mazur distance* between E and F by

$$d_{B-M}(E, F) = inf\left\{ ||T||\,||T^{-1}|| : \quad T: E \to F \text{ is an isomorphism}\right\}.$$

For Banach spaces E and F, $L(E,F)$ ($K(E,F)$) denotes the Banach space of all continuous (compact) linear maps from E into F. If $E = F$ then we write $L(E)$ or $K(E)$, respectively.

If X is locally compact Hausdorff space then $C_0(X, E)$ ($C_0(X)$, $C_0^R(X)$) denotes the space of all E-valued ($\mathbb{C}$-valued, $\mathbb{R}$-valued) continuous functions on X vanishing at infinity provided with the supremum norm. We delete the subscript 0 if X is compact. For $x \in X$, δ_x is the probability measure with the mass concentrated at x, we identify it with an element of the dual space $C_0(X)^*$. An isometry T from $C_0(X)$ into $C_0(Y)$ is said to be isotonic if for all $f, g \in C(X)$ we have $f \geq g$ if and only if $Tf \geq Tg$.

By c and c_0 we mean the space of all convergent sequences and the space of all sequences convergent to zero with the usual sup norm.

3. Banach-Stone Theorem.

A fundamental result in the isometry theory of function spaces is the classical Banach-Stone theorem. The theorem was proved in 1932 by Banach [B1] for isometries from $C^R(X)$ onto $C^R(Y)$ under the assumption that X and Y are metrizable. In 1937 Stone [S3] removed the metrizability condition and in 1947 Arens and Kelley [A3] extended the result to the spaces of complex valued functions. The proof was subsequently simplified by Eilenberg [E1], Behrends [B4] and others. The theorem stated in its present form is as follows.

Theorem 3.1. Let T be an isometry from $C_0(X)$ onto $C_0(Y)$. Then there is a homeomorphism $\tau : Y \to X$ and unimodular continuous map $u : Y \to \mathbb{C}$ such that

$$(1) \qquad (Tf)(y) = u(y)\, f(\tau(y)) \qquad f \in C_0(X), \quad y \in Y.$$

A number of proofs of this theorem are available in the literature. The main idea in all these proofs is to show that there exists a family of objects associated with $C_0(X)$ which are invariant under isometries and which can be indexed by X and then to prove that this correspondence also determines the topology of X. Extreme linear functionals, *T-sets* and *M ideals* are some of the objects associated with $C_0(X)$ having the desired properties. Using these objects three different proofs of the theorem are given in [B4].

The Banach-Stone theorem has been generalized in various directions, either by replacing the space $C_0(X)$ by its subspaces, with the uniform norm or with a Banach function norm, or by a space of vector valued continuous functions, as well as by taking T to be an isomorphism with small bound rather than an isometry.

4. Isometries between subspaces of spaces of continuous functions.

4.1. Injective isometries.

The Banach-Stone theorem asserts that every isometry of $C(X)$ **onto** $C(Y)$ is canonical, that is of the form (1). Here we discuss whether this can be extended to isometries from $C(X)$ **into** $C(Y)$. First we have to define what we mean by a canonical injective isometry. This means, roughly speaking, what is the easiest way to define an isometry from $C(X)$ into $C(Y)$. To this end let Y_0 be a closed subset of Y and let ϕ be a continuous map from Y_0 onto X. We define $T_1: C(X) \to C(Y_0)$ by $T_1(f) = f \circ \phi$. If Y is metrizable then there exists a norm one linear extension E: $C(Y_0) \to C(Y)$ ([S1], p. 365), and so $E \circ T_1$ gives an isometry from $C(X)$ into $C(Y)$. Thus we say that an isometry T from $C(X)$ into $C(Y)$ is canonical if there exists a subset Y_0 of Y and a continuous map ϕ from Y_0 onto X such that

$$(Tf)\,(y) \;=\; u(y)\,f(\phi(y)), \qquad\qquad f \in C(X),\; y \in Y_0$$

where u is a unimodular continuous function on Y_0.

Injective isometries of function spaces have been investigated since 1960 when Geba and Semadeni [G1] proved that any isotonic isometry from $C(X)$ into $C(Y)$ is canonical. In 1965 Holsztyński [H1] generalised this result by showing that any isometry T from $C(X)$ into $C(Y)$ is canonical. In 1975 Novinger [N2] extended the Holsztyński's result and described isometries of certain subspaces of $C(X)$ into $C(Y)$.

Theorem 4.1.1. Let A be a linear subspace of $C(X)$ which separates points of X and contains the constant functions. Suppose T is a linear isometry from A into $C_0(Y)$ and let $B = T(A)$. Then there are continuous functions u and ϕ such that $u\colon Ch(B) \to \Gamma$, $\phi\colon Ch(B) \to Ch(A)$ and such that

$$(Tf)\,(y) \;=\; u(y)\,f(\phi(y)) \qquad\qquad f \in A, \quad y \in Ch(B).$$

Here $Ch(A)$, the *Choquet boundary* of A, is defined as $\{x \in X\colon \delta_x \in ext(A^*)\}$.

If we put $A = C(X)$ then $Ch(A) = X$ and the Holsztyński's result follows. If, in addition, T is surjective then $B = C_0(Y)$, $Ch(B) = Y$ and ϕ is a homeomorphism from Y onto X, this gives the Banach-Stone theorem.

4.2. Surjective isometries between subspaces and subalgebras of $C(X)$.

The Banach-Stone theorem states that in the category of all $C(X)$ spaces the topology of X is uniquely determined by the geometry of the Banach space $C(X)$. In 1948 Mayer [M1] showed that certain type of closed linear subspaces of

$C(X)$ also determined the topology of X. We call a closed linear subspace A of $C_0(X)$ completely regular if for each $x_0 \in X$ and each neighborhood V of x_0 there is a function $f \in A$ such that $1 = ||f||_\infty = f(x_0) > sup\{ |f(x)| : x \in X \setminus V\}$. Mayer proved that if A and B are completely regular subspaces of $C^R(X)$ and $C^R(Y)$, respectively, and if T is an isometry from A onto B then X and Y are homeomorphic.

By the Banach-Stone theorem any isometry from $C(X)$ onto $C(Y)$ is a composition of two isometries: the first one is an **algebra isomorphism** $f \to f \circ \phi$, and the second one is a multiplication by a unimodular function. On the other hand any algebra isomorphism of $C(X)$ onto $C(Y)$ is isometric. Thus we conclude that $C(X)$ and $C(Y)$ are linearly isometric if and only if they are isometric in the category of Banach algebras. In 1959 Nagasawa [N1] proved that the same is true for function algebras, that is closed subalgebras, with unit, of $C(X)$-spaces.

Theorem 4.2.1. Two function algebras are algebraically isomorphic if and only if they are isometric when regarded as Banach spaces.

5. Isometries between Banach function spaces and algebras.

In the previous sections we discussed isometries of subspaces or subalgebras of $C_0(X)$, equipped with the *sup* norm. In this section we present some results characterizing isometries of certain well known subspaces of $C_0(X)$ equipped with different norms. We consider spaces of differentiable functions, of absolutely continuous functions and spaces of Lipschitz functions under various norms.

5.1 Spaces of differentiable functions.

The study of isometries of spaces of differentiable functions was initiated in 1965 by Cambern [C1]. He considered the space $C^1([0,1])$ of all continuously differentiable functions defined on the unit segment $[0,1]$, with the norm defined by

$$||f||_c = sup\left\{ |f(x)| + |f'(x)| : \ x \in [0,1] \right\}.$$

Cambern proved that any isometry of $C^1([0,1])$ onto itself is induced by an isometry of the unit segment.

Theorem 5.1.1. Let T be an isometry from $C^1([0,1])$ onto itself. Then

(2) $$(Tf)(x) = \lambda f(\phi(x)) \qquad f \in C^1([0,1]), \quad x \in [0,1],$$

where $\lambda \in \Gamma$ and ϕ is either id or $1 - id$.

The method introduced by Cambern in his proof became later standard in this type of problems. First the space $C^1([0,1])$ is isometrically embedded into a space $C(W)$, where $W = ext\left(\left(C^1([0,1])\right)^*\right)$ is equipped with the weak $*$ topology. Since T^* is an isometry and is weak $*$ continuous it is a homeomorphism of W onto itself. Using functions from $C^1([0,1])$ that peak at a given point $w \in W$ we associate each functional from $ext\left(\left(C^1([0,1])\right)^*\right)$ with a point from the unit segment. A number of different functionals can be associated we the same point and so $ext\left(\left(C^1([0,1])\right)^*\right)$ can be seen as a cartesian product of $[0,1]$ and a compact set K. T^* is a linear homeomorphism of $[0,1]$ onto itself. Since certain linear relations between extreme functionals corresponding to the same point from the unit segment are different than relations between functionals corresponding to different points from $[0,1]$, T^* induces a homeomorphism ϕ of the unit segment. The specific form of ϕ is then obtained by showing that ϕ is the image of a function from $C^1([0,1])$.

Following the same techniques, Pathak [P1] proved that a surjective isometry of a space $C^{(n)}([0,1])$ of n times continuously differentiable functions is of the form (2). The result was also extended, by Cambern and Pathak [C26], to cover isometries of other C^1-spaces.

Theorem 5.1.2. Let X and Y be locally compact subsets of the real line with no isolated points and let T be an isometry of $C_0^1(X)$ onto $C_0^1(Y)$. Then

$$(3) \qquad (Tf)(y) = u(\phi(y))\, f(\phi(y)), \qquad f \in C_0^1(X),\ y \in Y,$$

where u is a unimodular differentiable function on X with $u' = 0$ and ϕ is a C^1-diffeomorphism of Y onto X with $|\phi'| = 1$.

The results discussed above apply to the isometries of spaces of differentiable functions with the norm defined by $||f||_c = sup\ \{\ |f(x)| + |f'(x)| : x \in X\ \}$. Vasavada [V1], Pathak and Vasavada [P3] and Rao and Roy [R1] proved very similar results for isometries of C^1-type spaces equipped with a $\sum$-norm or M-norm defined by

$$||f||_{\sum} = ||f||_\infty + ||f'||_\infty \quad \text{and} \quad ||f||_M = max\{||f||_\infty, ||f'||_\infty\},$$

respectively.

5.2 Spaces of absolutely continuous functions.

Isometries of the space $AC([0,1])$ of absolutely continuous functions on the unit segment, with the $\sum$-norm, were investigated by Cambern [C1] in 1965 and by Rao and Roy [R1] in 1971. Using different method they proved that any isometry of $AC([0,1])$ onto itself is canonical, that is of the form (2). In 1982 Pathak [P2] extended this result as follows.

Theorem 5.2.1. Let X and Y be compact subsets of the real line. Then any isometry T from $AC(X)$ onto $AC(Y)$, provided with the $\sum$-norms, has the form (2). The homeomorphism ϕ of Y onto X is equal to $T(id)$, where id is the identity map of X onto itself.

5.3. Spaces of Lipschitz functions.

For a compact metric space (X,d) and a real number $0 < \alpha \leq 1$, we define Lipschitz spaces on X by

$$Lip_{\alpha}(X,d) = \left\{ f \in C(X) : \ \|f\|_{d^{\alpha}} = \sup_{x,y \in X} \frac{|f(x)-f(y)|}{d^{\alpha}(x, y)} < \infty \right\}$$

and

$$lip_{\alpha}(X,d) = \left\{ f \in Lip_{\alpha}(X,d) : \lim_{d(x,y) \to 0} \frac{f(x) - f(y)}{d^{\alpha}(x, y)} = 0 \right\}.$$

We write $Lip(X,d)$ and $lip(X,d)$ for $\alpha = 1$. By M-norm and $\sum$-norm on these spaces we mean the norms defined by

$$\|f\|_M = max\{\|f\|_{\infty}, \|f\|_{d_{\alpha}}\} \qquad \text{and} \qquad \|f\|_{\sum} = \|f\|_{\infty} + \|f\|_{d_{\alpha}},$$

respectively. By $Lip\alpha$ ($lip\,\alpha$) we denote the subspace of $Lip_{\alpha}(\mathbb{R})$ ($\mathrm{lip}_{\alpha}(\mathbb{R})$), consisting of all functions of period 1.

Isometries of spaces of Lipschitz functions have been studied since 1961 when De Leeuw [D1] described surjective isometries of the space $lip\,\alpha$ with M-norm. He proved that any surjective isometry T from $lip\,\alpha$ onto itself is canonical, this is of the form

$$(Tf)(x) = \lambda f(\phi(x)) \qquad f \in lip\,\alpha, \quad x \in \mathbb{R},$$

where $\lambda = \pm 1$ and ϕ is an isometry of the real line. This result had been extended by Roy [R2], Novinger [N2], Patil and Vasavada [P4] and Vasavada [V1] to cover isometries of other $Lip(X,d)$ spaces, all provided with M-norm. In 1971 Rao and Roy [R1] proved that any isometry of the space $Lip([0,1])$, with the $\sum$-norm, onto itself is also canonical.

In [J12] authors prove that for any compact metric space (X,d) and (Y,δ) any isometry from $\mathrm{Lip}_{\alpha}(X,d)$ onto $Lip_{\alpha}(Y,\delta)$, both being equipped with M-norm or $\sum$-norm, is cannonical.

5.4 Isometries between general Banach function spaces.

In the previous three sections we considered isometries of Banach spaces of differentiable, absolutely continuous and Lipshitz functions under various norms. Here we discuss a more general situation and present abstract paterns which provide easy proofs for all the previously mentioned results.

Let A be a subspace of $C(X)$, which separates points of X, and let T_A be a linear map from A into a Banach space E. We consider three norms on A:

$$||f||_M = max\{ ||f||_\infty, ||T_A f|| \}, \quad f \in A \tag{M}$$

$$||f||_\Sigma = ||f||_\infty + ||T_A f||, \quad f \in A \tag{Σ}$$

and

$$||f||_c = \sup_{x \in X} \{ |f(x)| + |T_A f(x)| \}, \quad f \in A \tag{C}$$

In the last case we assume that $E = C(X)$. We call A an *M-subspace* , Σ-*subspace* or *C-subspace* of $C(X)$ depending on which of the above formulae is used to define the norm on A.

Note that the space $C^1(X)$, of continuously differentiable functions on X, is defined by the map $T_A : C^1(X) \to C(X)$, $T_A(f) = f'$; the space $AC(X)$ of absolutely continuous function is defined by $T_A : AC(X) \to L^1(X)$, $T_A(f) = f'$ and the space $Lip_\alpha(X,d)$, of Lipshitz functions on X is defined by the map

$$T_A\colon\ Lip_\alpha(X,d) \to C(K), \qquad T_A(f)\,(x,y) = \frac{f(x) - f(y)}{d^\alpha(x,\, y)},$$

where K is a compactification of $X \times X \setminus \{(x,x)\colon x \in X\}$.

Let B be a subspace of $C(Y)$, which separates points of Y. Assume that the norm on B is given by a map T_B, from B into a Banach space F, via the same formula as the norm of A. An isometry T from A onto B is called canonical if

$$(Tf)(y) = u(y)\, f(\phi(y)), \qquad f \in A, \quad y \in Y,$$

where ϕ is a homeomorphism of Y onto X and u is a unimodular function defined on Y.

In their recent paper [J12] authors gave elementary technical schemes to verify if an isometry is canonical.

__Theorem__ __5.4.1.__ Let A and B be M-subspaces of $C(X)$ and $C(Y)$, respectively. Put $\tilde{X} = \{\alpha\delta_x \in extA^*:\ x \in X,\ \alpha \in \Gamma\}$, $\tilde{Y} = \{\alpha\delta_y \in extB^*:\ y \in Y,\ \alpha \in \Gamma\}$. Then an isometry T from A onto B is canonical if and only if $T^*(\tilde{Y}) = \tilde{X}$.

__Theorem__ __5.4.2.__ Let A and B be $\sum$-subspaces of $C(X)$ and $C(Y)$, respectively. Then an isometry T from A onto B is canonical, if and only if for any $\alpha_i\delta_{y_i} + G_i \circ T_B \in extB^*$, $i = 1,2,$ the following two implications hold.

(a) $y_1 = y_2$ if $x_1 = x_2$ and

(b) $G_1 \circ T_B$ and $G_2 \circ T_B$ are proportional if $F_1 \circ T_A$ and $F_2 \circ T_A$ are proportional,

where $x_i \in X$, $F_i \in extE^*$ are such that $T^*(\alpha_i\delta_{y_i} + G_i \circ T_B) = \beta_i\delta_{x_i} + F_i \circ T_A$, for some scalars β_i, $i = 1,2.$

__Theorem__ __5.4.3.__ Let A and B be C-subspaces of $C(X)$ and $C(Y)$, respectively. Then an isometry T from A onto B is canonical if and only if the following two conditions hold.

(i) for any $\alpha_i\delta_{y_i} + \beta_i\delta_{y_i} \circ T_B \in ext\ B^*$, $i = 1, 2$ we have

$$y_1 = y_2 \quad \text{iff} \quad x_1 = x_2$$

where $x_i \in X$ is such that

$$T^*(\alpha_i\delta_{y_i} + \beta_i\delta_{y_i} \circ T_B) = \alpha_i'\delta_{x_i} + \beta_i'\delta_{x_i} \circ T_A,$$

for some scalars α_i', β_i', $i = 1, 2$;

(ii) $T^*(\{\alpha\delta_y\colon y \in Y,\ \alpha \in \Gamma\}) \cap \{\alpha\delta_x \circ T_A\colon x \in X,\ \alpha \in \Gamma\} = \emptyset.$

The necessary and sufficient conditions listed in the above theorems appear to be strong. However they can be easily verified for all the classical function spaces considered in 5.1 - 5.3. For that it is usually necessary to have at least a partial description of the extreme functionals in the unit ball of the dual spaces. Hence in some cases it is easier to apply the following theorem which follows from the results of [J4].

Theorem 5.4.4. Let A be a complex subspace of $C(X)$ such that

(i) A is sup-norm dense in $C(X)$,

(ii) the norm on A is given by a map $T_A\colon A \to E_A$ via the formula (M) or (Σ),

(iii) A contains the constant function $\mathbf{1}$ and $T_A(\mathbf{1}) = 0$.

Further, let B be a complex subspace of $C(Y)$ which satisfies analogous assumptions (i) – (iii). Then any isometry from A onto B, such that $T(\mathbf{1}) = \mathbf{1}$, is of the form $Tf = f \circ \phi$ for $f \in A$, where ϕ is a homeomorphism from Y onto X.

In the case of the subspaces A and B of $C(X)$ and $C(Y)$ equipped with Σ-norms, yet another set of sufficient conditions is given in [J12] under which any isometry T from A onto B is canonical.

5.5. Isometries of semisimple commutative Banach algebras.

Any semisimple commutative Banach algebra A can be seen as a subalgebra of a $C(X)$ space, where X is the maximal ideal space of A. The Nagasawa theorem, disccussed in section 4.2, states that any surjective isometry between uniform algebras A and B, which maps unit element of the first algebra onto the unit element of the latter one, is multiplicative so it is given by a homeomorphism of the maximal ideal spaces of these algebras. Whether the Nagasawa theorem can be extended to other Banach algebras depends not only on the algebraic structures of A and B, but also, and in fact mostly, on the norms on these algebras. On any Banach algebra we can define a number of equivalent norms; some of them have the Nagasawa property, some do not. In [J4] Jarosz proved that any semisimple, commutative Banach algebra with unit has a *natural norm*, and that the Nagasawa theorem extends to algebras equipped with the natural norms. For most of the classical semisimple, commutative Banach algebras the natural norm coincides with the original norm of the algebra.

Theorem 5.5.1. Let A and B be complex, semisimple, commutative Banach algebras equipped with natural norms. Then any isometry T from A onto B which preserves the units of algebras is multiplicative.

6. Isometries of spaces of vector valued functions.

6.1. Isometries of spaces of E-valued continuous functions.

Let X, Y be locally compact Hausdorff spaces and E, F be Banach spaces. An isometry T from $C_0(X,E)$ onto $C_0(Y,F)$ is called canonical if there is a homeomorphism ϕ from Y onto X and a continuous map U from Y into $I(E,F)$, the set of all isometries from E onto F, such that

$$(Tf)(y) = U(y)f(\phi(y)), \qquad f \in C_0(X,E), \quad y \in Y.$$

A Banach space E has the *Banach-Stone property* if the existence of an isometry from $C_0(X,E)$ onto $C_0(Y,E)$ implies that X and Y are homeomorphic. E has the *strong Banach-Stone property* if every isometry of

$C_0(X,E)$ onto $C_0(Y,E)$ is canonical. The Banach-Stone theorem asserts that the scalar field has the strong Banach-Stone property.

Not all Banach spaces have the Banach-Stone property. If we put $E = C(K)$ then $C_0(X,E) = C_0(X \times K)$ so any isometry from $C_0(X,E)$ onto $C_0(Y,E)$ is given by a homeomorphism of $Y \times K$ onto $X \times K$, but not necessarily by a homeomorphism of Y onto X. Roughly speaking a Banach space has the Banach-Stone property or the strong Banach-Stone property if it has very little $C(K)$-structure. These properties were studied by Jerison [J13], Lau [L3] for strictly convex Banach spaces, by Sundareson [S4] for real cylindrical spaces, by Cambern [C8], [C10], [C11] for finite dimensional Banach spaces and reflexive Banach spaces, by Behrends [B3], [B5], [B10] for Banach spaces with small centralizers. A systematic account of many of these results was given by Behrends in [B4]. Here we present only the most far reaching and most recent results.

A linear map S from a Banach space E into itself is called a multiplier if any $e^* \in extE^*$ is an eigenvector of S^*, that is if $S^*(e^*) = a_S(e^*)e^*$, where a_S is a scalar valued function on $extE^*$. We denote by $Mult(E)$ the algebra of all multipliers on E. This algebra can be identified with a uniform algebra of continuous functions on $extE^*$. We denote by $Z(E)$ the centralizer of E, that is, the maximal self-adjoint subalgebra of $Mult(E)$. We call $Z(E)$ trivial if it consists of multiples of the identity operator only. Multipliers and centralizers play very important role in investigating the Banach-Stone property.

<u>Theorem</u> <u>6.1.1</u>. A Banach space with trivial centralizer has the strong Banach-Stone property.

The class of spaces with trivial centralizer includes strictly convex spaces, smooth spaces, and a number of other classical Banach spaces so the above theorem, due to Behrends [B4], covers a wide range of Banach spaces.

6.2. Isometries of injective tensor products of Banach spaces.

A Banach space $C_0(X,E)$ can be naturally identified with the injective tensor product $C_0(X) \otimes E$. So the question whether an isometry from $C_0(X,E)$ onto $C_0(Y,E)$ induces a homeomorphism from Y onto X is a special case of a question whether an isometry from $A \otimes E$ onto $B \otimes E$ induces an isometry between the Banach spaces A and B. The answer is "yes" if E satisfies certain geometrical properties, e.g. if E^* is strictly convex [J6]. The following theorem [J6] provides a complete description of isometries between $A \otimes E$ and $B \otimes E$ in this case.

Theorem 6.2.1. For any real Banach space A the following are equivalent.

(i) $dim(Z(A)) = 1$.

(ii) for any real Banach space E with E^* strictly convex, every isometry of $A \otimes E$ onto itself is canonical.

(iii) for any Hilbert space H every isometry from $A \otimes H$ onto itself is canonical.

(iv) for the two-dimensional real Hilbert space H_2 every isometry from $A \otimes H_2$ onto itself is canonical.

Here we say that an isometry T from $A \otimes E$ onto $B \otimes F$ is canonical if it has one of the following two forms.

(i) $T(a \otimes e) = T_1(a) \otimes T_2(e)$, for all $a \in A$, $e \in E$

where $T_1 : A \to B$, $T_2 : E \to F$ are surjective isometries.

(ii) There is a Banach space Z such that A is isometric with $Z \otimes F$ and B is isometric with $Z \otimes E$, and under this identification T is of the form

$$T(z \otimes f \otimes e) = z \otimes e \otimes f, \qquad \text{for all } z \in Z,\ f \in F,\ e \in E.$$

We remark that, for those who prefer the terminology and flavor of operator theory, Theorem 6.2.1 may be formulated in terms of spaces of compact linear maps, as $A^* \otimes E$ is isometric with the space of compact maps from A into E, provided E has the approximation property. More results on isometries and small bound isomorphisms of operator algebras are presented in [J5, J6, J7].

6.3. Isometries of modules of vector-valued functions.

Let K be a compact Hausdorff space, let $(E_k)_{k \in K}$ be a family of Banach spaces and let E be a closed subspace of the product $\prod\{E_k\colon\ k \in K\}$. E is called a $C(K)$-module, if

(i) $fe \in E$ for $f \in C(K)$ and $e \in E$,

(ii) $k \to \|e(k)\|$ is upper semicontinuous for every $e \in E$,

(iii) $\{ e(k) : e \in E \} = E_k$ for every $k \in K$,

(iv) $\{ k \in K\colon\ E_k \neq \{0\} \}$ is dense in K.

A $C(K)$-module E can be thought of as a space of functions on K where the values of the functions at different points of K lie (possibly) in different spaces. The following result of Behrends [B4] generalized Theorem 6.1.1.

Theorem 6.3.1. Let E and F be a $C(K)$-module and a $C(L)$-module, respectively. Assume that for any $k \in K$ and any $l \in L$ the spaces E_k and F_l have trivial centralizers. Then for any isometry $T\colon E \to F$ there is a homeomorphism $\phi\colon L \to K$ and a family of isometries $u_l : E_{\phi(l)} \to F_l$ such that $(Tx)(l) = u_l(x(\phi(l)))$ for $x \in E$ and $l \in L$.

The concept of $C(K)$-modules can be further extended if we replace the $C(K)$ space by a subalgebra A of $C(K)$. A result similar to the above theorem can be proved for isometries of A-modules [B6].

6.4. Isometries of spaces of vector valued weak continuous functions.

If X is a compact Hausdorff space and E^* a Banach dual, we denote by $C(X, (E^*, \sigma^*))$ the Banach space of continuous functions from X to E^* when the latter space is given its weak $*$ topology. This space arises quite naturally within a variety of mathematical contexts. In [C17] it is shown that the characterization of the bidual of $C(X)$ originally obtained by Kakutani [K1], and studied by Arens [A2] and Kaplan [K2], can be formulated for spaces of norm-continuous vector functions via the introduction of $C(X, (E^*, \sigma^*))$. The dual of the Bochner space $L^1(\mu, E)$ is always of the form $C(X, (E^*, \sigma^*))$ [C18] (whereas $L^{\infty}(\mu, E^*)$ fulfills this role only with an assumption regarding the Radon-Nikodym property [D3, p. 98]). $C(X, (E^*, \sigma^*))$ provides the dual of a space of vector measures [C18] in a manner which parallels the duality obtained for spaces of scalar measures by Gordon [G2]. And the results of Diximier and Grothendieck [D4, G9] characterizing those spaces $C(X)$ which are Banach duals have vector analogues which involve $C(X, (E^*, \sigma^*))$ [C19].

Cambern and Jarosz [C22] showed that if E_1^*, E_2^* have trivial centralizers and satisfy a topological condition introduced by Namioka and Phelps, then, given an isometry T mapping $C(X_1, (E_1^*, \sigma^*))$ onto $C(X_2, (E_2^*, \sigma^*))$ and given $F \in C(X_1, (E_1^*, \sigma^*))$, there exists a dense G_δ in X_2 on which

(4) $$(TF)(x) = U(x)\, F \circ \phi(x),$$

where ϕ is a homeomorphism of X_2 onto X_1 and $U(.)$ an operator-valued function independent of F. If the E_i^* are separable then the $U(x)$ are surjective isometries and $x \longrightarrow U(x)$ is continuous. When the X_i are metric the representation (4) holds on all of X_2. This result has been extended [C21] to cover small bound isomorphisms of spaces of weak continuous functions. If E_i^*, $i = 1, 2,$ belong to a class of Banach duals satisfying a condition involving the space of multipliers on E_i^*, then the existence of an isomorphism T mapping $C(X_1, (E_1^*, \sigma^*))$ onto $C(X_1, (E_1^*, \sigma^*))$ with $||T||\ ||T^{-1}||$ small implies that X_1 and X_2 are homeomorphic. Ultraproducts of Banach spaces and the notion of ε-multipliers play key roles in obtaining this result.

6.5 Isometries of spaces of vector valued analytic functions.

Let E be a finite-dimensional complex Banach space. Then H_E^p stands for the Banach space of all $F\colon \mathbb{D} \to E$ such that $<F, e^*>$ belongs to the Hardy class H^p for all $e^* \in E$. The norm on H_E^p is given by

$$||F||_p = \left\{ \frac{1}{2\pi} \int_{-\pi}^{\pi} ||F(e^{it})||^p \, dt \right\}^{1/p}, \ p < \infty$$

$$||F||_\infty = \ ess\ sup||F(e^{it})|| \ \left(= \sup_{z \in D} ||F(z)||\right).$$

(We use the same symbol F to denote the corresponding L_E^p element on the unit circle). When E is a Hilbert space we write H for E.

The isometries of H^∞ were determined by de Leeuw, Rudin and Wermer [L4] and quite independently by Nagasawa [N1]. Their results were generalized to the context of H_H^∞ in [C6]. In [D3] the isometries of H^1 were also described.

Theorem 6.5.1. Let E be a Banach space with $Mult(E) = \mathbb{C}$, and let T be an isometry of H_E^∞ onto itself. Then T is of the form

$$(TF)(z) = JF(\tau(z)), \qquad F \in H_E^\infty, \ z \in \mathbb{D},$$

where J is a constant isometry of E onto itself and τ is a conformal homeomorphism of the disc.

Theorem 6.5.2. Let $T\colon H^1_H \to H^1_H$ be a surjective isometry. Then T is of the form

$$TF(z) = UF(\tau(z))\tau'(z), \qquad F \in H^1_H, \quad z \in \mathbb{D},$$

where $U\colon H \to H$ is a fixed unitary operator and τ is a conformal map of $\mathbb{D}$ onto $\mathbb{D}$.

7. Isomorphisms with small bound.

In this section we consider the second problem mentioned in the introduction:

Let A be a Banach space of functions defined on X and B be a function space on Y. Assume that there is a linear isomorphism T from A onto B such that both $||T||$ and $||T^{-1}||$ are close to one. Can we conclude that X and Y are homeomorphic and that T is close to an isometry from A onto B?

The assumption, that T is an almost isometry, that is that $||T||$ and $||T^{-1}||$ are close to one, is essential. It is well-known that for any two compact, metric, uncountable sets X and Y the spaces $C(X)$ and $C(Y)$ are isomorphic.

7.1 Small bound isomorphisms of spaces of continuous functions.

The first result in this direction was proved in 1965-66 independently by Cambern and Amir. Amir [A1] proved that if T is a linear isomorphism from

$C(X)$ onto $C(Y)$ with $||T||\,||T^{-1}|| < 2$, then X and Y are homeomorphic. Cambern [C2,C3] showed the same for isomorphisms from $C_0(X)$ onto $C_0(Y)$.

Theorem 7.1.1. There exists an isomorphism T from $C_0(X)$ onto $C_0(Y)$ with $||T||\,||T^{-1}|| < 2$ if and only if X and Y are homeomorphic.

The idea of the proof is as follows. The elements of $ext(C_0(X)^*)$ are precisely the measures of the form $\lambda\mu_x$ with $x \in X$ and $\lambda \in \Gamma$. Thus, if T is an isometry from $C_0(X)$ onto $C_0(Y)$ then T^* maps $ext(C_0(Y)^*)$ onto $ext(C_0(X)^*)$, and for each $y \in Y$ we get a unique $x \in X$ such that $T^*\mu_y = \lambda\mu_x$ for some $\lambda \in \Gamma$. Now, if T is just a linear isomorphism this is no longer true. What Cambern shows is that if $||T^{-1}|| = 1$ and $||T|| < 2$ then for each $y \in Y$, $T^*(\mu_y)$ is close to a unique point of $ext(C_0(X)^*)$. This gives a point map from Y onto X which is then proved to be a homeomorphism.

There is a number of simply examples of non-homeomorphic compact Hausdorff spaces X and Y, such that $C(X)$ and $C(Y)$ are isomorphic under a map T with $||T||\,||T^{-1}|| = 3$. The following one was given by Amir [A1].

Let $X = [0,1]$, $Y = \Gamma \cup \{0\}$. For $f \in C(X)$ define Tf on Y by

$$(Tf)(0) = f(1) - f(0),$$

and

$$(Tf)(e^{2\pi ix}) = f(x) + \left(\tfrac{1}{2} - x\right)(f(1) - f(0)) \qquad \text{for } x \in \Gamma.$$

T is a linear isomorphism from $C(X)$ onto $C(Y)$ with $||T|| = 2$ and $||T^{-1}|| = \frac{3}{2}$, so that $||T||\,||T^{-1}|| = 3$.

The problem whether there exist non-homeomorphic compact Hausdorff spaces X and Y and an isomorphism of $C(X)$ onto $C(Y)$ with $2 \leq ||T||\,||T^{-1}|| < 3$ was open for a number of years. In 1968 Cambern [C4] showed that for any isomorphism T from c onto c_0 we have $||T||\,||T^{-1}|| \geq 3$. Two years later Gordon [G3] extended this result by showing that if X and Y are compact, countable metric spaces, where Y contains no subset homeomorphic to X, then $||T||\,||T^{-1}|| \geq 3$, for any isomorphism T from

$C(X)$ onto $C(Y)$. The problem was finally solved in 1975 by Cohen [C30]. He gave a surprisingly simple example of two non-homeomorphic compact metric sets X and Y and an isomorphism T from $C(X)$ onto $C(Y)$ such that $||T|| \, ||T^{-1}|| = 2$.

Amir-Cambern theorem has been generalized to cover various subspaces of $C(X)$-spaces, this includes extremely regular spaces [C29] and function algebras [J5]. A closed linear subspace A of $C_0(X)$ is called *extremely regular* if for any $x_0 \in X$, any real number ε with $0 < \varepsilon < 1$ and any neighborhood V of x_0 there is a function $f \in A$ such that $1 = ||f||_\infty = f(x_0) > \varepsilon > |f(x)|$ for every $x \in X \setminus V$.

Theorem 7.1.2. Let A and B be extremely regular subspaces of $C_0(X)$ and $C_0(Y)$, respectively. If there is a linear isomorphism T from A onto B with $||T|| \, ||T^{-1}|| < 2$, then X and Y are homeomorphic.

Theorem 7.1.3. Let A and B be function algebras. If there is a linear isomorphism T from A onto B with $||T|| \, ||T^{-1}|| < 2$, then Choquet boundaries of A and B are homeomorphic.

7.2. Injective Isomorphisms of Spaces of Continuous Functions.

Here we present a common generalization of the results of the previous section and of section 4.1.

Theorem 7.2.1. Let T be a linear map from an extremely regular subspace A of $C_0(X)$ into $C_0(Y)$. Let $0 < \varepsilon < 1$ and assume that $||f|| \leq ||Tf|| \leq (1 + \varepsilon)||f||$ for all $f \in C(X)$, then

(i) there is a continuous function ϕ from a subset of Y onto X,

(ii) if, in addition, X is metrizable and compact then there is an isometry Φ from $C(X)$ into $C(Y)$ such that $||\Phi - T|| \leq 3\varepsilon$.

The above theorem was proved, in the case of compact metric spaces, by Benyamini in 1981 [B12]. He also gave an example showing that (ii) does not hold for non-metric spaces. In [J3] the first part of this result was extended for nonmetrizable spaces.

7.3 Isomorphisms of spaces of differentiable functions.

Theorem 7.3.1. Let X and Y be locally compact subsets of the real line without isolated points. Assume that T is a linear map from $C_0^1(X)$ onto $C_0^1(Y)$ with $||T||\,||T^{-1}|| < 2$ and $||T||_\infty\,||T^{-1}||_\infty < \infty$. Then X and Y are homeomorphic.

The spaces $C_0^1(X)$ and $C_0^1(Y)$ in the above theorem can be equipped with the C-norm or with the M-norm as defined in section 5.1. In the first case the result was proved by Cambern and Pathak [C27] and in the second case by Pathak and Vasavada [P3]. In [C27] an example is given showing that 2 is the best number possible in both cases. The question whether the additional assumption, that T and T^{-1} are sup-norm continuous, is essential remains open.

7.4. Isomorphisms of spaces of continuous vector-valued functions.

A Banach space E has the *isomorphic Banach-Stone property* (*IBSP*) if there exists $\varepsilon > 0$ such that:

if X and Y are locally compact Hausdorff spaces and T is a linear isomorphism from $C_0(X,E)$ onto $C_0(Y,E)$ such that $||T||\,||T^{-1}|| <$

$1+\varepsilon$ then X and Y are homeomorphic.

A Banach space E has the *strong isomorphic Banach-Stone property* (*SIBSP*) if for every $\delta > 0$ there is an $\varepsilon > 0$ such that:

if X and Y are locally compact Hausdorff spaces and T is a linear isomorphism from $C_0(X,E)$ onto $C_0(Y,E)$ such that $||T||\,||T^{-1}|| < 1+\varepsilon$ then there is a homeomorphism φ: $X \to Y$ and a family $\{T_k\}_{k \in K}$ of linear isomorphisms of E such that $||T_k||\,||T_k^{-1}|| \leq 1+\delta$ and

$||(Tf)(\varphi(x)) - T_k(f(x))|| \leq \delta||f||$ for all $x \in X$, $f \in C_0(X,E)$.

The first result concerning isomorphic Banach-Stone property is due to Cambern [C9]. In 1976 he proved that a finite dimensional Hilbert space has *IBSP* with $\varepsilon = \sqrt{2} - 1$. Example given by Cohen [C30] clearly shows that ε cannot be greater than 1 for any Banach space, but the exact value of ε for a finite dimensional Hilbert space is not known.

In [J2] the Cambern's result has been extended to cover a wider class of Banach spaces including all spaces with uniformly convex duals, with ε depending on the geometry of the dual space. It was also shown that spaces with trivial centralizers may not have *IBSP*, so the Banach-Stone property and the isomorphic Banach-Stone property are not equivalent.

In the last few years the *IBSP* and the *SIBSP* have been investigated in a number of papers by Behrends, Cambern, Greim, Jarosz and others. Many theorems in this area are rather technical so we present here only few recent results. Most of them have been extended to cover injective isomorphisms and isomorphisms between more general function spaces. Roughly speaking a Banach space B has *IBSP* or *SIBSP* if it is far from any space having a nontrivial $C(X)$ structure. The following two parameters measure this by comparing two dimensional subspaces of B with the two dimensional $C(X)$ space.

$\lambda_0(E) = inf\left\{ d_{B-M}(E_2,\ell_2^1) : E_2 \text{ is a two-dimensional subspace of } E \right\}$,

where ℓ_2^1 is a 2-dimensional L^1-space,

$$\lambda(E) = inf\left\{ max\{ \| e_1+\lambda e_2 \| : \ |\lambda| = 1\} : \ e_1, e_2 \in E, \ \| e_1 \| = \| e_2 \| = 1 \right\}.$$

Any Banach space E such that $\lambda_0(E^*) > 1$ [B9] or $\lambda(E) > 1$ [J8] has the *SIBSP,* in particular any uniformly smooth or uniformly convex Banach space has this property. For an infinite dimensional Hilbert space the best ε for the *IBSP* is equal to $\sqrt{2} - 1$ [J8], and it is not known if there is a Banach space, other than the scalar field, with ε greater than $\sqrt{2} - 1$.

7.5. Non-linear Banach-Stone Theorem

In this section we discuss non-linear homeomorphisms between function spaces. We call a map $T: A \to B$ *ε-bi-Lipshitz* if

$$(1-\varepsilon) \|f - g\| \leq \| Tf - Tg \| \leq (1+\varepsilon)\|f - g\|, \qquad f, g \in A.$$

Bi-Lipshitz maps, uniform homeomorphisms and homeomorphisms between Banach spaces have been investigated intensively for years. It is well-known that any two separable infinite dimensional Banach spaces are homeomorphic, that a Banach space uniformly homeomorphic with a Hilbert space is lineary isomorphic with a Hilbert space and that two bi-Lipshitz equivalent $C(X)$ spaces need not be linear isomorphic. Much less is known about about quantitative nature of non-linear maps. The following results, proved in 1984 [J10], gives a nonlinear version of the Banach-Stone theorem.

Theorem. 7.5.1. Let T be an ε-bi-Lipshitz map from $C_0(X)$ onto $C_0(Y)$, with $\varepsilon < \varepsilon_0$ (absolute constant) then there is a homeomorphism $\phi: Y \to X$ such that

$$\| \ |T(f) \circ \phi| - |f| \ \| \leq \varepsilon' \|f\|, \qquad f \in A,$$

where $\varepsilon' \to 0$ as $\varepsilon \to 0$.

References

[A1] D. Amir, "On isomorphisms of continuous function spaces", Israel J. Math 3, 1966, 205-210.

[A2] R. Arens, "Operations induced in function classes", Monatsh. math. 55, 1951, 1-19.

[A3] ___ and J.L. Kelley, "Characterization of spaces of continuous functions over a commpact Hausdorff space," Trans. Amer. Math. Soc. 62, 1947, 499-508.

[B1] S. Banach, Theorie des operations lineares, Warszawa, 1932.

[B2] E. Behrends, " L^p-Structur in Banachraumen", Studia Math. 55, 1976, 71-85.

[B3] E. Behrends, "On the Banach-Stone theorem," Math. Annalen 233, 1978, 261-272.

[B4] ___, M-Structure and the Banach-Stone Theorem, Lecture Notes in Math. 736, Springer-Verlag, 1979.

[B5] ___, "T sets in function modules and an application to theorems of Banach-Stone type," Revue Roumania de Math. Pures et appl. XXV (6) 1980, 833-834.

[B6] ___, "Multiplier representation and an application to the problem whether $A \otimes_{\epsilon} X$ determines A and/or X", Math Scand. 52, 1983, 117-144.

[B7] ___, "Isomorphic Banach-Stone theorems and Isomorphisms which are closed to Isometries", Pacific J. Math. 133, 1988, 229-250.

[B8] ___, "Small-into isomorphisms between spaces continuous functions", Proc. Amer. Math. Soc. 85, 1981, 479-485.

[B9] ___ and M. Cambern, "An isomorphic Banach-Stone theorem", Studia Math. 90, 1988, 15-26.

[B10] ——— and U. Schmidt-Bichler, "M-structure and the Banach-Stone theorem", Studia Math. LXIX (1) 1980, 33-40.

[B11] ——— et al., L^p-structure in Real Banach Spaces, Lecture Notes in Math. 613, Springer-Verlag, 1977.

[B12] Y. Benyamini, "Near isometries in the class of L^1-preduals", Israel J. Math., 20, 1975, 275-281.

[C1] M. Cambern, "Isometries of certain Banach algebras," Studia Math. 25, 1965, 217-225.

[C2] ———, "A generalized Banach-Stone theorem", Proc. Amer. Math. Soc. 17, 1966, 396-400.

[C3] ———, "On isomorphisms with small bound", Proc. Amer. Math. Soc. 18, 1967, 1062-1066.

[C4] ———, "On mappings of sequence spaces", Studia Math. 30, 1968, 73-77.

[C5] ———, "Isomorphisms of $C_0(Y)$ onto $C(X)$ ", Pacific J. Math. 35, 1970, 307-312.

[C6] ———, "The isometries of $H^\infty(K)$ ", Proc. Amer. Math. Soc., 36, 1972, 173-178.

[C7] ———, "The isometries of $L^p(X,K)$ ", Pacific J. Math. 55, 1974, 9-17.

[C8] ———, "On mappings of spaces of functions with values in a Banach space", Duke Math. J. 42, 1975, 91-98.

[C9] ———, "Isomorphisms of spaces of continuous vector-valued functions", Ill. J. Math. 20, 1976, 1-11.

[C10] ———, "The Banach-Stone property and the weak Banach-Stone property in three dimensional spaces", Proc. Amer. Math. Soc., 67, 1977, 55-61.

[C11] ———, "Reflexive spaces with Banach-Stone property", Revne Roumania de Math. Pure et Appl., XXI (7), Bucarest, 1005-1010.

[C12] ———, "Holsztynski theorem for spaces of continuous vector-valued functions", Studia Math. LXIII, 1978, 213-217.

[C13] ———, "Isometries of measurable functions", Bull. Australian Math.

Soc., 24, 1981, 13-26.

[C14] ___, "Isometries of spaces of norm-continuous functions", Pacific J. Math., 116, 1985, 243-254.

[C15] ___, "A Banach-Stone theorem for spaces of weak * continuous functions", Proc. Royal Soc. Edinburgh. Soc., 101A, 1985, 203-206.

[C16] ___, "Near Isometries of spaces of weak * continuous functions with application to Bochner spaces", Studia Math. 85, 1987, 149-156.

[C17] ___ and P. Greim, "The bidual of $C(X, E)$ ", Proc. Amer. Math. Soc. 85, 1982, 53-58.

[C18] ___, ___, "The dual of a space of vector measures", Math. Z., 180, 1982, 373-378.

[C19] ___, ___, "Spaces of continuous vector functions as duals", Canad. Math. Bull 31, 1988, 70-78.

[C20] ___ and K. Jarosz, "Ultraproducts, ε-multipliers, and isomorphisms", Proc. Amer. Math. Soc., 105, 1989, 929-937.

[C21] ___, ___, "Isometries of spaces of weak * continuous functions", Proc. Amer. Math. Soc., 106, 1989, 707-712.

[C22] ___, ___, "The isometries of H_H^1 ", Proc. Amer. Math. Soc., 107, 1989, 205-214.

[C23] ___, ___, " Multipliers and isometries in H_E^1 ", Bull. London Math. Soc., to appear.

[C24] M. Cambern, K. Jarosz and G. Wodinski, "Almost L^p-projections and L^p-isomorphisms", Proc. Royal Soc. Edinburgh 113A, 1989, 13-25.

[C25] M. Cambern and V.D. Pathak, "Isometries of spaces of differentiable functions", Math. Japonica 26, 1981, 253-260.

[C26] ___, "Isometries of spaces of differentiable functions", Revne Roumaine de Math. Pure et Appl. XXII (7), 1982, 737-743.

[C27] N. L. Carothers and B. Turett, "Isometries on L_p ", Trans. Amer. Math. Soc. 297, 1986, 95-103.

[C28] B. Cengiz, "A generalization of Banach-Stone theorem", Proc. Amer.

Math. Soc. 40, 1973, 426-430.

[C29] H.B. Cohen, "A bound-two isomorphism between $C(X)$ Banach spaces", Proc. Amer. Math. Soc. 50, 1975, 215-217.

[C30] ———, "A second-dual method for $C(X)$ isomorphisms", J. Funct. Anal. 23, 1976, 107-118.

[D1] K. DeLeeuw, "Banach spaces of Lipschitz functions", Studia Math. 21, 1961, 55-66.

[D2] K. DeLeeuw, W. Rudin and J. Wermer, "The isometries of some function spaces", Proc. Amer. Math. Soc. 11, 1960, 694-698.

[D3] J. Diestel and J. J. Uhl, Jr., Vector measures, Math. Surveys, No. 15, Amer. Math. Soc., Providence, 1977.

[D4] J. Diximier, "Sur certains espaces consideres par M. H. Stone", Summa Brasil. Math 2, 1951, 151-152.

[E1] S. Eilenberg, "Banach space methods in topology", Ann. of Math. (2) 43, 1942, 568-579.

[E2] M. El-Gebeily and J. Wolfe, "Isometries of the disc algebra", Proc. Amer. Math. Soc. 93, 1985.

[F1] R.J. Fleming and J.E. Jamison, "Isometries of certain Banach spaces", J. London Math. Soc. 9, 1984, 121-127.

[F2] F. Forelli, "The Isometries of H^p", Can. J. Math., 16, 1964, 721-728.

[G1] K. Geba and Z. Semadeni, "Space of continuous functions (v)", Studia Math. 19, 1960, 303-320.

[G2] H. Gordon, "The maximal ideal space of a ring of measurable functions", Amer. J. Math. 88, 1966, 827-843.

[G3] ———, "On the distance coefficient between isomorphic function spaces", Israel J. Math. 8, 1970, 391-397.

[G4] P. Greim, "Banach-Stone theorems for non-separably valued Bochner L^∞-spaces", Rend. Circ. Mat. Palermo (2), Suppl. 2, 1982, 123-129

[G5] ———, "Hilbert spaces have the Banch-Stone property for Bochner spaces", Bull. Australian Math. Soc., 27, 1983, 121-128.

[G6] ___, "Isometries and L^p-structure of separably valued Bochner L^p-spaces", In: Measure Theory and Its Applications, Proc. Conf. Sherbrooke 1982, Lecture Notes in Math. 1033, Springer 1983, 209-218.

[G7] ___, "Banach spaces with the L^1-Banach-Stone property", Trans. Amer. Math. Soc. 287, 1985, 819-828.

[G8] ___ and J.E. Jamison, "Hilbert spaces have the strong Banach-Stone property for Bochner spaces", Math. Z. 196, 1987, 511-515.

[G9] A. Grothendieck, "Une caraterisation vectorielle metrique des espaces L^1 ", Canad. J. Math. 7, 1955, 552-561.

[H1] W. Holsztyński, "Continuous mappings induced by isometries of spaces of continuous function", Studia Math. XXVI, 1966, 133-136.

[J1] J. E. Jamison and I. Loomis, "Isometries of Orlicz spares of vector valued functions", Math. Z. 193, 1986, 363-371.

[J2] K. Jarosz, "A generalization of the Banach-Stone theorem", Studia Math. LXXIII, 1982, 33-39.

[J3] ___, "Into isomorphisms of spaces of continuous functions", Proc. Amer. Math. Soc. 90, 1984, 373-377.

[J4] ___, "Isometries in semisimple commutative Banach algebras", Proc. Amer. Math. Soc. 94, 1985, 65-71.

[J5] ___, Perturbations of Banach Algebras, Lecture Notes in Math. 1120, Springer-Verlag, 1985.

[J6] ___, "Isometries between injective tensor products of Banach spaces". Pacific J. Math. 121, 1986, 383-395.

[J7] ___, "Small Isomorphisms between operator algebras", Proc. Edinburgh Math. Soc. 28, 1985, 121-131.

[J8] ___, "Small Isomorphisms of $C(X,E)$ spaces", Pacific J. Math., 138, 1989, 295-315.

[J9] ___, "$H^\infty(\mathbb{D})$ is stable", J. London Math. Soc., 37, 1988, 490-498.

[J10] ___, "Nonlinear generalizations of the Banach-Stone theorem", Studia Math. 93, 1989, 97-107.

[J11] ___,"Ultraproducts and small bound perturbations" , Pacific J. Math., 148, 1991, 81-88.

[J12] K. Jarosz and V.D. Pathak, "Isometries between function spaces", Trans. Amer. Math. Soc., 305, 1988, 193-206.

[J13] M. Jerison, "The space of bounded maps into a Banach space", Ann. of Math. 52, 1950, 309-327.

[K1] S. Kakutani, "Concrete representations of abstract (M)-sapces", Ann. of Math. 42, 1941, 994-1024.

[K2] S. Kaplan, "On the second dual of the space of continuous functions", Trans. Amer. Math. Soc. 86, 1957, 70-90.

[K3] R. Khalil, "Isometries of $L^p \times L^p$ ", Tamkang J. Math. 16, 1989, 77-85.

[L1] N. Lal and S. Merril III, "Isometries of H^p spaces of torus", Proc. Amer. Math. Soc. 31, 1972, 465-471.

[L2] J. Lamperti, "On the isometries of certain function spaces", Pacific J. Math. 8, 1958, 459-466.

[L3] K.S. Lau, "A representation theorem for isometreies of $C(X,E)$", Pacific J. Math. 60, 1975, 229-232.

[L4] ___, "On the isometries of reflexive Orlicz spaces", Ann. Inst. Fourier, Grenoble, 18, 1963, 99-109.

[M1] S.B. Mayer, "Banach spaces of continuous functions", Ann. of Math. 49, 1948, 132-140.

[N1] M. Nagasawa, "Isomorphisms between commutative Banach algebras with applications to ring of Analytic functions", Kodia Math. Sem. Rep. 11, 1959, 182-188.

[N2] W.P. Novinger, "Linear isometries of subspaces of spaces of continuous functions", Studia Math. LIII, 1975, 273-276.

[P1] V.D. Pathak, "Isometries of $C^{(n)}[0,1]$", Pacific J. Math. (1) 96, 1981, 211-222.

[P2] ___, "Linear isometries of Absolutely Continuous functions", Can. J. Math. XXXIV, 1982, 298-306.

[P3] V.D. Pathak and M.H. Vasavada, "Isometries and isomorphism of Banach spaces of differentiable functions", preprint.

[P4] D.J. Patil and M.H. Vasavada, "Linear isometries of Banach space of Lipschitz functions", unpublished.

[P5] P.K. Prased, "Isometries of $C^1([0,1]^m)$ ", S.P. University Technical Report, V.V. Nagar, India.

[R1] V. Rao and A.K. Roy, "Linear Isometries of some function spaces", Pacific J. Math. 38, 1971, 177-192.

[R2] R. Rochberg, "Deformation of uniform algebras on Riemann surfaces", Pacific J. Math. 121, 1986, 135-181.

[R3] A.K. Roy, "Extreme points and linear isometries of the Banach spaces of Lipschitz functions", Can. J. Math., 20, 1968, 1150-1164.

[R4] W. Rudin, "L^p-isometries and equimeasurability", Indiana Univ. Math. J., 25, 1976, 215-228.

[S1] Z. Semadeni, Banach spaces of continuous functions, Vol. 1, PWN, Warsaw.

[S2] A.R. Sourour, "The isometries of $L^p(\Omega,X)$ ", J. Funct. Anal. 30, 1978, 276-285.

[S3] M.H. Stone, "Applications of the theory of Boolean rings in topology", Trans. Amer. Math. Soc. 41, 1937, 375-481.

[S4] K. Sundareson, "Spaces of continuous functions into a Banach space", Studia Math. 48, 1973, 15-22.

[T1] K.W. Tam, "Isometries of certain function spaces", Pacific J. Math., 31, 1969, 233-246.

[V1] M.H. Vasavada, Ph.D. Thesis, University of Wisconsin, 1969.

[W1] J.D.M. Wright and M. Youngson, "On isometries of Jordan algebras", J. London Math. Soc. 17, 1978, 339-344.

On the Theorems of Pick and von Neumann

KEITH LEWIS Department of Mathematics, Brown University, Providence, Rhode Island

JOHN WERMER Department of Mathematics, Brown University, Providence, Rhode Island

1. Let $\mathfrak{B}$ be the unit ball in the space of bounded analytic functions on the open unit disk D. Fix distinct points z_j, $1 \le j \le n$, in D.

Put $z = (z_1, \ldots, z_n) \in \mathbb{C}^n$.

<u>Problem</u>: Given an n-tuple $\{w_j\}$ of complex numbers in $|w| < 1$. When does $\exists f \in \mathfrak{B}$ such that $f(z_j) = w_j$, $1 \le j \le n$?

This problem was solved by G. Pick in 1916. Let us say that $w = \{w_j\}$ <u>satisfies Pick's condition</u>, <u>relative to</u> z, if

$$\sum_{j,k=1}^{n} \frac{w_j \bar{w}_k}{1 - z_j \bar{z}_k} t_j \bar{t}_k \le \sum_{j,k=1}^{n} \frac{t_j \bar{t}_k}{1 - z_j \bar{z}_k}$$

for every $t = (t_1, \ldots, t_n) \in \mathbb{C}^n$.

<u>Pick's Theorem</u>: $\exists f \in \mathfrak{B}$ <u>such that</u> $f(z_j) = w_j$, $1 \le j \le n$, <u>if and only if</u> $w = \{w_j\}$ <u>satisfies Pick's condition</u>.

If Pick's condition holds for w, we may choose f in the disk algebra. (See Garnett, [G], p. 7.)

The relation between Pick's theorem and operator theory on Hilbert space was established by Donald Sarason in his important paper, [S].

Here we shall consider related questions.

Fix $z_1,\dots,z_n$ in D. Define the inner product $(,)$ on $\mathbb{C}^n$ by

$$(c,c') = \sum_{j,k=1}^{n} \frac{c_j \bar{c'}_k}{1-z_j\bar{z}_k}, \quad c,c' \in \mathbb{C}^n .$$

Direct calculation shows that $(c,c) > 0$ unless $c = 0$, and so $(,)$ makes $\mathbb{C}^n$ into a Hilbert space H.

For $w = \{w_j\}$ in $\mathbb{C}^n$, we define the linear tranformation T_w on H by:

$$T_w c = \{w_j c_j\}, \quad c \in H .$$

For t in $\mathbb{C}^n$, then,

$$\|T_w t\|^2 = \sum_{j,k=1}^{n} \frac{w_j t_j \bar{w}_k \bar{t}_k}{1-z_j\bar{z}_k} .$$

Hence Pick's condition on w states

$$\|T_w t\|^2 \le \|t\|^2 \text{ for each } t \text{ in } H .$$

In other words, we have

<u>Result 1</u>: w satisfies Pick's condition if and only if $\|T_w\| \le 1$.

Direct computation gives, for $t \in H$:

$$\|T_z t\|^2 - \|t\|^2 = -\left|\sum_j t_j\right|^2 \le 0 ,$$

whence we have

<u>Result 2</u>: $\|T_z\| \le 1$.

Let now P be a polynomial. Put $w = \{P(z_j)\}$ in $\mathbb{C}^n$. We have

<u>Result 3</u>: $T_w t = P(T_z)\, t$, for $t \in H$.

We now recall Von Neumann's theorem on contractions, which he proved in 1951.

<u>Von Neumann's Theorem</u>: <u>Let</u> T <u>be a contraction on a Hilbert space, i.e.,</u> T <u>is a linear transformation of norm</u> ≤ 1. <u>If</u> P <u>is any polynomial in</u> z, <u>then</u>

$$(*) \quad \|P(T)\| \le \max_{|z|\le 1} |P(z)| .$$

If f is an element of the disk algebra $A(D)$, we choose polynomials P_n with $P_n \to f$ in $A(D)$ and $\|P_n\| \le \|f\|$ for all n. Then

$$\|P_n(T) - P_m(T)\| \le \max_{|z|\le 1} |P_n(z) - P_m(z)| \to 0$$

as $n,m \to \infty$, so $\lim_{n\to\infty} P_n(T)$ exists. We denote the limit $f(T)$; it depends only on f. Because of $(*)$ we get

$$\|f(T)\| \le \|f\| = \max_{|z|\le 1} |f(z)| .$$

<u>Claim</u>: <u>Von Neumann's Theorem implies the Necessity of Pick's condition</u>.

<u>Proof</u>: (suggested by Brian Cole).

Fix $z_1,\dots,z_n$ in D. Let $f \in \mathfrak{B}$ and put $w_j = f(z_j)$, $1 \le j \le n$. We must show that $w = \{w_j\}$ satisfies Pick's condition. It suffices to consider the case when f is a polynomial P, with $\sup_{|z|\le 1} |P(z)| \le 1$. Put $w_j = P(z_j)$, $1 \le j \le n$. Then

$$\|T_w\| = \|P(T_z)\| \le 1$$

by Von Neumann's Theorem and Results 2 and 3 above, and so w satisfies Pick's condition, by Result 1.

2. <u>Commuting Contractions</u>.

Let H be a Hilbert space and let $T_1,\dots,T_n$ be an n-tuple of commuting contractions on H, i.e. $T_j T_k = T_k T_j$ for all j,k. The analogue of Von Neumann's inequality here is the following:

We denote by Δ^n the closed unit polydisk in $\mathbb{C}^n$, i.e. $\Delta^n = \{(z_1 \dots, z_n) \,|\, |z_j| \le 1, 1 \le j \le n\}$. Let P be a polynomial in $z_1,\dots,z_n$. Then

$$(**) \quad \|P(T_1,\dots,T_n)\| \le \max_{z\in\Delta^n} |P(z)| = \|P\|_{\Delta^n} .$$

<u>When is this true</u>? In 1973, N. Varopoulos [V] and in 1975, A. M. Davie and Crabb [D-C], gave counterexamples in the general case. They found a commuting triple of contractions on a finite dimensional Hilbert space such that $(**)$ fails. In their examples, the contractions were nilpotent.

We are interested in the following case: Fix an invertible $k \times k$ matrix A with complex number entries. Let $\mathfrak{A}$ denote the algebra of all $k \times k$ matrices

$$T = A W A^{-1}$$

where W is a diagonal matrix. We regard such T as linear transformations on $\mathbb{C}^k$ with the standard inner product:

$$(c, c') = \sum_{j=1}^{k} c_j \bar{c}'_j .$$

$\mathfrak{A}$ is a commutative algebra of linear transformations on $\mathbb{C}^k$. Put $X_j = A E_j$, $j = 1, \ldots, k$, where $E_j = (0, 0, \ldots, 1, \ldots 0)$, with 1 in the j'th place. $\mathfrak{A}$ is the algebra of all those linear transformations of $\mathbb{C}^k$ which have each X_j, $j = 1, \ldots, k$, as eigenvector.

Question 1: Does inequality (∗∗) hold for all n-tuples of contractions $T_1, \ldots, T_n$ in $\mathfrak{A}$?

Note: John Holbrook has told us (November 1990) that he and a coworker have found a counterexample to (∗∗) for $n = 3$ by perturbing one of the known nilpotent counterexamples.

3. The Case: $k = 2$.

We consider $\mathfrak{A}$ now for the case: $k = 2$.

For $w = (w_1, w_2) \in \mathbb{C}^2$, we put

$$T_w = A \begin{pmatrix} w_1 & 0 \\ 0 & w_2 \end{pmatrix} A^{-1} .$$

Let φ_1, φ_2 be linearly independent unit eigenvectors for $\mathfrak{A}$. Put

$$a = (\varphi_1, \varphi_2) .$$

We assume $a \neq 0$. If $a = 0$, each T_w is normal.

Fix w with $|w_1| < 1$, $|w_2| < 1$, and put $T = T_w$. Then

$$\|T\| \leq 1 \Leftrightarrow I - T^* T \geq 0$$

$\Leftrightarrow$ the matrix $((c_{ij}))$, $1 \leq i, j \leq 2$, with

$$c_{ij} = ((I - T^* T)\, \varphi_i, \varphi_j))$$

is positive semi-definite. This matrix is

$$\begin{pmatrix} 1-|w_1|^2 & a-aw_1\overline{w}_2 \\ \overline{a}-\overline{a}w_2\overline{w}_1 & 1-|w_2|^2 \end{pmatrix}$$

and since $1 - |w_j|^2 > 0$, $j = 1,2$, it is positive semi-definite if and only if the determinant

$$(1 - |w_1|^2)(1 - |w_2|^2) - |a|^2|1 - w_1\overline{w}_2|^2 \geq 0 \text{ , or}$$

$$(1 - |w_1|^2)(1 - |w_2|^2) \geq |a|^2|1 - w_1\overline{w}_2|^2 .$$

This $\Leftrightarrow$

$$\frac{(1-|w_1|^2)(1-|w_2|^2)}{|1-w_1\overline{w}_2|^2} \geq |a|^2 \text{ , or}$$

$$1 - \left|\frac{w_1-w_2}{1-w_1\overline{w}_2}\right|^2 \geq |a|^2 \text{ , or}$$

(1) $1 - |a|^2 \geq \left|\dfrac{w_1-w_2}{1-w_1\overline{w}_2}\right|^2 .$

We have proved: $\|T\| \leq 1 \Leftrightarrow$ (1) holds. In other words, we have

<u>Lemma 1</u>: Put

$$\mathfrak{D}_A = \{w \in \mathbb{C}^2 \,|\, \|T_w\| \leq 1 , |w_1| < 1 , |w_2| < 1\} .$$

<u>Then</u>

$$\mathfrak{D}_A = \{w \in \mathbb{C}^2 \,\Big|\, \left|\frac{w_1-w_2}{1-\overline{w}_1 w_2}\right|^2 \leq 1 - |a|^2 , |w_1| < 1 , |w_2| < 1\} .$$

Choose z_1, z_2 in the open unit disk such that

$$1 - |a|^2 = \left|\frac{z_1-z_2}{1-\bar{z}_1 z_2}\right|^2$$

Put

$$S = T_z \text{ , where } z_1 = (z_1, z_2) .$$

By Lemma 1, $\|S\| \leq 1$.

<u>Theorem 1</u>: <u>If</u> $T \in \mathfrak{A}$ <u>and</u> $\|T\| < 1$, <u>then</u> $\exists f \in$ <u>the disk algebra</u> $A(D)$ <u>with</u> $\|f\| \leq 1$ <u>such that</u> $T = f(S)$.

<u>Proof</u>: By Lemma 1, $T = T_w$

with $1 - |a|^2 \geq \left|\frac{w_1 - w_2}{1 - \bar{w}_1 w_2}\right|^2$.

By choice of $z_1 z_2$, then, we have

(2) $\left|\frac{z_1 - z_2}{1 - \bar{z}_1 z_2}\right| \geq \left|\frac{w_1 - w_2}{1 - \bar{w}_1 w_2}\right|$.

(2) is equivalent to Pick's condition in the case of pairs of points. (See .e.g. [G], p. 7). Pick's theorem then yields $f \in A(D)$, the disk algebra, with $\|f\| \leq 1$ and $f(z_j) = w_j$, $j = 1,2$. Then $f(S)\varphi_j = w_j\varphi_j$, $j = 1,2$, where φ_1, φ_2 are simultaneous eigenvectors for $\mathfrak{A}$. Hence $f(S) = T_w = T$.

Theorem 2: If $T_1, \ldots, T_n$ are n contractions in $\mathfrak{A}$, then inequality (**) holds.

Note: Theorem 2 was proved by S. W. Drury in [D]. We shall use Theorem 1. (See also Holbrook [H], p. 121.)

Proof: Let P be a polynomial in n variables. Without loss of generality, $\|P\|_{\Delta_n} \leq 1$.

First assume $\|T_j\| < 1$, $1 \leq j \leq n$. By Theorem 1, $\exists f_j \in A(D)$, $\|f_j\| \leq 1$, such that $f_j(S) = T_j$, $1 \leq j \leq n$. Define $Q \in A(D)$ by

$$Q(\lambda) = P(f_1(\lambda), \ldots, f_n(\lambda)) .$$

Then $\|Q\| \leq 1$. By Von Neumann's Theorem, $\|Q(S)\| \leq 1$.
Hence $\|P(T_1, \ldots, T_n)\| = \|P(f_1(S), \ldots, f_n(S)\| = \|Q(S)\| \leq 1$.
So (**) holds in this case.

If $\|T_j\| = 1$ for some j , we replace each T_i by $r\,T_i$, where r is a scalar, $0 < r < 1$. We use the previous result, let $r \to 1$, and get (**) in general.

Theorem 3: Let $\mathfrak{A}$ be as above. Then $\mathfrak{A}$ is isometrically isomorphic to a quotient algebra of the disk algebra.

Proof: Choose z_1, z_2 in D and choose S as above. Let I denote the ideal in $A(D)$,

$$I = \{g \in A(D) \,|\, g(z_1) = g(z_2) = 0\} .$$

The quotient algebra $A(D)/I$ is given the quotient norm.

Fix $w \in \mathbb{C}^2$. Choose $f \in A(D)$ such that $f(z_j) = w_j$, $j = 1,2$. Let $[f]$ denote the coset of f in $A(D)/I$. Then $[f]$ depends only on w. We define the map $\tau : \mathfrak{A} \to A(D)/I$ by

$$\tau(T_w) = [f].$$

Then τ is an algebraic isomorphism of $\mathfrak{A}$ onto $A(D)/I$.

Choose $T_w \in \mathfrak{A}$ with $\|T_w\| < 1$. By Theorem 1, $\exists f \in A(D)$, $\|f\| \le 1$, such that $T_w = f(S)$. Hence $w_j = f(z_j)$, $j = 1, 2$ and so $\tau(T_w) = [f]$ and

$$\|\tau(T_w)\| = \|[f]\| \le \|f\| \le 1.$$

Conversely, fix $T_w \in A$ such that $\|\tau(T_w)\| < 1$. Then $\tau(T_w) = [g]$ for some g in $A(D)$ with $\|g\| < 1$; $g(z_j) = w_j$, $j = 1,2$, so $T_w = g(S)$. By von Neumann's Theorem, then, $\|T_w\| = \|g(S)\| \le \|g\| < 1$.

So $\|T_w\| < 1 \Rightarrow \|\tau(T_w)\| \le 1$ and

$$\|\tau(T_w)\| < 1 \Rightarrow \|T_w\| < 1.$$

Hence τ preserves norm, and so τ is the desired map from $\mathfrak{A}$ to $A(D)/I$.

4. Q-Algebras and Commuting Contractions.

A Q-Algebra (introduced by N. Varopoulos) is defined as a commutative Banach algebra which is isometrically isomorphic to the quotient algebra of a uniform algebra by a closed ideal. Theorem 3, above, shows that our algebra $\mathfrak{A}$ on $\mathbb{C}^k$ is a Q-algebra for the case $k = 2$.

Question 2: Fix $k \ge 2$ and let $\mathfrak{A}$ be an algebra as in Question 1. Is $\mathfrak{A}$ a Q-algebra?

An unpublished result of Brian Cole states that Q-algebras are characterized by inequality (**), i.e. a commutative Banach algebra $\mathfrak{L}$ is a Q-algebra if and only if for each n-tuple of elements $x_1, \dots, x_n \in \mathfrak{L}$ with $\|x_j\| \le 1$, all j, we have

$$\|P(x_1, \dots, x_n)\| \le \|P\|_{\Delta^n}$$

for every polynomial P in n variables, n = 1,2,... . Hence Questions 1 and 2 are equivalent.

REFERENCES

[G] J. Garnett, Bounded Analytic Functions, Academic Press (1981).

[S] D. Sarason, Generalized Interpolation in H^∞, Trans AMS 127 (1967), pp. 179-203.

[V] N. Varapoulos, Sur une inégalité de Von Neumann, C. R. Acad. Sci. Paris, Sér A. B. 277 (1973), A19-A22.

[C-D] M. J. Crabb and A. M. Davie, Von Neumann's inequality for Hilbert space operators, Bull. London Math. Soc. 7 (1975), pp. 49-50.

[D] S. W. Drury, Remarks on Von Neumann's Inequality, Springer Lecture Notes in Mathematics 995 (1983), pp. 14-32.

[H] J. Holbrook, Von Neumann's Inequality and the Poisson Radius for Operators, Bull. Acad. Polon. Sci., Sér. Math., Astr. et Phys., Vol. 22 (1974), pp. 121-127.

Some Geometric Properties Related to Uniform Convexity of Banach Spaces

Bor–Luh Lin
Department of Mathematics
The University of Iowa
Iowa City, IA 52242

Wenyao Zhang
Department of Mathematics
The University of Iowa
Iowa City, IA 52242

In 1955, two generalizations of uniformly convex (UR) Banach spaces were introduced. Lovaglia [L] studied the locally uniformly convex (LUR) Banach spaces. Fan and Glicksberg [FG1,2] extended the 2R Banach spaces, introduced by V. Smulian, and studied the fully k–convex (kR) Banach spaces. Later, two more generalizations of uniform convexity were introduced for Banach spaces. In 1979, Sullivan [S] studied the k–uniformly rotund (k–UR) Banach spaces and in 1980, Huff [H] studied the nearly uniformly convex (NUC) Banach spaces. It is known that every k–UR space is NUC [Y] and every strictly convex k–UR space is (k+1)R [LY]. Recently, the locally fully k–convex (LkR) Banach spaces are defined in [NW] and it is proved that $LUR \Rightarrow L2R \Rightarrow \cdots \Rightarrow LkR \Rightarrow L(k+1)R$ and every strictly convex locally k–uniformly rotund Banach space is L(k+1)R. Furthermore, Kutzarova [K] introduces k–β Banach spaces and k–nearly uniformly convex (k–NUC) Banach spaces and shows that these are two classes of Banach spaces that lie strictly in between the classes of k–UR and kR spaces. The relationships between kβ and k–NUC spaces are also determined in [K]. The class of locally k–β and locally k–NUC Banach spaces are identical and has been studied in [KL]. By definitions, it follows that the NUC space is exactly the limit space of k–NUC when $k \longrightarrow \infty$. In this paper, we study the corresponding spaces of k–UR, Lk–UR, kR, LkR and L k–NUC when $k \longrightarrow \infty$. The authors wish to thank Denka Kutzarova for sending them the preprint of her paper [K].

1. For a Banach space X, let S_X (resp. B_X) be the unit sphere (resp. ball) of X. Recalled that for an integer k, $k \geq 2$, X is kR (resp. LkR) if for any sequence $\{x_n\}$ in B_X, $\lim_{n_1,\dots,n_k \to \infty} \|x_{n_1} + \cdots + x_{n_k}\| = k$ (resp. for all x in S_X, $\lim_{n_1,\dots,n_k \to \infty} \|x + x_{n_1} + \cdots + x_{n_k}\| = k+1$), then the sequence $\{x_n\}$ is convergent (resp. to x) in X.

Definition. Let X be a Banach space. X is said to be a ωR space if for any sequence $\{x_n\}$ in B_X, $\lim_{n_1,\dots,n_k \to \infty} \|x_{n_1} + \cdots + x_{n_k}\| = k$ for all k in $\mathbb{N}$, then $\{x_n\}$ is convergent in X.

X is said to be a LωR (resp. w LωR) space if for any sequence $\{x_n\}$ in B_X and for any x in S_X, $\lim_{n_1,\dots,n_k \to \infty} \|x + x_{n_1} + \cdots + x_{n_k}\| = k+1$ for all k in $\mathbb{N}$, then $\lim_n \|x_n - x\| = 0$ (resp., $w\text{-}\lim_n x_n = x$).

For any sequence $\{x_n\}$ in B_X, by triangle inequality, it is easy to see that if $\lim_{n_1,\dots,n_k \to \infty} \|x_{n_1} + \cdots + x_{n_{k+1}}\| = k+1$, then $\lim_{n_1,\dots,n_k \to \infty} \|x_{n_1} + \cdots + x_{n_k}\| = k$. Hence $UR \Rightarrow 2R \Rightarrow \cdots \Rightarrow kR \Rightarrow (k+1)R \Rightarrow \cdots \Rightarrow \omega R$ and $LUR \Rightarrow L2R \Rightarrow \cdots \Rightarrow LkR \Rightarrow \cdots \Rightarrow L\omega R$. It is clear that $\omega R \Rightarrow L\omega R \Rightarrow wL\omega R$.

Proposition 1. *If a Banach space* X *is* ωR *then* X *is reflexive.*

Proof. Let x^* be any element in X^* with norm one. Choose a sequence $\{x_n\}$ in B_X such that $\lim_n x^*(x_n) = 1$. Then for any k in $\mathbb{N}$, $\lim_{n_1,\dots,n_k\to\infty} \|x_{n_1} + \cdots + x_{n_k}\| = k$. Hence $\{x_n\}$ converges to some element x in X. It is clear that $\|x\| = 1 = x^*(x)$. By James' theorem, X is reflexive. □

Proposition 2. *Let* X *be a Banach space.*

(i) *If* X *is* $w\,L\omega R$ *then* X *is strictly convex;*

(ii) *If* X *is* $L\omega R$, *then* X *has the property* (G), *that is, every point in* S_X *is a denting point of* B_X. *In fact, in this case, every point in* S_X *is a strongly exposed point of* B_X.

Proof. (i) Suppose there exist $x \neq y$ and $\lambda x + (1-\lambda)y \in S_X$ for all $\lambda \in [0,1]$. Let $x_n = y$ for all n. Then for all $n_1,\dots,n_k$, $\|x + x_{n_1} + \cdots + x_{n_k}\| = k+1$ for any k in $\mathbb{N}$. If X is $w\,L\omega R$, then $y = w\text{-}\lim_n x_n = x$ which is impossible.

(ii). Let x be any element in S_X. Choose x^* in X^*, $\|x^*\| = 1 = x^*(x)$. For any sequence $\{x_n\}$ in X, if $\lim_n x^*(x_n) = x^*(x)$, then

$\lim_{n_1,\dots,n_k\to\infty} \|x + x_{n_1} + \cdots + x_{n_k}\| = k+1$ for any k in $\mathbb{N}$. If X is $L\omega R$, then $\lim_n \|x_n - x\| = 0$. Hence x is a strongly exposed point of B_X. □

Corollary 3. Let X be a Banach space. Then X is $L\omega R$ (resp., $w\,L\omega R$) if and only if X is strictly convex and for any x in S_X and any sequence $\{x_n\}$ in B_X, $\lim_{n_1,\dots,n_k\to\infty} \|x + x_{n_1} + \cdots + x_{n_k}\| = k+1$ for all k in $\mathbb{N}$ imply that there is a subsequence of $\{x_n\}$ convergent (resp. weakly) to some element with norm one in X.

Theorem 4. *Let X be a Banach space. If X^* is LωR (resp. w LωR) then $(S_{X^*}, w^*) = (S_{X^*}, \|\cdot\|)$ (resp. $(S_{X^*}, w^*) = (S_{X^*}, w)$), that is, the weak* topology and the norm (resp. weak) topology coincide on S_{X^*}.*

Proof. Suppose there is a net $\{x^*_\alpha\}$ and an element x^* in S_{X^*} such that $w^*\text{–}\lim x^*_\alpha = x^*$ but $\lim\|x^*_\alpha - x^*\| \neq 0$. Then there is an $\epsilon > 0$ such that for each α there is $\beta > \alpha$ and $\|x^*_\beta - x^*\| > \epsilon$. Choose a sequence $\{x_n\}$ in S_X such that $x^*(x_n) > 1 - \frac{1}{n}$, $n \in \mathbb{N}$. For each $n \in \mathbb{N}$, let $V_n = \{y^* : y^* \in S_{X^*}, |(y^* - x^*)(x_i)| < \frac{1}{i},$ $1 \leq i \leq n\}$. Since $w^*\text{–}\lim x^*_\alpha = x^*$, for each $n \in \mathbb{N}$, there is α_n such that $x^*_\alpha \in V_n$ for all $\alpha > \alpha_n$. Choose $\beta_n > \alpha_n$ and $\|x^*_{\beta_n} - x^*\| > \epsilon$. Then $x^*_{\beta_n} \in V_n$ for all $n \in \mathbb{N}$. Let $k \in \mathbb{N}$ be fixed. For any integer N, if $n_1, \ldots, n_k \geq N$, then

$$\|x^* + x^*_{\beta_{n_1}} + \cdots + x^*_{\beta_{n_k}}\| \geq (x^* + x^*_{\beta_{n_1}} + \cdots + x^*_{\beta_{n_k}})(x_N)$$

$$\geq (k+1)x^*(x_N) - \frac{k}{N} > (k+1)(1 - \frac{1}{N}) - \frac{k}{N} = (k+1)(1 - \frac{1}{N}).$$

Hence $\lim_{n_1, \ldots, n_k \to \infty} \|x^* + x^*_{\beta_{n_1}} + \cdots + x^*_{\beta_{n_k}}\| = k+1$. If X is LωR, then $\lim_n \|x^*_{\beta_n} - x^*\| = 0$ which is a contradiction.

The case when X is w LωR can be proved similarly. □

Corollary 5. Let X be a Banach space. Then

(i) If X^* is LωR (resp. w LωR) then X^* has the weak* asymptotic–norming property II (resp. III). Hence if X^* is w LωR, then X^* has the Radon–Nikodym property;

(ii) If X^* is LωR (resp. w LωR) then X is Frechet differentiable (resp. very smooth). Hence if X^{**} if w LωR, then X is reflexive;

(iii) If X is reflexive and is LωR, in particular, when X is ωR, then for any bounded closed convex set M in X, the metric projection on M, $P_M(x) = \{y : y \in M, \|x-y\| = d(x,M)\}$, $x \in X$ is continuous on X.

Proof. (i) Recall that a set Φ in B_X is called a norming set of X^* if $\|x^*\| = \sup_{x \in \Phi} x^*(x)$ for all x^* in X^*. X^* is said to have the w^*ANP–II (resp. w^*ANP–III) if there is a norming set Φ in B_X for X^* such that for any sequence $\{x_n^*\}$ in S_{X^*} with the property that for any $\epsilon > 0$, there is N in $\mathbb{N}$ and x in Φ such that $x_n^*(x) > 1-\epsilon$ for all $n \geq N$, then $\{x_n^*\}$ has a convergent subsequence (resp. $\bigcap_{n=1}^{\infty} \overline{co}\{x_k^* : k \geq n\} \neq \phi$).

(i) follows from Theorem 4 and Theorem 3.1 in [HL] and the fact that ANP implies RNP [JH].

(ii) See e.g. [D].

(iii) If X is reflexive and LωR, then, by (ii), X^* is Frechet differentiable. By [O], it follows that P_M is continuous on X for all bounded closed convex sets M in X. □

2. For $x_1, \ldots, x_{k+1}$ in X, let

$$V(x_1,\ldots,x_{k+1}) = \sup\left\{ \begin{vmatrix} 1 & \cdots & 1 \\ f_1(x_1) & \cdots & f_1(x_{k+1}) \\ \cdots & \cdots & \cdots \\ f_k(x_1) & \cdots & f_k(x_{k+1}) \end{vmatrix} : \begin{matrix} f_i \in B_{X^*} \\ i = 1, \ldots, k \end{matrix} \right\}$$

and for $i = 1,\dots,k$, let $d_i = d(x_{i+1}, \mathrm{aff}[x_1,\dots,x_i])$ where $\mathrm{aff}[x_1,\dots,x_i]$ is the affine span of $\{x_1,\dots,x_i\}$. Define $d(x_1,\dots,x_{k+1}) = \min\{d_1,\dots,d_k\}$. In [GS], it is proved that $\prod_{i=1}^{k} d_i \le V(x_1,\dots,x_{k+1}) \le k^{k/2} \prod_{i=1}^{k} d_i$. Hence $V(x_1,\dots,x_{k+1}) \to 0$ if and only if $d(x_1,\dots,x_{k+1}) \to 0$. Thus a Banach space X is k–UR (resp. Lk–UR) if for every $\epsilon > 0$ (resp. and for every x in S_X), there exists $\delta = \delta(\epsilon) > 0$ (resp. $\delta = \delta(\epsilon,x) > 0$) such that for all $x_1,\dots,x_{k+1}$ in B_X with $\frac{1}{k+1}\|\sum_{i=1}^{k+1} x_i\| \ge 1-\delta$ $\Big($resp. $\frac{1}{k+1}\|x + \sum_{i=1}^{k} x_i\| \ge 1-\delta\Big)$ then $d(x_1,\dots,x_{k+1}) < \epsilon$ (resp. $d(x,x_1,\dots,x_k) < \epsilon$). This leads to the following definition.

<u>Definition</u>. Let X be a Banach space. X is said to be ωUR (resp. L ωUR) if for all triangular sequence $\{x_i^{(n)} : 1 \le i \le n,\ n \in \mathbb{N}\}$ in B_X with $\lim_n \Big(n - \|\sum_{i=1}^{n} x_i^{(n)}\|\Big) = 0$ $\Big($resp. $\lim_n \Big((n+1) - \|x + \sum_{i=1}^{n} x_i^{(n)}\| = 0\Big)$ then $\lim_n d(x_1^{(n)},\dots,x_n^{(n)}) = 0$ $\Big($resp. $\lim_n d\big(x,x_1^{(n)},\dots,x_n^{(n)}\big) = 0\Big)$.

It is clear that for any $k \in \mathbb{N}$, every k–UR (resp. L k–UR) space is ωUR (resp. L ωUR).

<u>Proposition 6</u>. *If a Banach space X is ωUR then X is superreflexive.*

<u>Proof</u>. Suppose X is not superreflexive. Then for each $n \in \mathbb{N}$, there are $x_1^{(n)},\dots,x_n^{(n)}$ in B_X and $f_1^{(n)},\dots,f_n^{(n)}$ in B_{X^*} such that

$$f_i^{(n)}\left(x_j^{(n)}\right) = \begin{cases} 1-\frac{1}{n^2} & \text{if } j \geq i \\ 0 & \text{if } j < i. \end{cases} \qquad i,j = 1,2,\ldots,n$$

Then $\|x_1^{(n)} + \cdots + x_n^{(n)}\| = n - \frac{1}{n}$, $n \in \mathbb{N}$ but $d\left(x_{i+1}^{(n)}, \text{aff}[x_1^{(n)},\ldots,x_i^{(n)}]\right) \geq 1 - \frac{1}{n^2} \geq \frac{1}{2}$ for all $i = 1,\ldots,n-1$, $n \in \mathbb{N}$. Hence X is not ωUR. □

Let $\{x_n\}$ be a sequence in a Banach space X. The separation constant of $\{x_n\}$ is defined to be $\text{sep}\{x_n\} = \inf\{\|x_n - x_m\| : n \neq m\}$.

<u>Lemma 7</u>. Let $\{x_n\}$ be a sequence in the unit ball of a Banach space X. If $\text{sep}\{x_n\} > \epsilon > 0$ then for each $k \in \mathbb{N}$, there exist $n_1,\ldots,n_k$ such that $d\left(x_1, x_{n_1},\ldots,x_{n_k}\right) \geq \frac{\epsilon}{4}$.

<u>Proof</u>. We proceed by induction. For $k = 1$, it is obvious.

Suppose that $n_1,\ldots,n_k$ have been chosen such that $d\left(x_1, x_{n_1},\ldots,x_{n_{k+1}}\right) \geq \frac{\epsilon}{4}$. Let F be the subspace spanned by $x_1, x_{n_1},\ldots,x_{n_{k+1}}$. If there is $n_{k+1} > n_k$ such that $d(x_{n_{k+1}}, F) \geq \frac{\epsilon}{4}$. Then $d(x_1, x_{n_1},\ldots,x_{n_{k+1}}) \geq d(x_{n_{k+1}}, F) \geq \frac{\epsilon}{4}$. Suppose for all $n > n_k$, $d(x_n, F) < \frac{\epsilon}{4}$. Then for each $n > n_k$, there is $y_n \in F$ such that $\|x_n - y_n\| < \frac{\epsilon}{4}$. Hence $\|y_n\| \leq \|x_n\| + \|x_n - y_n\| \leq 1 + \frac{\epsilon}{4} \leq 2$. However, F is finite dimension, there is a finite $\frac{\epsilon}{4}$-net A in $2B_F$. Thus for every $n > n_k$, there is an element y in A such that $\|x_n - y\| \leq \|x_n - y_n\| + \|y_n - y\| < \frac{\epsilon}{4} + \frac{\epsilon}{4} = \frac{\epsilon}{2}$. Since A is finite, there are $n \neq m$ in N and some y in A such that $\|x_n - y\| < \frac{\epsilon}{2}$ and $\|x_m - y\| < \frac{\epsilon}{2}$. But then $\|x_n - x_m\| < \epsilon$ which is a contradiction. □

Theorem 8. *Let* X *be a strictly convex Banach space. If* X *is* ωUR *(resp.* L ωUR*) then* X *is* ωR *(resp.* $L\omega R$*).*

Proof. We prove the case when X is L ωUR only. The global case is proved similarly.

Let x be an element in S_X and let $\{x_n\}$ be a sequence in B_X such that $\lim_{n_1,\dots,n_k \to \infty} \|x + x_{n_1} + \cdots + x_{n_k}\| = k+1$ for all k in $\mathbb{N}$. If $\{x_n\}$ does not converge to x, by taking a subsequence if necessary, we may assume that there is an $\epsilon > 0$ such that $\text{sep}\{x,x_n\} > \epsilon$. By Lemma 7, for each k, there are $x_1^{(k)},\dots,x_k^{(k)}$ in $\{x_n\}$ such that $\|x + x_1^{(k)} + \cdots + x_k^{(k)}\| \geq k + 1 - \frac{1}{k+1}$ and $d(x,x_1^{(k)},\dots,x_k^{(k)}) \geq \frac{\epsilon}{4}$ which is impossible when X is L ωUR. □

A Banach space X is nearly uniformly convex [H] if for every $\epsilon > 0$, there is a $\delta > 0$ such that for any sequence $\{x_n\}$ in B_X with $\text{sep}\{x_n\} < \epsilon$, then $\text{co}(\{x_n\}) \cap (1-\delta)B_X \neq \phi$. We, now, localize the concept of nearly uniformly convexity and introduce the following.

Definition. A Banach space X is said to be locally nearly uniformly convex (LNUC) if for every $\epsilon > 0$ and for every x in B_X, there is a $\delta = \delta(x,\epsilon) > 0$ such that for any sequence $\{x_n\}$ in B_X with $\text{sep}\{x_n\} > \epsilon$ then $\text{co}(\{x,x_n\}) \cap (1-\delta)B_X \neq \phi$.

It is clear that every NUC space is LNUC.

Theorem 9. *Let* X *be a Banach space. If* X *is* ωUR *(resp.* $L\omega UR$*), then* X *is NUC (resp. LNUC).*

Proof. Again, the proof for the two cases are similar. We prove the local case only.

Suppose X is not LNUC. Then there is an element x in B_X and $\epsilon > 0$ such that for each $n \in \mathbb{N}$, there is a sequence $\{x_i^{(n)}\}$ in B_X with $sep\{x_i^{(n)}\} > 2\epsilon$ but

$$co(\{x,x_i^{(n)}\}) \cap \left(1 - \frac{1}{n(n+1)}\right)B_X = \phi. \qquad (*)$$

Since $sep(\{x_n\}) > 2\epsilon$, by taking a subsequence if necessary, we may assume that $sep(\{x,x_n\}) > \epsilon$. By Lemma 7, for each $n \in \mathbb{N}$, there are $\{y_1^{(n)},\dots,y_n^{(n)}\}$ in $\{x_n\}$ such that $d(x,y_1^{(n)},\dots,y_n^{(n)}) \geq \frac{\epsilon}{4}$. By $(*)$, we have that $\frac{1}{n+1}\|x + y_1^{(n)} + \cdots + y_n^{(n)}\| > 1 - \frac{1}{n(n+1)}$ for all n. Hence $\lim_n\left((n+1) - \|x + y_1^{(n)} + \cdots + y_n^{(n)}\|\right) = 0$. However $\lim_n d(x,y_1^{(n)},\dots,y_n^{(n)}) \geq \frac{\epsilon}{4}$ which is a contradiction to the hypothesis that X is LωUR. □

We conjecture that every strictly convex NUC (resp. LNUC) Banach space is ωR (resp. LωR). Although we are not able to prove the conjecture. In the following, we shall prove some properties of LNUC spaces to substantial our conjecture.

Theorem 10. *Let X be a Banach space.* (i) *If X is LNUC then X has the Kadec property, that is,* $(S_X,w) = (S_X,\|\cdot\|)$. (ii) *If* X^* *is LNUC, then* $(S_{X^*},w^*) = (S_{X^*},\|\cdot\|)$.

Proof. (i) Suppose there are an element x_0 in S_X and an $\epsilon > 0$ such that $diam(V \cap S_X) \geq 2\epsilon$ for all weak neighborhood V of x_0. Let x_0^* be an element in S_{X^*} such that $x_0^*(x_0) = 1$. Let $V_0 = \{x : x \in S_X, |x_0^*(x-x_0)| < 1/2\}$. Then there is an element x_1 in V_0 such that $\|x_1-x_0\| \geq \epsilon$. Choose $x_1^* \in S_{X^*}$ such that $x_1^*(x_1-x_0) \geq \epsilon$.

We continue by induction, suppose that there are $x_1,\dots,x_k$ in S_X, $x_1^*,\dots,x_k^* \in S_{X^*}$ such that $x_i^*(x_i-x_0) \geq \epsilon$ and $\|x_i-x_y\| \geq \frac{\epsilon}{2}$ for all $i \neq j$, $i,j = 1,2,\dots,k$. Let $V_k = \{x : x \in S_X, |x_0^*(x-x_0)| < \frac{1}{2^k}, |x_i^*(x-x_0)| < \frac{\epsilon}{2}, i = 1,2,\dots,k\}$. Then there is $x_{k+1} \in V_k$ such that $\|x_{k+1}-x_0\| \geq \epsilon$. Then $x_0^*(x_{k+1}) > x_0^*(x_0) - \frac{1}{2^k} = 1 - \frac{1}{2^k}$ and $\|x_{k+1}-x_i\| \geq x_i^*(x_i-x_{k+1}) = x_i^*(x_i-x_0) - x_i^*(x_{k+1}-x_0) \geq \epsilon - \frac{\epsilon}{2} = \frac{\epsilon}{2}$ for all $i = 1,2,\dots,k$. In this way, $\{x_n\}$ is a sequence in B_X and $\operatorname{sep}(x_0,x_n) \geq \frac{\epsilon}{2}$. However, for any $\delta > 0$, choose N such that $\frac{1}{2^N} < \delta$.

For any $\lambda_i \geq 0$, $i = 0,1,\dots,k$, $\sum_{i=1}^{k} \lambda_i = 1$, if $n_1,\dots,n_k > N$ then $\|\lambda_0 x_0 + \lambda_1 x_{n_1} + \cdots + \lambda_k x_{n_k}\| \geq x_0^*(\lambda_0 x_0 + \cdots + \lambda_k x_{n_k}) \geq 1 - \frac{1}{2^N} > 1-\delta$. Hence $\operatorname{co}(\{x_0,x_n : n \in \mathbb{N}\}) \cap (1-\delta)B_X = \phi$. Thus X fails to be LNUC.

The proof for (ii) is similar. □

<u>Corollary 11</u>. Let X be a Banach space.

(i) If X^* is LNUC then X^* has the RNP;

(ii) If X^{**} is LNUC then X is reflexive.

The following diagrams show the relations between the convexity properties considered in this paper.

$$UR \Rightarrow \cdots \Rightarrow kUR \quad \Rightarrow (k+1)UR \Rightarrow \cdots \Rightarrow \omega UR \Rightarrow \text{super reflexivity}$$
$$\Downarrow \qquad\qquad \Downarrow (+R) \qquad\qquad\qquad \Downarrow (+R) \searrow NUC$$
$$\qquad\qquad\qquad\qquad\qquad\qquad\qquad\qquad\qquad\qquad \Downarrow$$
$$2R \Rightarrow \cdots \Rightarrow (k+1)R \Rightarrow (k+2)R \Rightarrow \cdots \Rightarrow \omega R \Rightarrow \text{reflexivity.}$$

$$LUR \Rightarrow \cdots \Rightarrow L\,kUR \quad \Rightarrow L(k+1)UR \Rightarrow \cdots \Rightarrow L\,\omega UR \Rightarrow \qquad LNUC$$
$$\qquad\qquad\qquad\qquad\qquad\qquad\qquad\qquad\qquad\qquad\qquad (+R) \qquad \Downarrow$$
$$w\,LUR \quad L2R \Rightarrow \cdots \Rightarrow L(k+1)R \Rightarrow L(k+2)R \Rightarrow \cdots \Rightarrow L\,\omega R \Rightarrow (G) \Rightarrow (K) \Rightarrow (KK)$$
$$\searrow \quad \Downarrow \qquad\qquad\qquad\qquad\qquad\qquad\qquad\qquad\qquad\qquad \Downarrow \qquad \Downarrow$$
$$w\,L2R \Rightarrow \cdots \Rightarrow w\,L(k+1)R \Rightarrow w\,L(k+2)R \Rightarrow \cdots \Rightarrow w\,L\omega R \Rightarrow (R)$$

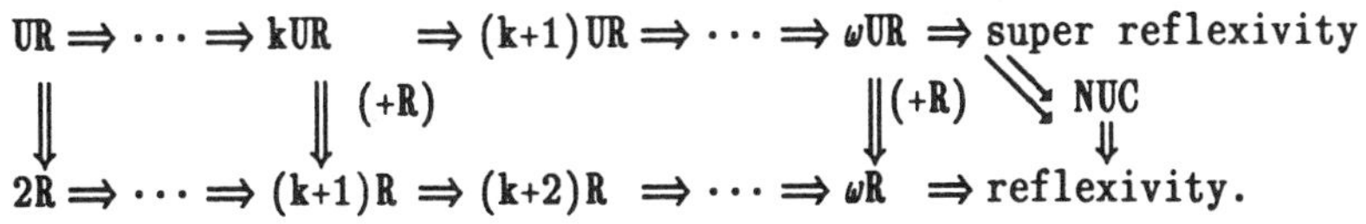

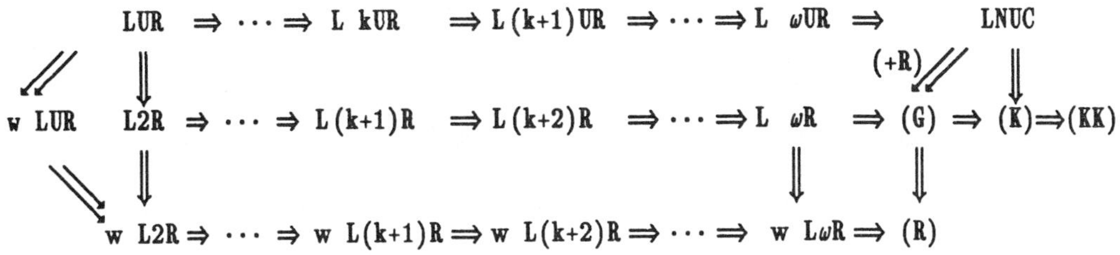

References

[BS] J. Bernal and F. Sullivan, Multi–dimensional volumes, superreflexivity and normal structure in Banach spaces, III. *J. Math.* **27**(1983), 501–513.

[D] J. Diestel, Geometry of Banach spaces, *Lecture Notes in Math.*, Springer–Verlag, **485**(1975).

[FG1] Ky Fan and I. Glicksberg, Fully convex normed linear spaces, *Proc. Nat. Acad. Sci. U.S.A.* **41**(1955), 947–953.

[FG2] Ky Fan and I. Glicksberg, Some geometric properties of the spheres in a normed linear space, *Duke Math. J.* **25**(1958), 553–568.

[GS] R. Geremia and F. Sullivan, Multi–dimensional volumes and moduli of convexity in Banach spaces, *Ann. Mat. Pura Appl.* **127**(1981), 231–251.

[HL] Zhibao Hu and Bor–Luh Lin, On the asymptotic–norming property of Banach spaces (to appear).

[H] R. Huff, Banach spaces which are nearly uniformly convex, *Rocky Mountain J. Math.* **19**(1980), 734–749.

[K] D. Kutzarova, k–β and k–nearly uniformly convex Banach spaces, (to appear).

[KL] D. Kutzarova and Bor–Luh Lin, Locally k–nearly uniformly convex Banach spaces (to appear).

[LLT] Bor–Luh Lin, Pei–Kee Lin and S.L. Troyanski, A characterization of denting point of a closed bounded convex set, Longhorn Notes, The University of Texas at Austin, (1985/86), 99–101.

[LY] Bor–Luh Lin and Xintai Yu, On the k–uniformly rotund and the fully convex Banach spaces, *J. Math. Anal. Appl.* **110**(1985), 407–410.

[L] A.R. Lovaglia, Locally uniformly convex Banach spaces, *Trans. Amer. Math. Soc.* **78**(1955), 225–238.

[NW] Nan Chao–Xun and Wang Jian–Hua, On the Lk–UR and L–kR spaces, *Math. Proc. Camb. Phil. Soc.* **104**(1988), 521–526.

[O] E.V. Oshman, Continuously criterion for metric projections in a Banach space, *Math. Notes* **10**(1971), 697–701.

[S] F. Sullivan, A generalization of uniformly rotund Banach spaces, *Canad. J. Math.* **31**(1979), 628–636.

[Y] Xintai Yu, k–UR spaces are NUC spaces, *Ke Xue Tong Pao* **24**(1983), 1473–1475 (Chinese).

Toeplitz Operators on Weighted Hardy Spaces

B. A. Lotto
Department of Mathematics
University of California, Davis
Davis CA 95616

The classical Hardy space H^2 can be viewed as the closure of the analytic polynomials in L^2 of the unit circle. One generalization of H^2 can be obtained by taking the closure of the analytic polynomials in $L^2(\mu)$, where μ is a positive Borel measure on the unit circle. This space, a weighted Hardy space, is denoted by $H^2(\mu)$.

For ϕ belonging to L^∞, the Toeplitz operator T_ϕ with symbol ϕ is defined on H^2 by the formula $T_\phi f = P(\phi f)$, where P is the bounded projection from L^2 onto H^2 that truncates an arbitrary trigonometric polynomial $\sum_{j=-n}^{n} c_j z^j$ to the analytic polynomial $\sum_{j=0}^{n} c_j z^j$. We could try to generalize Toeplitz operators to weighted Hardy spaces by using the same formula as in the classical case; in general, however, P is not bounded from $L^2(\mu)$ onto $H^2(\mu)$, so this is not possible.

A different possibility for generalization stems from the observation that when m is in H^∞, $T_{\bar{m}}$ maps analytic polynomials to analytic polynomials. Hence we can regard $T_{\bar{m}}$ as a densely defined operator on $H^2(\mu)$ and ask the question: When is $T_{\bar{m}}$ bounded on $H^2(\mu)$?

The purpose of this paper is to connect the operator $T_{\bar{m}}$ on $H^2(\mu)$ with a multiplication operator on a Hilbert space of analytic functions.

For w in the unit disk $\mathbf{D}$, the function $k_w(z) = (1-\bar{w}z)^{-1}$ is the reproducing kernel function at w for H^2; that is, we have the equality $\langle f, k_w \rangle_2 = f(w)$ for every f in H^2. Note that k_w is continuous on $\partial\mathbf{D}$.

Lemma 1 *The span of* $\{k_w : w \in \mathbf{D}\}$ *is dense in* $H^2(\mu)$.

For w in $\mathbf{D}$, k_w is defined on $\partial\mathbf{D}$ and is continuous there, so k_w belongs to $L^2(\mu)$. Also, the Taylor series for k_w converges uniformly on $\overline{\mathbf{D}}$ and hence on $\partial\mathbf{D}$. Since uniform convergence implies $L^2(\mu)$ convergence, it follows that k_w is in $H^2(\mu)$.

To show that the span of $\{k_w : w \in \mathbf{D}\}$ is dense in $H^2(\mu)$, it will suffice to prove the stronger fact that z^n belongs to the uniform closure of the span of the k_w's. We proceed by induction. That $z^0 = 1 = k_0$ belongs to this closure is immediate, so fix $n > 0$ and suppose that z^k belongs to the uniform closure of the span of the k_w's for all $0 \leq k < n$. Since

$$\frac{1}{\bar{w}^n}(k_w(z) - 1 - \bar{w}z - \ldots - \bar{w}^{n-1}z^{n-1})$$

tends to z^n uniformly on $\partial\mathbf{D}$ as w tends to 0, the lemma is proved.

Lemma 2 *Suppose that m belongs to H^∞. The operator $T_{\bar{m}}$ is bounded on $H^2(\mu)$ if and only if the map $k_w \mapsto \overline{m(w)}k_w$ extends to a continuous linear operator on $H^2(\mu)$.*

Note that the collection $\{k_w : w \in \mathbf{D}\}$ is a linearly independent subset of $H^2(\mu)$, so the linear extension of the map $k_w \mapsto \overline{m(w)}k_w$ to the span of $\{k_w : w \in \mathbf{D}\}$ is well-defined.

Suppose first that $T_{\bar{m}}$ is bounded on $H^2(\mu)$. Then $T_{\bar{m}}$ extends to a continuous linear operator on all of $H^2(\mu)$. We will also call this extension $T_{\bar{m}}$.

Fix w in $\mathbf{D}$. Since

$$k_w(z) = \frac{1}{1 - \bar{w}z} = \sum_{n=0}^{\infty} \bar{w}^n z^n$$

converges uniformly on $\partial\mathbf{D}$, the series also converges in $H^2(\mu)$. Hence

$$T_{\bar{m}}k_w = \sum_{n=0}^{\infty} \bar{w}^n T_{\bar{m}} z^n$$

with the series converging in $H^2(\mu)$. A calculation (using the identity $T_{\bar{m}}z^n = \sum_{j=0}^{n} \overline{\hat{m}(n-j)}z^j$, where $\hat{m}(k)$ is the coefficient of z^k in the Taylor expansion of $m(z)$ about the origin) shows that this series converges to $\overline{m(w)}k_w$, absolutely and uniformly for z in $\partial\mathbf{D}$ and $|w| \leq r < 1$, and therefore in $H^2(\mu)$ for each fixed w in $\mathbf{D}$. Hence $T_{\bar{m}}k_w = \overline{m(w)}k_w$ and $T_{\bar{m}}$ is itself the continuous linear extension to $H^2(\mu)$ of the map $k_w \mapsto \overline{m(w)}k_w$.

Now suppose conversely that the map $k_w \mapsto \overline{m(w)}k_w$ extends to a continuous linear operator T on $H^2(\mu)$. Since $k_w(z) = \sum_{n=0}^{\infty} \bar{w}^n z^n$ in $H^2(\mu)$, we have that $Tk_w = \sum_{n=0}^{\infty} \bar{w}^n T z^n$. But $Tk_w = \overline{m(w)}k_w = \sum_{n=0}^{\infty} \bar{w}^n T_{\bar{m}} z^n$ in $H^2(\mu)$ as well. Since these two power series (in $\bar{w}$) represent the same conjugate analytic function in $\mathbf{D}$, their coefficients must be the same, so $Tz^n = T_{\bar{m}} z^n$ for all $n \geq 0$.

Let b belong to the unit ball of H^∞ and define the $\mathcal{H}(b)$ to be the range of the operator $(1 - T_b T_{\bar{b}})^{1/2}$. We make $\mathcal{H}(b)$ into a Hilbert space by giving it the norm that makes $(1 - T_b T_{\bar{b}})^{1/2}$ into a coisometry of H^2 onto $\mathcal{H}(b)$; that is, if f in H^2 is orthogonal to the kernel of $(1 - T_b T_{\bar{b}})^{1/2}$, then we have $\|(1 - T_b T_{\bar{b}})^{1/2} f\|_{\mathcal{H}(b)} = \|f\|_2$. (We exclude the trivial case where b identically equals a constant of modulus 1, in which case $\mathcal{H}(b)$ contains only the zero function.) The space $\mathcal{H}(b)$ is called a de Branges space after L. de Branges, who, with J. Rovnyak, first studied these spaces [5].

Clearly $\mathcal{H}(b)$ is contained in H^2 and, since the operator $(1 - T_b T_{\bar{b}})^{1/2}$ is a contraction, the inclusion map from $\mathcal{H}(b)$ into H^2 is also a contraction. In particular, evaluation at any w in $\mathbf{D}$ is a continuous linear functional in $\mathcal{H}(b)$, so there is a reproducing kernel function $k_w^b(z)$ in $\mathcal{H}(b)$.

Since b maps $\mathbf{D}$ into itself, the function $(1+b)/(1-b)$ has positive real part in $\mathbf{D}$. We therefore have the Herglotz representation

$$\frac{1+b(z)}{1-b(z)} = \int_{\partial \mathbf{D}} \frac{e^{i\theta} + z}{e^{i\theta} - z}\, d\mu(e^{i\theta}) + ic$$

where μ is a positive Borel measure on $\partial\mathbf{D}$ and c is a real constant. Note that if we start with a positive Borel measure μ on $\partial\mathbf{D}$ we can define b by this formula (by setting c equal to zero) and b will be in the unit ball of H^∞ (and not identically equal to a constant of modulus 1). We will assume below that b and μ are always related in this manner.

The following theorem connecting $H^2(\mu)$ and $\mathcal{H}(b)$, in the special case where μ is a singular measure (equivalently, when b is an inner function), is due to D. N. Clark [3]. A vector-valued version containing this one as a special case can be found in J. A. Ball and A. Lubin [2]; another vector-valued result encompassing this one appears in D. Alpay and H. Dym [1]. A proof of the result in this form can be found in [9].

Theorem 3 *There is a unitary map of $H^2(\mu)$ onto $\mathcal{H}(b)$ that sends k_w to $(1 - \overline{b(w)})^{-1} k_w^b$ for every w in $\mathbf{D}$.*

If m is in H^∞, we say that m is a multiplier of $\mathcal{H}(b)$ if mf is in $\mathcal{H}(b)$ whenever f is.

Lemma 4 *Suppose that m belongs to H^∞. Then m is a multiplier of $\mathcal{H}(b)$ if and only if the map $k_w^b \mapsto \overline{m(w)}k_w^b$ extends to a continuous linear operator on $\mathcal{H}(b)$.*

This is a standard calculation. Suppose that m is a multiplier of $\mathcal{H}(b)$ and let M_m denote the operator "multiplication by m" on $\mathcal{H}(b)$. A closed graph theorem argument shows that M_m is bounded on $\mathcal{H}(b)$.

Fix w in $\mathbf{D}$. Then we have

$$\langle M_m^* k_w^b, f \rangle_b = \langle k_w^b, mf \rangle_b = \overline{m(w)f(w)} = \overline{m(w)}\langle k_w^b, f \rangle_b = \langle \overline{m(w)}k_w^b, f \rangle_b$$

for every f in $\mathcal{H}(b)$. (The notation $\langle \cdot, \cdot \rangle_b$ refers to the inner product in $\mathcal{H}(b)$.) Hence $M_m^* k_w^b = \overline{m(w)}k_w^b$ and M_m^* is the desired extension.

Conversely, suppose that M is a bounded linear operator such that $Mk_w^b = \overline{m(w)}k_w^b$ for every w in $\mathbf{D}$. A similar calculation now shows that $M^*f = mf$ for every f in $\mathcal{H}(b)$. In particular, mf is in $\mathcal{H}(b)$ whenever f in $\mathcal{H}(b)$, so m is a multiplier of $\mathcal{H}(b)$.

Corollary 5 *Suppose m belongs to H^∞. Then $T_{\bar{m}}$ is bounded on $H^2(\mu)$ if and only if m is a multiplier of $\mathcal{H}(b)$.*

Since the unitary operator of Theorem 3 takes k_w in $H^2(\mu)$ to a constant multiple of k_w^b in $\mathcal{H}(b)$, the map $k_w \mapsto \overline{m(w)}k_w$ extends to a bounded linear operator on $H^2(\mu)$ if and only if the map $k_w^b \mapsto \overline{m(w)}k_w^b$ extends to a bounded linear operator on $\mathcal{H}(b)$. The corollary now follows from Lemmas 2 and 4.

To give an example of how the above results can be used, let us consider the case $m(z) = z$. We will denote $T_{\bar{m}}$ in this case by S^*. (The operator S denotes the unilateral shift on H^2 defined by the formula $Sf(z) = zf(z)$.) If p is an analytic polynomial, then we have the formula

$$S^*p(z) = \bar{z}(p(z) - p(0)),$$

valid for all z in $\partial\mathbf{D}$. Since multiplication by $\bar{z}$ is a unitary operator on $L^2(\mu)$, it follows that S^* is bounded on $H^2(\mu)$ if and only if the functional $p \mapsto p(0)$ of evaluation at 0 is bounded on $H^2(\mu)$. A theorem of Szegö [7,

p. 49] says that this is the case if and only if the Radon-Nikodym derivative of the absolutely continuous part of μ with respect to (normalized) Lebesgue measure is log-integrable on $\partial\mathbf{D}$, or equivalently if and only if $H^2(\mu) \neq L^2(\mu)$.

An easy calculation shows that this Radon-Nikodym derivative equals $(1-|b|^2)/|1-b|^2$. Since $\log|1-b|^2$ is always integrable (being the logarithm of the modulus of the H^∞ function $1-b$), we see that S^* is bounded on $H^2(\mu)$ if and only if $\int \log(1-|b|^2) > -\infty$. By the above results, this condition also characterizes those functions b for which $m(z) = z$ is a multiplier of $\mathcal{H}(b)$. This is a well-known result, and can be found (with a different proof) in [5] and [10]. (We note that the condition $\int \log(1-|b|^2) > -\infty$ is exactly the condition that b not be an extreme point of the unit ball of H^∞ [6]. The structure of $\mathcal{H}(b)$ depends strongly on whether b is an extreme point.)

Since the collection of multipliers of $\mathcal{H}(b)$ is an algebra, every analytic polynomial is a multiplier of $\mathcal{H}(b)$ if z is. On the other hand, the formula $T_{\overline{S^*m}}p = T_{\bar{m}}(zp)$, valid for all analytic polynomials p and m in H^∞, shows that if $T_{\bar{m}}$ is bounded on $H^2(\mu)$ then so is $T_{\overline{S^*m}}$. Hence if there is any nonconstant analytic polynomial m such that $T_{\bar{m}}$ is bounded on $H^2(\mu)$, then so is S^*, and $H^2(\mu) \neq L^2(\mu)$. So if $H^2(\mu) = L^2(\mu)$, the only analytic polynomials that give rise to bounded coanalytic Toeplitz operators are the constants.

The question then arises: if $H^2(\mu) = L^2(\mu)$, are there any nontrivial H^∞ functions m such that $T_{\bar{m}}$ is bounded on $H^2(\mu)$? The answer is yes, and follows from results about the multipliers of $\mathcal{H}(b)$ in [8].

More results about bounded coanalytic Toeplitz operators on $H^2(\mu)$ can be found in [4].

I would like to thank the referee for his suggestions in improving the proofs of Lemmas 1 and 2.

References

[1] D. Alpay and H. Dym. Hilbert spaces of analytic functions, inverse scattering, and operator models, I. *Integral Equations and Operator Theory*, 7, 1984.

[2] J. A. Ball and A. Lubin. On a class of contractive perturbations of restricted shifts. *Pacific J. Math.*, 63:309–323, 1976.

[3] D. N. Clark. One dimensional perturbations of restricted shifts. *J. Analyse Math.*, 25:169–191, 1972.

[4] B. Mark Davis and John E. McCarthy. Multipliers of de Branges' spaces. Preprint.

[5] L. de Branges and J. Rovnyak. *Square Summable Power Series.* Holt, Rinehart, and Winston, New York, 1966.

[6] K. de Leeuw and W. Rudin. Extreme point and extremum problems in H^1. *Pacific J. Math.*, 8, 1958.

[7] Kenneth Hoffman. *Banach Spaces of Analytic Functions.* Prentice-Hall, Englewood Cliffs, New Jersey, 1962.

[8] B. A. Lotto and D. Sarason. Multiplicative structure of de Branges's spaces. *Revista Mathematica Iberoamericana*, to appear.

[9] D. Sarason. Exposed points in H^1, I. In *Operator Theory: Advances and Applications, vol. 41*, pages 485–496, Birkhäuser Verlag, Basel, 1989.

[10] D. Sarason. Shift-invariant spaces from the Brangesian point of view. In *The Bieberbach Conjecture—Proceedings of the Symposium on the Occasion of the Proof*, American Mathematical Society, Providence, 1986.

Weighted Bergman Spaces

JOHN E. McCARTHY Department of Mathematics, Indiana University, Bloomington, Indiana 47405

0. INTRODUCTION

The intention of this article is to describe a particular example; it is a simple example, but I hope it is sufficiently appealing to induce the reader to think about the questions it raises. The reader is warned that this is not a research article, but rather an illustrative one.

For μ a compactly supported (positive) measure on $\mathbb{C}$, let $P^2(\mu)$ denote the closure of the polynomials in $L^2(\mu)$. If μ is Lebesgue measure on the circle (hereinafter referred to as σ) $P^2(\sigma)$ is the classical Hardy space of square summable power series; if μ is area measure on the unit disk (this measure will be called A), $P^2(A)$ is normally called the Bergman space.

In **[Hel]** Henry Helson studied the union of all the spaces $P^2(\mu)$ as μ ranges over those measures mutually absolutely continuous with respect to σ that satisfy $P^2(\mu) \cap L^\infty(\mu) = H^\infty$, where H^∞ is the space of bounded analytic functions on the unit disk (as the measures are supported on the unit circle, we really mean radial limits of bounded analytic functions; but, as usual, we will slur this distinction). Endowing this space with the inductive limit topology H (so a neighbourhood base at zero is given by those balanced convex sets whose intersection with every $P^2(\mu)$ is a neighbourhood of zero in $P^2(\mu)$), he identified the dual with the common range of every co-analytic Toeplitz operator on $P^2(\sigma)$ (for a given bounded function m, the Toeplitz operator with symbol m on $P^2(\mu)$, written T_m^μ, or T_m if μ is understood, is the operator of multiplication by m followed by orthogonal projection from $L^2(\mu)$

onto $P^2(\mu)$; if $\bar{m}$ is in H^∞, T_m is called *co-analytic*).

In **[McC1]** it was shown, using this, that a function f is in the common range of all co-analytic Toeplitz operators on $P^2(\sigma)$ if and only if its Fourier coefficients $\hat{f}(n)$ are $O(e^{-c\sqrt{n}})$ for some positive c. Moreover, it was proved in **[McC2]** that the space

$$\bigcup\{P^2(\mu) : \mu \equiv \sigma \quad \textit{and} \quad P^2(\mu) \cap L^\infty(\mu) = H^\infty\}$$

with the topology H is a metrizable topological algebra in which Fourier series converge.

We consider analogues of these results for the Bergman space $P^2(A)$. The theory turns out to be much simpler in this case, because one can prove straight off that Fourier series converge in the inductive limit topology, and this greatly facilitates calculation of the dual.

In our inductive limits, we always look at the union of the $P^2(\mu)$-spaces as μ ranges over those measures mutually absolutely continuous with respect to some fixed measure μ_0, and satisfying $P^2(\mu) \cap L^\infty(\mu) = P^2(\mu_0) \cap L^\infty(\mu_0)$. The reason for this latter restriction is that otherwise one would end up with all μ_0-measurable functions **[Br]**, and would lose whatever analytic structure one started with. For general facts about $P^2(\mu)$ see **[Co]**.

1. RADIALLY SYMMETRIC MEASURES

Throughout this section, μ_0 will be a fixed probability measure on the unit disk $\mathbb{D}$, that is radially symmetric (*i.e.* of the form $d\nu(r)d\sigma(\theta)$), satisfies $\mu_0(\mathbb{T}) = 0$ (*i.e.* $\nu(1) = 0$), but lives on the whole disk in the sense that the convex hull of its support is $\overline{\mathbb{D}}$ (*i.e.* $\nu([1-\epsilon, 1]) > 0$ for all $\epsilon > 0$). This guarantees that $P^2(\mu_0) \cap L^\infty(\mu_0) = H^\infty$ (*i.e.* the identity map on polynomials extends to an isometric isomorphism and weak-star homeomorphism between the two spaces; alternatively, any function in the former space agrees μ_0-a.e. with a function in the latter, and *vice versa*). The motivating example is $\mu_0 = A$, but no simplification ensues by restricting to this case, so we indulge in a little more generality.

First, let us prove our asymptotic Szegö theorem on how fast the Taylor coefficients can grow:

PROPOSITION 1. *Let μ be a measure absolutely continuous with respect to μ_0, and satisfying $P^2(\mu) \cap L^\infty(\mu) = H^\infty$; let $c > 0$ be given. Then there is a constant K such that*

$$sup\{|\hat{p}(n)| : p \quad a \quad polynomial, \int_{\mathbb{D}} |p|^2 d\mu \leq 1\} \leq Ke^{cn}.$$

Moreover, e^{cn} cannot be replaced by $e^{c_n n}$ for any sequence $\{c_n\}$ decreasing to zero.

PROOF: By **[Th]**, if all the bounded functions in $P^2(\mu)$ are analytic on $\mathbb{D}$, then every function in $P^2(\mu)$ is analytic there, and so has a power series with radius of convergence at least one. Therefore the family of linear functionals Γ_k, given by

$$\Gamma_k(f) = \hat{f}(k)e^{-ck}$$

is pointwise bounded on $P^2(\mu)$, and hence equicontinuous. Therefore the inequality holds.

Now if c_n decreases to zero, the power series of the function $f(z) = \sum_{n=0}^{\infty} ne^{c_n n} z^n$ has radius of convergence one, so f is analytic on $\mathbb{D}$. Define

$$w(r) = \frac{1}{\sup_{|z|=r} |f(z)|^2 + 1}. \qquad (\star)$$

Then if $\mu = w(r)\mu_0$, f is in the unit ball of $P^2(\mu)$, so

$$\sup_{\int |p|^2 d\mu = 1} |\hat{p}(n)| \geq ne^{c_n n}.$$

□

Defining O^+ by

$$O^+ = \cup\{P^2(\mu) : \mu \equiv \mu_0, \quad \textit{and} \quad P^2(\mu) \cap L^\infty(\mu) = H^\infty\},$$

and endowing it with the inductive limit topology H, then we first observe that Taylor series converge in (O^+, H): for if f is any function in O^+, define $w(r)$ as in $(\star)$; let $d\mu(re^{i\theta}) = (1-r)w(r)d\mu_0(re^{i\theta})$; then the Taylor series for f converges in $P^2(\mu)$, and hence in (O^+, H).

Because Taylor series converge, the dual of (O^+, H) consists of those sequences that, when multiplied term by term by the Taylor coefficients of a function in O^+, give a convergent series. It follows from Proposition 1 that those are the sequences that decay exponentially:

Proposition 2. *Let Γ be a linear functional, defined on polynomials by $\Gamma(p) = \sum_{n=0}^{N} a_n \gamma_n$, where $p(z) = \sum_{n=0}^{N} a_n z^n$. Then Γ extends to be continuous on (O^+, H) if, and only if, for some $c > 0$, $\gamma_n = O(e^{-cn})$.*

Proof: (Sufficiency) Suppose $\gamma_n = O(e^{-cn})$. By Proposition 1, applied to $c/2$, the family of linear functionals $\{\Gamma_k\}$, where $\Gamma_k(\sum_n a_n z^n) = \sum_{n=0}^{k} a_n \gamma_n$, is pointwise bounded, and each Γ_k is continuous because it is continuous on each $P^2(\mu)$.

But O^+, as the inductive limit of Banach spaces, is barrelled (*i.e.* every balanced convex absorbing closed set is a neighbourhood of zero), which is equivalent to having the property that every pointwise bounded family of continuous linear functionals is equicontinuous [**Wi**, 9-3.4]. So $\{\Gamma_k\}$ is equicontinuous, and Γ, as the pointwise limit, is continuous.

(Necessity): Suppose Γ is continuous. Because Fourier series converge,

$$\Gamma(f(z) \sim \sum_{n=0}^{\infty} a_n z^n) = \lim_{N\to\infty} \sum_{n=0}^{N} a_n \gamma_n,$$

and this series must converge for all f in O^+.

If $\gamma_n \neq O(e^{-cn})$ for any $c > 0$, then $|\gamma_{n_k}| \geq ke^{-n_k/k}$ for some subsequence $\{\gamma_{n_k}\}$. Let $f(z) = \sum_{k=0}^{\infty} e^{n_k/k} z^{n_k}$. The power series of f has radius of convergence one, so f is analytic on $\mathbb{D}$, and hence is in O^+; but $\sum_{k=0}^{K} \gamma_{n_k} e^{n_k/k}$ does not converge. □

From Proposition 2 it also follows that the inductive limit topology on O^+ agrees with the topology of uniform convergence on compacta: for both topologies give rise to the same continuous linear functionals, and the one is barrelled and the other bornological, so both are Mackey (*i.e.* the finest locally convex topology with a given dual), and hence must coincide (see [**Wi**] for explanations of these terms). Actually, because we can use weights that decay as rapidly as we like at the boundary, it's easy to prove directly that the two topologies coincide.

However, this simple approach breaks down if the boundary has mass. Even the measure $A + \sigma$ is hard to handle. We are led to ask the following questions:

QUESTION 1. *Let μ be a compactly supported measure on $\mathbb{C}$, which satisfies $P^2(\mu) \neq L^2(\mu)$. Let Ω be the set of mutually absolutely continuous measures ν that satisfy*

$$P^2(\nu) \cap L^\infty(\nu) = P^2(\mu) \cap L^\infty(\mu).$$

Is the union of all $P^2(\nu)$, as ν ranges over Ω, a metrizable topological space in the inductive limit topology?

The reason for the restriction $P^2(\mu) \neq L^2(\mu)$ is that otherwise the only continuous linear functionals on $\cup_{\nu\in\Omega} P^2(\nu)$ are evaluation at the atoms of μ; if μ is not purely atomic, H is therefore not Hausdorff.

It seems, however, that for function spaces, the locally convex inductive limit is often a bad choice. Instead, consider the full inductive limit topology, where a set is open if and only if its intersection with every $P^2(\nu)$ is open. This will not be any finer on O^+ than the locally convex inductive limit topology (basically because the underlying topology, that of uniform convergence on compacta, is locally convex). It does, however, make a big difference in the case of Lebesgue measure on the circle: the new topology comes from a complete metric, that can be explicitly described -see [**McC1**]. It also makes a difference if $P^2(\mu) = L^2(\mu)$. The full inductive limit topology then becomes the (metrizable) topology of convergence in measure. Thus perhaps we should ask:

QUESTION 2. *Let μ be a compactly supported measure on $\mathbb{C}$. Let Ω be as above. Is the union of all $P^2(\nu)$, as ν ranges over Ω, a metrizable topological space in the full inductive limit topology?*

Locally convex inductive limits seem to work better with sequence spaces; and an analytic function can be thought of as a sequence of Taylor coefficients. In particular, the following question has an affirmative answer for σ only if the locally convex inductive limit is used:

QUESTION 3. *Let μ be a measure on the unit disk, satisfying $P^2(\mu) \cap L^\infty(\mu) = H^\infty$. Let Ω be as above. Do Taylor series converge in the inductive limit topology on $\cup_{\nu\in\Omega} P^2(\nu)$?*

It is easy to see that for specific spaces, *e.g.* $H^2(|e^{i\theta} - 1|^2)$, Taylor series do not necessarily converge. Those weighted Hardy spaces in which Fourier series always converge have been classified by Helson and Szegö **[HS]**; but the inductive limit topology, certainly in spaces that are rotation invariant, seems to smear things out enough to allow convergence.

For more information on inductive limits, see the books of Köthe **[Kö]** and Wilansky **[Wi]**, and the article **[He2]**.

2. COMMON RANGE OF CO-ANALYTIC TOEPLITZ OPERATORS

Because the decay condition on Fourier coefficients that places a function in the common range of all Toeplitz operators on $P^2(\sigma)$ is so stringent, the same condition will work for any radial measure μ whose moments $\gamma_n := \int |z^n|^2 d\mu$ satisfy $1/\gamma_n = O(exp[o(\sqrt{n})])$. In particular, it works for the Bergman space:

PROPOSITION 3. *The function f is in the range of every co-analytic Toeplitz operator on the Bergman space if and only if there exists a constant $c > 0$ such that $\hat{f}(n) = O(e^{-c\sqrt{n}})$.*

PROOF: Obviously the range of $T^\sigma_{\bar{m}}$ is contained in the range of $T^A_{\bar{m}}$ for every m in H^∞.

But f is in the range of $T^A_{\bar{m}}$ implies

$$|\frac{1}{\pi}\int_{\mathbb{D}} p\bar{f}dA|^2 \leq C\frac{1}{\pi}\int_{\mathbb{D}} |p|^2|m|^2 dA \leq C\frac{1}{2\pi}\int_{\mathbb{T}} |p|^2|m|^2 d\sigma,$$

for all polynomials p, and so

$$\sum_{n=0}^{\infty} \frac{\hat{f}(n)}{n+1} z^n$$

is in the range of $T^\sigma_{\bar{m}}$. Therefore to lie in the intersection of the ranges of all the $T^A_{\bar{m}}$, a function must decay like $e^{-c\sqrt{n}}$. □

The reason for the discrepancy with Proposition 2, which does not appear in the case of σ, is that f is in the range of every co-analytic Toeplitz operator if and only if it is in the dual of

$$\cup\{P^2(\mu) : \mu \equiv A, \quad and \quad \frac{d\mu}{dA} = |m|, \quad some \quad m \in H^\infty\},$$

which is a much smaller space, and hence has a larger dual, than

$$\cup\{P^2(\mu) : \mu \equiv A, \quad and \quad P^2(\mu) \cap L^\infty(\mu) = H^\infty\}.$$

QUESTION 4. *Let μ be a measure on the disk, satisfying $P^2(\mu) \cap L^\infty(\mu) = H^\infty$. When is a function in the range of every co-analytic Toeplitz operator on $P^2(\mu)$?*

This is probably very hard. The case where μ is radial (but with rapidly decreasing moments) seems more tractable.

REFERENCES

Br J. Bram "Subnormal operators," *Duke Math. Jour.* 22 [1955] 75-94

Co J.B. Conway "Subnormal Operators," Pitman, Boston, 1981

He1 H. Helson "Large analytic functions II," in *Analysis and partial differential equations,* editor Cora Sadosky, Marcel Dekker, Basel, 1990

He2 H. Helson "Large analytic functions III," to appear

HS H. Helson and G. Szegö "A problem in prediction theory," *Ann. Mat. Pura Appl* 51 [1960] 107 - 138

Kö G. Köthe "Topological vector spaces I," Springer-Verlag, Heidelberg, 1969; and "Topological vector spaces II," Springer-Verlag, Heidelberg, 1979

McC1 J.E. McCarthy "Common range of co-analytic Toeplitz operators ," to appear

McC2 J.E. McCarthy "Topologies on the Smirnov class," to appear

Th J. E. Thomson "Approximation in the mean by polynomials," to appear

Wi A. Wilansky "Modern methods in topological vector spaces," McGraw-Hill, New York, 1978

Banach Algebras, Decomposable Convolution Operators, and a Spectral Mapping Property

MICHAEL M. NEUMANN Department of Mathematics and Statistics, Mississippi State University, Mississippi State, Mississippi

1. INTRODUCTION

In this note, we are concerned with certain spectral properties of convolution operators given by measures on locally compact abelian groups. Some of the long–standing problems for these operators arise naturally from harmonic analysis and automatic continuity theory; it seems appropriate to point out that these problems are non–trivial even in the case of the circle group or the real line. Four of these classical problems and the solutions obtained so far will be briefly reviewed in Section 2.

We then present a uniform approach in the framework of Banach algebra theory, which we believe to give some new insight into these problems. The basic idea is to investigate a certain algebra of continuous functions on a compact, but not necessarily Hausdorff topological space. More specifically, we will focus on those elements of a commutative semi–simple Banach algebra, whose Gelfand transform is continuous with respect to the hull–kernel topology on the maximal ideal space. It turns out that this continuity property characterizes the decomposability of the corresponding multiplication operator in the sense of Foiaş [14]. Moreover, there are interesting spectral consequences for representations and applications both to regular and to non–regular Banach algebras. All these aspects

Research partially supported by NSF Grant DMS 90–96108

will be discussed in Section 3. Some of these results are closely related to and actually an extension of certain parts of our recent paper [22], where the interested reader will find some further results and examples.

In Section 4, the general theory will be applied to the measure algebra M(G) and to certain natural subalgebras thereof, where G is an arbitrary locally compact abelian group. These algebras are typical examples of non–regular semi–simple Banach algebras, whenever the underlying group is non–discrete, and an important aspect will be to determine the greatest regular subalgebra for these Banach algebras. We will be able to obtain some new solutions for the classical problems mentioned in Section 2, but certain questions remain. Moreover, our approach leads naturally to some new problems, which will be stated at the end of this paper. It seems likely that the methods presented here will have various further applications, for instance to more general types of multipliers. Some of these aspects are pursued in [21]. The list of references at the end of this note is far from being complete and contains only papers which are immediately connected with the topics discussed here.

The results of this paper were presented at the Conference on Function Spaces at Southern Illinois University at Edwardsville in April 1990. It is a pleasure to acknowledge the stimulating atmosphere of this conference and to express our warmest thanks to the organizers. Our research was supported by the U.S. National Science Foundation (NSF) under Grant No. DMS 90–96108; this support is acknowledged with thanks.

2. SOME MOTIVATING PROBLEMS

We start by recalling some basic notions from abstract harmonic analysis. Given a locally compact abelian group G, let M(G) denote the Banach algebra of all regular complex Borel measures on G, and for $1 \leq p < \infty$ let $L_p(G)$ denote the usual Lebesgue space of p–integrable functions with respect to the Haar measure on G. In the following, we will have to consider various natural closed subalgebras of the measure algebra M(G). The algebras of all absolutely continuous and all discrete measures on G will be denoted by $M_a(G)$ and $M_d(G)$, respectively. The algebra $M_a(G)$ will be canonically identified with $L_1(G)$; and $M_a(G) + M_d(G)$ is precisely the subalgebra of all those measures on G whose continuous part is even absolutely continuous. Finally, let $M_o(G)$ denote the sub–algebra consisting of all $\mu \in M(G)$ for which the corresponding Fourier–Stieltjes transform $\hat{\mu}$ on the dual group Γ vanishes at ∞. It is well–known that the maximal ideal space

$\Delta(L_1(G))$ of the group algebra $L_1(G)$ can be identified with the dual group Γ and that Γ can be canonically embedded in the maximal ideal space $\Delta(M(G))$ of the measure algebra $M(G)$; for further information on Fourier analysis on groups and the theory of multipliers we refer to the monographs of Rudin [25] and Larsen [19] as standard references.

For a measure $\mu \in M(G)$ we are interested in the spectral properties of the corresponding convolution operator $T_\mu: M(G) \to M(G)$ given by $T_\mu(\nu) := \mu * \nu$ for all ν in $M(G)$. It is well-known and easily seen that

$$\sigma(T_\mu | L_1(G)) = \sigma(T_\mu) \supseteq \hat{\mu}(\Gamma)^- \qquad \text{for each } \mu \in M(G),$$

where σ denotes the spectrum; and it is clear that

$$\hat{\mu}(\Gamma)^- = \hat{\mu}(\Gamma) \cup \{0\} \qquad \text{for each } \mu \in M_0(G).$$

Moreover, it is an elementary fact that

$$\sigma(T_\mu | L_1(G)) = \sigma(T_\mu) = \hat{\mu}(\Gamma)^- = \hat{\mu}(\Gamma) \cup \{0\} \quad \text{for each } \mu \in M_a(G) \cong L_1(G).$$

On the other hand, it has been observed by Zafran [29] that for every non-discrete locally compact abelian group G there exists some $\mu \in M_0(G)$ such that the preceding formula fails to hold. This phenomenon is due to the fact that the Banach algebra $M_0(G)$ is not symmetric and that its maximal ideal space is, in a sense, much larger than Γ. With the terminology of Zafran [29], [30] we are led to the following question.

Problem 1 (Zafran): *Which measures $\mu \in M(G)$ have a "natural" spectrum*

$$\sigma(T_\mu) = \hat{\mu}(\Gamma)^- ?$$

It seems that the systematic investigation of such measures and of the corresponding subalgebra of $M_0(G)$ for an arbitrary locally compact group G was originated by Zafran in [29], but it should also be noted that this problem and related questions of a similar type have a long tradition in classical Fourier analysis for the case $G = \mathbb{R}$; see for instance the work of Beurling [8], Hartman [16], and Wiener – Pitt [28]. The classical result of Beurling [8] yields, in modern terms, that if a measure $\mu \in M_a(\mathbb{R}) + M_d(\mathbb{R})$ satisfies $|\mu^\wedge| \geq \delta > 0$ on $\mathbb{R}$ for some $\delta > 0$, then the reciprocal of $\mu^\wedge$ is the Fourier–Stieltjes transform of some measure in $M_a(\mathbb{R}) + M_d(\mathbb{R})$. Of course, this result shows in the classical case $G = \mathbb{R}$ that every measure $\mu \in M_a(G) + M_d(G)$ has a natural spectrum. Actually, it follows from more recent work of D'Antoni, Longo, and Zsidó [12] that the same is true for an arbitrary locally compact abelian group G, even in a somewhat more general context. Stronger results were obtained by Arveson [7] and also by Albrecht [2], but the class of measures with natural spectrum is still far from being understood completely. In the following sections, we will relate problem 1 to some other problems from spectral theory and present some new solutions.

Let us mention in passing that a similar problem arises for general classes of multiplier transformations on the spaces $L_p(G)$. On the one hand, it is known from the work of Hörmander [17] and Zafran [30] that

$$\sigma(T_\mu ; L_p(G)) = \hat{\mu}(\Gamma)^- \qquad \text{for all } \mu \in M_0(G) \text{ and all } 1 < p < \infty,$$

where T_μ is considered as a convolution operator on $L_p(G)$. On the other hand, this identity ceases to be true for arbitrary measures $\mu \in M(G)$ on a non–discrete group G. Moreover, for a more general class of multipliers on $L_p(G)$ with $1 < p < \infty$ and $p \neq 2$, Zafran [30] obtained interesting examples of multipliers with non–natural spectrum in the case that G is either the n–torus or the n–dimensinal Euclidean space. The question whether the spectrum is natural or not is also of considerable importance for the non–spectrality of certain differential and multiplier operators in the space $L_p(G)$, where G denotes again the n–dimensional Euclidean space. In this connection, we refer to the recent work of Albrecht – Ricker [4], where it is shown that elliptic linear partial differential operators with constant coefficients in the space $L_p(G)$ are always decomposable (as unbounded operators) in the sense of Foiaş, but rarely spectral in the sense of Dunford.

Another important aspect arises in the theory of group representations. Given a complex Banach space X, let $\mathcal{L}(X)$ denote the Banach algebra of all continuous linear operators on X, and consider a *weakly continuous representation* of a locally compact abelian group G by means of isometries on X, i.e. a mapping $U: G \to \mathcal{L}(X)$ with the following three properties:

(a) $U(s+t) = U(s)\,U(t)$ for all $s,t \in G$ and $U(0) = I$ (identity operator on X)

(b) $\| U(t)x \| = \| x \|$ for all $t \in G$ and $x \in X$

(c) $U(\cdot)x : G \to X,\ t \to U(t)x$ is weakly continuous for each $x \in X$

Following Arveson [6], one can smear the representation U to a representation Φ of the measure algebra M(G) by considering in a weak sense

$$\Phi(\mu)x := \int_G U(t)x\, d\mu(t) \qquad \text{for all } \mu \in M(G) \text{ and } x \in X.$$

Then Φ: $M(G) \to \mathcal{L}(X)$ is a norm–decreasing homomorphism, and the *Arveson spectrum* sp(U) of the representation U is defined to be the hull of the kernel of $\Phi | L_1(G)$, i.e. the set of all $\gamma \in \Gamma$ which satisfy $f^\wedge(\gamma) = 0$ for all $f \in L_1(G)$ with $\Phi(f) = 0$; see also Pedersen [23] for details and for the more general case of an integrable representation U. Note that for $X = L_1(G)$ and for the representation U: $G \to \mathcal{L}(X)$ given by translation on $L_1(G)$ we obtain the convolution operator $\Phi(\mu) = T_\mu$ for all $\mu \in M(G)$ and consequently $sp(U) = \Gamma$. Hence the following problem, which was posed and discussed by Arveson in section 2 of [7], contains problem 1 as a special case.

Problem 2 (Arveson): *Given a weakly continuous (or at least integrable) representation* U: $G \to \mathcal{L}(X)$, *which measures* $\mu \in M(G)$ *have the spectral mapping property*

$$\sigma(\Phi(\mu)) = \hat{\mu}\,(sp(U))^{-}\ ?$$

It is easily seen that the inclusion "$\supseteq$" holds in general; and in the non–discrete case it is known that there is always a measure $\mu \in M(G)$ for which this inclusion is strict. On the other hand, it was shown by Connes [11] in the context of automorphism groups of von Neumann algebras that the spectral mapping property holds for all Dirac measures and also for all absolutely continuous measures. Then D'Antoni, Longo, and Zsidó [12] proved the spectral mapping property for all measures $\mu \in M_a(G) + M_d(G)$. A generalization of their result and a thorough discussion of the spectral mapping property can be found in Arveson's paper [7]: it is shown that this property holds for all measures in the closure of $L_1(G) + A$, where A denotes an arbitrary closed regular subalgebra of M(G). We will obtain a further extension of this result by a completely different approach.

Another long–standing and apparently difficult problem on convolution operators arose naturally in the context of automatic continuity theory. In [18], p. 98, Barry Johnson proved that an arbitrary everywhere defined linear transformation on $L_1(\mathbb{R})$ that commutes with one non–trivial shift operator is necessarily continuous and asked the following question (originally for the case $G = \mathbb{R}$).

Problem 3 (Johnson): *Let* $\mu \in M(G)$ *be a measure such that its Fourier–Stieltjes transform* $\mu^\wedge$ *is non–constant on every non–empty open subset of* Γ. *Does it follow that every linear transformation* $S: L_1(G) \to L_1(G)$ *with the property* $S\,T_\mu = T_\mu\,S$ *is automatically continuous ?*

It is embarrassing to confess that we do not know of any counterexample; actually the construction of a counterexample seems to require some new methods in automatic continuity theory. On the positive side, it was shown by Laursen and the author in [20] that the answer to problem 3 is "yes" at least for every measure $\mu \in M_a(G) + M_d(G)$. Moreover, a positive solution to this problem is given for all those measures $\mu \in M(G)$ for which the corresponding convolution operator on $L_1(G)$ is super–decomposable [20], which provides an interesting link with the theory of decomposable operators in the sense of Foiaş [14].

Given a complex Banach space X, an operator $T \in \mathcal{L}(X)$ is called *decomposable* if for every open covering $\{U,V\}$ of the complex plane $\mathbb{C}$ there exist T–invariant closed linear subspaces Y and Z of X such that $\sigma(T|Y) \subseteq U$, $\sigma(T|Z) \subseteq V$ and $Y + Z = X$. It has been shown by Albrecht that this simple definition of operator decomposability is actually equivalent to the original definition due to Foiaş [14]; see [10] and [26] for a comprehensive discussion of decomposable operators (including the case of unbounded operators). Let us also recall from [20] that an operator $T \in \mathcal{L}(X)$ is said to be *super–decomposable* if for every open covering $\{U,V\}$ of $\mathbb{C}$ there exists some $R \in \mathcal{L}(X)$ commuting with T such that $\sigma(T|\overline{R(X)}) \subseteq U$ and $\sigma(T|\overline{(I-R)(X)}) \subseteq V$. Such operators have also been studied in [5] in the context of certain normal operator algebras and in [27] as operators which are decomposable with respect to the identity. An example of a decomposable, but not super–decomposable operator can be found in [3].

In [10], p. 218, Colojoară and Foiaş ask if the convolution operator T_μ is decomposable on $L_1(G)$ for every measure $\mu \in M(G)$. They were able to provide a positive answer for all $\mu \in M_a(G)$, but later it has been shown independently by Albrecht [2] and by Eschmeier [13] that counterexamples do exist whenever the group G is non–discrete. There remains, however, the following updated version of their question.

Problem 4 (Colojoară – Foaiş): *For which measures* $\mu \in M(G)$ *is the corresponding convolution operator* T_μ *decomposable on* $M(G)$ *and on* $L_1(G)$?

It has been shown in [20] that for every $\mu \in M_a(G) + M_d(G)$ the operator T_μ is actually super–decomposable both on $M(G)$ and on $L_1(G)$. Some related results can also be found in [1] , [2] , [13] ; but a measure theoretic characterization of all those measures μ in $M(G)$, for which T_μ is decomposable or even super–decomposable, is still missing. In the following sections, we will obtain some new results in this direction.

3. BANACH ALGEBRAS AND DECOMPOSABLE OPERATORS

Given a commutative complex Banach algebra A, let $\Delta(A)$ stand for the spectrum or maximal ideal space of A, i.e. the set of all non–trivial multiplicative linear functionals on A. For each $a \in A$, let $\hat{a}: \Delta(A) \to \mathbb{C}$ denote the corresponding Gelfand transform given by $\hat{a}(\varphi) := \varphi(a)$ for all $\varphi \in \Delta(A)$. On $\Delta(A)$ we will have to consider the hull–kernel topology, which is always weaker than the usual Gelfand topology and coincides with the Gelfand topology if and only if the algebra A is regular. Thus some of the Gelfand transforms $\hat{a}$ will not be hull–kernel continuous on $\Delta(A)$ unless A is regular. Recall that the hull–kernel topology on $\Delta(A)$ is determined by the Kuratowski closure operation $cl(E) := hul(ker(E)) := \{\psi \in \Delta(A): \psi(c) = 0$ for all $c \in A$ with $\varphi(c) = 0$ for all $\varphi \in E\}$ for each $E \subseteq \Delta(A)$ and that, in the unital case, the hull–kernel topology is Hausdorff if and only if A is regular. For the basic facts concerning the hull–kernel topology we refer to [9] and to [24].

Throughout this section, let A be a semi–simple commutative complex Banach algebra with or without an identity element, and let X denote a complex Banach space. The following theorem generalizes a main result from [22].

Theorem 1. *Let* A *be a semi–simple commutative Banach algebra over* $\mathbb{C}$, *and consider an algebraic homomorphism* $\Phi : A \to \mathcal{L}(X)$. *Then, for every* $a \in A$ *for which the Gelfand transform* $\hat{a}$ *is continuous with respect to the hull–kernel topology of* $\Delta(A)$, *the corresponding operator* $T := \Phi(a) \in \mathcal{L}(X)$ *is super–decomposable.*

Proof. (1) We first assume that A has an identity element $1 \in A$ and that $\Phi(1) = I$. In this case $\Delta(A)$ is compact in the hull–kernel topology. Now, given an arbitrary open covering $\{U,V\}$ of $\mathbb{C}$, let us choose a pair of open sets $G,H \subseteq \mathbb{C}$ such that $\overline{G} \subseteq U$, $\overline{H} \subseteq V$, and $G \cup H = \mathbb{C}$. Since the complements $\mathbb{C}\setminus G$ and $\mathbb{C}\setminus H$ are closed and disjoint and since $\hat{a}$ is hull–kernel continuous, it follows that the preimages $\hat{a}^{-1}(\mathbb{C}\setminus G)$ and $\hat{a}^{-1}(\mathbb{C}\setminus H)$ are disjoint hulls in the compact space $\Delta(A)$. Hence, by corollary 3.6.10 of [24], there exists some $r \in A$ such that $\hat{r} = 0$ on $\hat{a}^{-1}(\mathbb{C}\backslash G)$ and $\hat{r} = 1$ on $\hat{a}^{-1}(\mathbb{C}\backslash H)$. We claim that the operator $R := \Phi(r) \in \mathcal{L}(X)$ satisfies the conditions for the super–decomposability of T with respect to the given covering $\{U,V\}$. Clearly, R and T commute. In order to show that $\sigma(T|\overline{R(X)})$ is contained in U, let $\lambda \in \mathbb{C}\backslash U$ be arbitrarily given, and let δ denote the distance from λ to $\overline{G}$. Because of $\overline{G} \subseteq U$ and $\lambda \notin U$ we have $\delta > 0$. Since the estimate $|(a-\lambda 1)\hat{}\,| \geq \delta > 0$ holds on the hull $\hat{a}^{-1}(\overline{G})$, we may apply theorem 3.6.15 of [24] to obtain some $c \in A$ with the property that $((a-\lambda 1)c)\hat{} = 1$ on $\hat{a}^{-1}(\overline{G})$. Since $\hat{r} = 0$ on $\hat{a}^{-1}(\mathbb{C}\backslash G)$ by the choice of r, we conclude that $(\hat{a}-\lambda 1)\hat{c}\hat{r} = \hat{r}$ holds on the whole spectrum $\Delta(A)$, which implies $(a-\lambda 1)cr = r$ in view of the semi–simplicity of A. Consider now the operator $S := \Phi(c) \in \mathcal{L}(X)$ and apply the homomorphism Φ to the equation $(a-\lambda 1)cr = r$. It follows that $(T-\lambda I)S(Rx) = S(T-\lambda I)(Rx) = Rx$ holds for all $x \in X$, from which we conclude that $(T-\lambda I)S = S(T-\lambda I) = I$ on $\overline{R(X)}$. Since the space $\overline{R(X)}$ is certainly invariant under S, these identities reveal that λ is in the resolvent set for the restriction $T|\overline{R(X)}$, which completes the proof of the inclusion $\sigma(T|\overline{R(X)}) \subseteq U$. A similar argument shows that $\sigma(T|\overline{(I-R)(X)}) \subseteq V$. Indeed, given an arbitrary $\mu \in \mathbb{C}\backslash V$, it is clear that the distance ε from μ to $\overline{H}$ is strictly positive and satisfies $|(a-\mu 1)\hat{}\,| \geq \varepsilon > 0$ on the hull $\hat{a}^{-1}(\overline{H})$. Hence, again by theorem 3.6.15 of [24], there exists some $e \in A$ such that $((a-\mu 1)e)\hat{} = 1$ on $\hat{a}^{-1}(\overline{H})$. We conclude that $(\hat{a}-\mu 1)\hat{e}(1-\hat{r}) = 1-\hat{r}$ holds on the entire spectrum $\Delta(A)$, which implies $(a-\mu 1)e(1-r) = 1-r$ by the semi–simplicity of A. Now define $Q := \Phi(e) \in \mathcal{L}(X)$ and recall that $\Phi(1) = I$ by our present assumption. It follows that $(T-\mu I)Q(I-R)(x) = (I-R)(x)$ holds for all $x \in X$, which shows that the restriction $(T-\mu I)|\overline{(I-R)(X)}$ is invertible. Hence $\sigma(T|\overline{(I-R)(X)})$ is contained in V, which completes the proof of the theorem in the unital case.

(2) If the Banach algebra A has no identity element at all or if the identity $1 \in A$ satisfies $\Phi(1) \neq I$, then the assertion can be reduced to the situation of the preceding part of the proof by means of the following construction. In either case, we consider the standard unitization $A_e := A \oplus \mathbb{C}e$ of A, endowed with the usual structure of a Banach algebra, as well as the canonical extension Φ^* of Φ to A_e given by $\Phi^*(u+\lambda e) := \Phi(u) + \lambda I$ for all $u \in A$ and $\lambda \in \mathbb{C}$. It is well–known and easily seen that $\Delta(A_e) = \Delta(A) \cup \{\varphi_\infty\}$, where each $\varphi \in \Delta(A)$ is identified with its canonical extension φ^* from A to A_e, and $\varphi_\infty: A_e \to \mathbb{C}$ denotes the functional given by $\varphi_\infty(u+\lambda e) := \lambda$ for all $u \in A$ and $\lambda \in \mathbb{C}$, i.e. the canonical extension of the zero functional on A. It should be noted that this construction works even if A happens to have an identity element $1 \in A$, which however ceases to be the identity for the extension A_e. Now, the first part of the proof will establish that the

operator $T = \Phi(a) = \Phi^*(a)$ is super–decomposable, once $\hat{a}$ is seen to be hull–kernel continuous on the extended spectrum $\Delta(A_e)$. Thus, given an arbitrary non–empty closed subset F of $\mathbb{C}$, we have to show that the set E of all $\varphi \in \Delta(A_e)$ satisfying $\varphi(a) \in F$ is a hull in $\Delta(A_e)$. To this end, let us recall that the closure $cl(E)$ of E with respect to the hull–kernel topology of $\Delta(A_e)$ consists precisely of all those $\psi \in \Delta(A_e)$ which satisfy $\psi(c) = 0$ for all $c \in A_e$ with $\varphi(c) = 0$ for all $\varphi \in E$. Since $\hat{a}$ is assumed to be hull–kernel continuous on $\Delta(A)$, it is easily seen that $cl(E) \cap \Delta(A)$ is contained in E. To deal with the remaining case φ_∞, let us assume that $\varphi_\infty \notin E$. Then obviously $0 \notin F$ and hence $|\hat{a}| \geq \delta > 0$ on E, where $\delta := \inf\{|\lambda| : \lambda \in F\}$. Moreover, since in the case $\varphi_\infty \notin E$ we have $E \subseteq \Delta(A)$ and since $\hat{a}$ is hull–kernel continuous on $\Delta(A)$, we conclude that E is a hull in $\Delta(A)$. Hence, again by theorem 3.6.15 of [24], there exists some $b \in A$ such that $\varphi(ab) = 1$ holds for all $\varphi \in E$. Thus $\varphi(e-ab) = 0$ for all $\varphi \in E$, whereas of course $\varphi_\infty(e-ab) = 1$. Thus $\varphi_\infty \notin cl(E)$, which shows that $\varphi_\infty \in cl(E)$ implies that actually $\varphi_\infty \in E$. We conclude that $cl(E) = E$, which completes the proof of the theorem.

The following result shows that the hull–kernel continuity on $\Delta(A)$ is the appropriate assumption in theorem 1. For each $a \in A$, let $T_a: A \to A$ denote the corresponding multi–plication operator given by $T_a(x) := ax$ for all $x \in A$. Recall from [13] that an operator $T \in \mathcal{L}(X)$ is said to have the *weak 2–spectral decomposition property* (SDP), if for every open covering $\{U,V\}$ of $\mathbb{C}$ there exist T–invariant closed linear subspaces Y and Z of X such that $\sigma(T|Y) \subseteq U$, $\sigma(T|Z) \subseteq V$, and $Y + Z$ is dense in X. In general, this con–dition is much weaker than decomposability, but in the present context it turns out that the various decomposability concepts all coincide.

Theorem 2. *Let* A *be a semi–simple commutative Banach algebra. Then for each* $a \in A$ *the following assertions are equivalent :*

(a) *The Gelfand transform* $\hat{a}$ *is hull–kernel continuous on* $\Delta(A)$.

(b) *The multiplication operator* $T_a: A \to A$ *is super–decomposable.*

(c) $T_a: A \to A$ *is decomposable.*

(d) $T_a: A \to A$ *has the weak 2–SDP.*

Proof. Taking Φ to be the left regular representation of A on the Banach space $X := A$, we infer from theorem 1 that (a) implies (b). Since the implications (b) ⇒ (c) and (c) ⇒ (d) are obvious, it remains to show that (d) implies (a). For completeness, we include the corresponding argument from our paper [22]. Suppose that (d) holds, but not (a). Then there exists some closed subset F of $\mathbb{C}$ such that $E := \{\varphi \in \Delta(A): \varphi(a) \in F\}$ is not a hull in $\Delta(A)$. Let us fix some $\psi \in \mathrm{cl}(E) \setminus E$ and define $\lambda := \psi(a) \in \mathbb{C} \setminus F$. By the weak 2–SDP there exists a pair of T–invariant closed subspaces Y and Z of A such that $\sigma(T_a|Y) \subseteq \mathbb{C} \setminus \{\lambda\}$, $\sigma(T_a|Z) \subseteq \mathbb{C} \setminus F$, and $Y + Z$ is dense in A. From the first inclusion we conclude that for every $u \in Y$ there exists some $v \in Y$ such that $u = (a-\lambda)v$, which implies $\psi(u) = (\psi(a) - \lambda)\,\psi(v) = 0$ and therefore $\psi = 0$ on Y. On the other hand, given an arbitrary $\varphi \in E$, we have $\mu := \varphi(a) \in F$. Since $\sigma(T_a|Z) \subseteq \mathbb{C} \setminus F$, it follows that for each $u \in Z$ there exists some $v \in Z$ such that $u = (a-\mu)v$, which implies $\varphi(u) = (\varphi(a) - \mu)\,\varphi(v) = 0$ and hence $\varphi = 0$ on Z. From $\psi \in \mathrm{cl}(E)$ we conclude that also $\psi = 0$ on Z. Since ψ vanishes both on Y and Z and since $Y + Z$ is dense in A, it follows that $\psi = 0$ on A. This obvious contradiction completes the proof.

Theorem 3. *For every semi–simple commutative Banach algebra* A, *the following assertions are equivalent*:

(a) A *is regular.*

(b) *For each* $a \in A$, *the Gelfand transform* $\hat{a}$ *is hull–kernel continuous on* $\Delta(A)$.

(c) *For each* $a \in A$, *the multiplication operator* $T_a: A \to A$ *is super–decomposable.*

(d) *For each* $a \in A$, T_a *is decomposable.*

(e) *There exists a system of generators* B *in* A *such that* T_a *has the weak 2–SDP for each* $a \in B$.

Proof. In view of theorem 2, it remains to show that (e) implies (a), and in view of the discussion under (2) in the proof of theorem 1, it suffices to do this under the additional assumption that A has an identity element. If (e) holds, then $\hat{a}$ is hull–kernel continuous for each $a \in B$. Moreover, since B generates the Banach algebra A, it is clear that the family $\{\hat{a}: a \in B\}$ separates the points of $\Delta(A)$. We conclude that the hull–kernel topology on $\Delta(A)$ is Hausdorff and hence that A is regular.

The preceding result is a generalization of a theorem due to Frunză [15]. Note that condition (e) is particularly useful in the context of monothetic Banach algebras, where it suffices to study the decomposability properties of one generator. Some further examples, where the weak form of condition (e) turns out to be relevant, can be found in [22]. In the following, we will discuss some applications of the preceding theorems to typically non–regular algebras. For an arbitrary semi–simple commutative Banach algebra A, let

Dec(A) consist of all those $a \in A$ for which the multiplication operator $T_a: A \to A$ is decomposable. The spectral radius of an element $x \in A$ will be denoted by $\rho(x)$.

Theorem 4. (a) Dec(A) *is a closed subalgebra of* A.

(b) *If* $a_n \in \mathrm{Dec}(A)$ *and* $a \in A$ *such that* $\rho(a-a_n) \to 0$ *as* $n \to \infty$, *then also* $a \in \mathrm{Dec}(A)$.

(c) *If* $a \in \mathrm{Dec}(A)$ *and if* $B \subseteq A$ *is a closed ideal in* A, *then the restriction* $T_a|B$ *is super–decomposable on* B.

(d) *If* $a \in \mathrm{Dec}(A)$ *and if* $C \supseteq A$ *is a not necessarily commutative or semi–simple Banach algebra containing* A *as a subalgebra, then the canonical extension* S_a *given by* $S_a(x) := ax$ *for all* $x \in C$ *is super–decomposable on* C.

Proof. The assertions (a) and (b) are clear from theorem 2 via the hull–kernel continuity of the corresponding Gelfand transforms, whereas (c) and (d) follow from an obvious combination of theorem 2 with theorem 1, which has to be applied to the left regular representation of A on B and C, respectively.

Theorem 5. *Every semi–simple commutative Banach algebra* A *contains a greatest regular closed subalgebra, denoted by* Reg(A); *this subalgebra is contained in* Dec(A).

Proof. Let C denote the (possibly trivial) closed subalgebra generated by the union B of all regular closed subalgebras M of A. To show that C is regular, by theorem 3 it suffices to prove that for each $a \in B$ the multiplication operator T_a is decomposable on C. Now, given an arbitrary $a \in B$, we have $a \in M$ for some regular closed subalgebra M of A. Again by theorem 3, we conclude that T_a is decomposable on M and hence, by theorem 4, also on the superalgebras C and A. It follows that C is regular and that B is contained in Dec(A). Since B generates C and since Dec(A) is known to be a closed subalgebra, we conclude that $C \subseteq \mathrm{Dec}(A)$, which completes the proof.

Let us note that the last two results are related to the work of Apostol [5] and Albrecht [2] on decomposable multiplication operators and multipliers, but the technicalities are entirely different. For instance, assertion (b) of theorem 4 is certainly a special case of a well–known limit theorem due to Apostol, for which we refer to theorem IV.6.10 of [26], but we believe that the present approach sheds some new light on this result. The existence of a greatest regular subalgebra of a semi–simple commutative Banach algebra was apparently first discovered by Albrecht [2] as an application of the general theory of decomposable operators. We feel that our more elementary proof of theorem 5 should be of some interest. Let us also note that semi–simplicity is actually not essential for theorem 5: in [22] we have established the existence of a greatest regular subalgebra of an arbitrary commutative Banach algebra by a different approach, which uses nothing but elementary Banach algebra theory. Of course, the greatest regular subalgebra may well be trivial: for instance, the greatest regular subalgebra of the disc algebra is easily seen to consist only of the complex constants [22]. On the other hand, the greatest regular subalgebra seems to be a useful tool for certain problems in harmonic analysis.

4. APPLICATIONS TO HARMONIC ANALYSIS

Throughout this section, let G be a locally compact abelian group with dual group Γ. The relevance of the hull–kernel topology for the spectral mapping property of a measure $\mu \in M(G)$ is most easily seen by the following argument: if $\mu^\wedge$ is hull–kernel continuous on $\Delta(M(G))$, then obviously

$$\sigma(T_\mu | L_1(G)) = \sigma(T_\mu) = \hat{\mu}(\Delta(M(G))) = \hat{\mu}(\Gamma)^-,$$

since it is well–known and easily seen that Γ is dense in $\Delta(M(G))$ with respect to the hull–kernel topology on $\Delta(M(G))$ (but certainly not with respect to the Gelfand topology unless G is discrete). Moreover, from theorem 2, part (c) of theorem 4, and theorem 2 of [13] we obtain immediately the following result.

Theorem 6. *For an arbitrary* $\mu \in M(G)$, *consider the following assertions :*

(a) $\hat{\mu}$ *is hull–kernel continuous on* $\Delta(M(G))$.
(b) T_μ *is super–decomposable on* $M(G)$.
(c) T_μ *is decomposable on* $M(G)$.
(d) T_μ *is decomposable on* $L_1(G)$.
(e) T_μ *has the weak 2–SDP on* $L_1(G)$.
(f) μ *has a natural spectrum in the sense of* $\sigma(T_\mu) = \mu^\wedge(\Gamma)^-$.

Then the following implications hold : (a) $\Leftrightarrow$ (b) $\Leftrightarrow$ (c) $\Rightarrow$ (d) $\Rightarrow$ (e) $\Rightarrow$ (f) .

Some related material can be found in the work of Albrecht; in particular, it has been observed in [2] that for suitable groups G there exist measures $\mu \in M(G)$ with a natural spectrum, for which the corresponding convolution operator on $L_1(G)$ is not decomposable. Hence the spectral mapping property does not characterize decomposability in general. The following result provides a partial answer to problem 1, problem 4, and hence, by theorem 4.3 of [20], also to problem 3.

Theorem 7. Reg(M(G)) *and hence* Dec(M(G)) *contains the closed subalgebra generated by* $M_a(G)$, $M_d(G)$ *and* $L_1(H)$, *where* $H \subseteq G$ *is a closed subgroup of* G *and* $L_1(H)$ *is canonically identified with the space of all measures on* G, *which are concentrated on* H *and absolutely continuous with respect to the Haar measure on* H. *In particular, for each measure* $\mu \in M_a(G) + M_d(G) + L_1(H)$, *it follows that* T_μ *is super–decomposable on* M(G) *and* $L_1(G)$ *and has a natural spectrum* $\sigma(T_\mu) = \mu\hat{}(\Gamma)^-$.

The proof follows immediately from the preceding results. Note that all measures in the subalgebra $L_1(H)$ are singular with respect to the Haar measure of G whenever the subgroup H is non–trivial, which shows that Reg(M(G)) may contain lots of singular measures. Using standard techniques from abstract harmonic analysis, it is also easy to find further examples of regular subalgebras of the measure algebra M(G), for instance by considering all those locally compact group topologies on G which are finer than the given group topology on G. This indicates that it may be difficult to find a precise description of Reg(M(G)) in general. In the following, we restrict our attention to the apparently more tractable subalgebra $M_0(G)$.

Theorem 8. *For each* $\mu \in M_0(G)$, *the following assertions are equivalent:*

(a) $\hat{\mu} = 0$ *on* $\Delta(M(G)) \setminus \Gamma$.
(b) T_μ *is decomposable on* M(G).
(c) T_μ *is decomposable on* $M_0(G)$.
(d) T_μ *is decomposable on* $L_1(G)$.

Proof. To show that (a) implies (b), recall that $\{f\hat{}\,|\,\Gamma\colon f \in L_1(G)\}$ is a dense subalgebra of $C_0(\Gamma)$; see for instance theorem 1.2.4 of [25]. Since by assumption $\mu \in M_0(G)$, it follows that there exist $f_n \in L_1(G)$ such that $f_n\hat{} \to \mu\hat{}$ uniformly on Γ as $n \to \infty$. Since $\mu\hat{}$ and $f_n\hat{}$ vanish on $\Delta(M(G)) \setminus \Gamma$ for all $n \in \mathbb{N}$, we conclude that $f_n\hat{} \to \mu\hat{}$ uniformly on the entire spectrum $\Delta(M(G))$ and hence $\rho(\mu - f_n) \to 0$ as $n \to \infty$. Since each f_n in–

duces a decomposable convolution operator on M(G), we conclude from the limit result in theorem 4 that T_μ is decomposable on M(G). Finally, the implications (b) ⇒ (c) and (c) ⇒ (d) are clear from theorem 4, whereas (d) ⇒ (a) follows from corollary 3.3 of [2].

Theorem 9. *Assume that the group* G *is compact and let* $A := M_0(G)$. *Then we have:*

$$\mathrm{Reg}(A) = \mathrm{Dec}(A) = \{\mu \in A: \hat{\mu} = 0 \text{ on } \Delta(M(G)) \setminus \Gamma\} = \{\mu \in A: \sigma(T_\mu) = \hat{\mu}(\Gamma)^-\}.$$

Proof. Theorem 5 shows that $\mathrm{Reg}(A) \subseteq \mathrm{Dec}(A)$, whereas theorem 8 yields the second of the desired identities. Since every $\mu \in M(G)$ with $\mu\hat{} = 0$ on $\Delta(M(G)) \setminus \Gamma$ certainly has a natural spectrum, it remains to prove that $B := \{\mu \in A: \sigma(T_\mu) = \mu\hat{}(\Gamma)^-\}$ is contained in Reg(A). But this follows immediately from the work of Zafran, as pointed out to the author by Wolfgang Arendt (Besançon) in a private communication: it is shown in theorem 3.2 of [29] that B is a Banach algebra whose spectrum can be canonically identified with Γ. Since $L_1(G) \subseteq B$, it follows that B is regular and hence contained in Reg(A), which completes the proof.

We finally prove the decomposability and the spectral mapping property for a wide class of generalized convolution operators.

Theorem 10. *Let* G *be an arbitrary locally compact abelian group and consider a weakly continuous (or at least integrable) representation* $U: G \to \mathcal{L}(X)$ *on a complex Banach space* X. *Let* $\Phi: M(G) \to \mathcal{L}(X)$ *denote the corresponding representation of the measure algebra* M(G). *Then, for each* $\mu \in M(G)$ *for which* $\mu\hat{}$ *is hull–kernel continuous on* $\Delta(M(G))$, *the operator* $T := \Phi(\mu) \in \mathcal{L}(X)$ *is super–decomposable and has the spectral mapping property* $\sigma(T) = \mu\hat{}(\mathrm{sp}(U)^-$.

Proof. By theorem 1, we know that T is super–decomposable. Hence, by proposition 2.10 of [7], it remains to show that $\sigma(T)$ is contained in $\mu\hat{}(\mathrm{sp}(U)^-$. Given an arbitrary $\lambda \in \mathbb{C} \setminus \mu\hat{}(\mathrm{sp}(U)^-$, let us fix a pair of numbers $\varepsilon > \delta > 0$ such that $|\mu\hat{}(\gamma) - \lambda| \geq \varepsilon$ for all $\gamma \in \mathrm{sp}(U) \subseteq \Gamma$, and define $E := \{\varphi \in \Delta(M(G)): |\mu\hat{}(\varphi) - \lambda| \geq \delta\} \subseteq \Delta(M(G))$. Since $\mu\hat{}$ is hull–kernel continuous on $\Delta(M(G))$, the set E turns out to be a hull in $\Delta(M(G))$. Hence, by theorem 3.6.15 of [24], there exists some $\nu \in M(G)$ such that $(\mu\hat{} - \lambda)\, \nu\hat{} = 1$ on E and hence on some open subset of Γ containing sp(U). By theorem 2.5 of [7] we obtain $\Phi((\mu-\lambda)\nu) = I$, which shows that the operator $T-\lambda I \in \mathcal{L}(X)$ is invertible. Thus $\lambda \notin \sigma(T)$, which completes the proof of the theorem.

Let us note in passing that it is also possible to express the spectral maximal spaces of the operator $T = \Phi(\mu)$ in the sense of Foiaş [10], [14] in terms of the spectral subspaces introduced by Arveson [6], [7]. This shows an interesting connection between decomposable operators and group representations, which may find some applications. For details and further information we refer to [1] and to [21].

The preceding results provide some partial solutions to the problems stated in the second section. In particular, theorem 9 contains an answer to the question raised by Arveson in [7], p. 223. Of course, some challenges remain, and there are some new problems which naturally result from our approach. For instance, it would be interesting to know if Reg(A) can be strictly contained in Dec(A) for certain semi–simple commutative Banach algebras A. We do not know the answer to this question even in the case of the measure algebra M(G) for an arbitrary locally compact abelian group G, although some partial results, which cover the case $A = M_0(G)$, have recently been obtained in [21]. Moreover, it would certainly be of interest to have some better insight into the role of the hull–kernel topology in the context of the measure algebra M(G). And last, but not least, it would be interesting to find an accessible characterization of the greatest regular subalgebra of M(G) in measure–theoretic terms.

REFERENCES

1. E. Albrecht, Spectral decompositions for systems of commuting operators, *Proc. Royal Irish Acad.* **81A** (1981), 81–98.
2. E. Albrecht, Decomposable systems of operators in harmonic analysis, pp. 19–35 in *Toeplitz Centennial* (Birkhäuser – Verlag, Basel 1982).
3. E. Albrecht, J. Eschmeier, and M. M. Neumann, Some topics in the theory of decomposable operators, pp. 15–34 in *Operator Theory : Advances and Applications Vol. 17* (Birkhäuser – Verlag, Basel 1986).
4. E. Albrecht and W. J. Ricker, Local spectral properties of constant coefficient differential operators, 27 p., Preprint Universität des Saarlandes, Saarbrücken 1989.
5. C. Apostol, Decomposable multiplication operators, *Rev. Roumaine Math. Pures Appl.* **17** (1972), 323–333.

6. W. B. Arveson, On groups of automorphisms of operator algebras, *J. Functional Analysis* **15** (1974), 217–243.

7. W. B. Arveson, The harmonic analysis of automorphism groups, pp. 199–269 in *Proc. Sympos. Pure Math. Vol. 38 Part I* (Amer. Math. Society, Providence, R.I. 1982).

8. A. Beurling, Sur les intégrales de Fourier absolument convergentes et leurs application à une transformation fonctionnelle, pp. 345–366 in *Proc. Neuv. Congr. Math. Scand. Helsingfors 1938* (Mercator, Helsingfors 1939).

9. F. F. Bonsall and J. Duncan, *Complete normed algebras* (Springer – Verlag, New York 1973).

10. I. Colojoară and C. Foiaş, *Theory of generalized spectral operators* (Gordon and Breach, New York 1968).

11. A. Connes, Une classification des facteurs de type III, *Ann. Sci. Ecole Norm. Sup. Paris* (4) **6** (1973), 133–252.

12. C. D'Antoni, R. Longo, and L. Zsidó, A spectral mapping theorem for locally compact groups of operators, *Pacific J. Math.* **103** (1981), 17–24.

13. J. Eschmeier, Operator decomposability and weakly continuous representations of locally compact abelian groups, *J. Operator Theory* **7** (1982), 201–208.

14. C. Foiaş, Spectral maximal spaces and decomposable operators in Banach space, *Arch. Math.* **14** (1963), 341–349.

15. Ş. Frunză, A characterization of regular Banach algebras, *Rev. Roumaine Math. Pures Appl.* **18** (1973), 1057–1059.

16. S. Hartman, Beitrag zur Theorie des Maßringes mit Faltung, *Studia Math.* **18** (1959), 67–79.

17. L. Hörmander, Estimates for translation invariant operators in L^p spaces, *Acta Math.* **104** (1960), 93–140.

18. B. E. Johnson, Continuity of linear transformations commuting with continuous linear operartors, *Trans. Amer. Math. Soc.* **128** (1967), 88–102.

19. R. Larsen, *An introduction to the theory of multipliers* (Springer – Verlag, New York 1971).

20. K. B. Laursen and M. M. Neumann, Decomposable operators and automatic continuity, *J. Operator Theory* **15** (1986), 33–51.

21. K. B. Laursen and M. M. Neumann, Decomposable multipliers and applications to harmonic analysis, 38 p., Preprint University of Copenhagen and Mississippi State University 1990 (submitted).

22. M. M. Neumann, Commutative Banach algebras and decomposable operators, 17 p., Preprint Mississippi State University 1990 (submitted).

23. G. K. Pedersen, *C *– algebras and their automorphism groups* (Academic Press, London 1979).

24. C. E. Rickart, *General theory of Banach algebras* (Van Nostrand, Princeton, N.J. 1960; reprinted by R. E. Krieger Publ. Co., Huntington, NY 1974).

25. W. Rudin, *Fourier analysis on groups* (Interscience Publ., New York 1962).

26. F.–H. Vasilescu, *Analytic functional calculus and spectral decompositions* (Editura Academiei and D. Reidel Publ. Comp., Bucureşti and Dordrecht 1982).

27. S. Wang, Local resolvents and operators decomposable with respect to the identity (in Chinese), *Acta Math. Sinica* **26** (1983), 153–162.

28. N. Wiener and H. R. Pitt, On absolutely convergent Fourier–Stieltjes transforms, *Duke Math. J.* **4** (1938), 420–436.

29. M. Zafran, On the spectra of multipliers, *Pacific J. Math.* **47** (1973), 609–626.

30. M. Zafran, On the spectra of multiplier transformations on the L^p spaces, *Ann. of Math.* **103** (1976), 355–374.

Middle Hankel Operators on Bergman Space

Lizhong Peng* Peking University, Beijing, China and Genkai Zhang** University of Stockholm, Stockholm, Sweden

§1. Introduction and main results

Let $dm(z)$ be the Lebesque measure on the unit disk D of complex plane and let, for $-1 < \alpha < \infty$, $d\mu_\alpha(z)$ be the measure $\frac{\alpha+1}{\pi}(1-|z|^2)^\alpha dm(z)$. Thus $L^{\alpha,2}(D) = L^2(d\mu_\alpha)$ is the space of all measurable functions on the unit disk for which the norm

$$\|f\|_\alpha^2 = \frac{\alpha+1}{\pi}\int_D |f(z)|^2(1-|z|^2)^\alpha dm(z)$$

is finite. The Bergman space $A^{\alpha,2}(D)$ is the subspace of all analytic functions in $L^{\alpha,2}(D)$. The orthogonal projection of $L^{\alpha,2}(D)$ onto $A^{\alpha,2}(D)$ will be denoted by P^α. The subspace of all anti-analytic functions in $L^{\alpha,2}(D)$ will be denoted by $\overline{A}^{\alpha,2}(D)$ and the corresponding projection will be denoted by $\overline{P}^\alpha$.

Let b be an analytic function on D. The big Hankel operator H_b with symbol b is defined by

$$H_b(f) = (I - P^\alpha)(\bar{b}f), \tag{1.1}$$

and the small Hankel operator

$$\widetilde{H}_b(f) = \overline{P}^\alpha(\bar{b}f). \tag{1.2}$$

Thus H_b maps $A^{\alpha,2}(D)$ into $(A^{\alpha,2}(D))^\perp$ if it is bounded, while $\widetilde{H}_b$ maps $A^{\alpha,2}(D)$ into $\overline{A}^{\alpha,2}(D)$ if it is bounded. In fact one can study the big Hankel operators and the small

*Research supported by NFR of Sweden and NNSF of P.R. China.
**Research supported as a doktorändtjanst in the University of Stockholm.

Hankel operators of one Bergman space into another, i.e. defines H_b from $A^{\beta,2}(D)$ into $(A^{\alpha,2}(D))^{\perp}$ and $\widetilde{H}_b$ from $A^{\beta,2}(D)$ into $\overline{A}^{\alpha,2}(D)$ (see Janson [4]).

We denote the Schatten-von Neumann class of operators from one Hilbert space H_1 into another H_2 by $S_p(H_1, H_2)$. In particular, $S_p^{\beta\alpha} = S_p(A^{\beta,2}(D), L^{\alpha,2}(D))$, and for the simplicity $S_p = S_p^{\alpha\alpha}$.

We recall the following two theorems.

THEOREM A. *Let $\alpha, \beta > -1$ and $0 < p \leq \infty$.*

(i) If $\frac{1}{p} < 1 + \frac{1}{2}(\alpha - \beta)$, then $H_b \in S_p^{\beta\alpha}$ iff $b \in B_p^{\frac{1}{p}+\frac{(\beta-\alpha)}{2}}$.

(ii) If $\frac{1}{p} \geq 1 + \frac{1}{2}(\alpha - \beta)$, then except in the case $p = \infty$ and $\beta = \alpha + 2$, $H_b \in S_p^{\beta\alpha}$ only if b is constant (and thus $H_b = 0$).

(iii) If $\beta < \alpha + 2$, then H_b is compact from $A^{\beta,2}(D)$ into $L^{\alpha,2}(D)$ iff $b \in b_{\infty}^{\frac{(\beta-\alpha)}{2}}$.

(iv) If $\beta \geq \alpha + 2$, then H_b is never compact from $A^{\beta,2}(D)$ into $L^{\alpha,2}(D)$ unless b is constant.

THEOREM B. *Let $\alpha, \beta > -1$ and $0 < p \leq \infty$.*

(i) $\widetilde{H}_b \in S_p^{\beta\alpha}$ iff $b \in B_p^{\frac{1}{p}+\frac{(\beta-\alpha)}{2}}$.

(ii) $\widetilde{H}_b$ is compact from $A^{\beta,2}(D)$ into $\overline{A}^{\alpha,2}(D)$ iff $b \in b_{\infty}^{(\beta-\alpha)/2}$.

Remark. For the definitions of the Besov spaces B_p^s and b_{∞}^s, see e.g. Janson [4] and the references there.

The results of Theorem B are due to Peetre [6], Peller [7,8,9], Rochberg [13] and Semmes [14]. The results of Theorem A are due to Axler [2] ($p = \infty$ and $\alpha = 0$), Arazy, Fisher and Peetre [1] ($\alpha = \beta$) and Janson [4], see Janson [4].

The phenomenon in (ii) of Theorem A is called the cut-off. Theorem A says that the big Hankel operator H_b has the cut-off at $p_0 = 1 (\alpha = \beta)$, and Theorem B says that the small Hankel operator $\widetilde{H}_b$ has no the cut-off at all. Rochberg proposed following problem:

Does there exist "middle" Hankel operator? (personal communication.)

Janson and Rochberg [5] have found a "middle" Hankel operator, which is close to the small Hankel operator. In fact it is easy to find a "middle" Hankel operator which is close to the big Hankel operator (see below). The aim of this paper is to find a strict "middle" Hankel operator so that it has the cut-off phenomenon at some $p_0 \in (0, 1)$.

To compare the size of operators, we introduce the following concept.

DEFINITION 1.1. *Let T_1, T_2 be two operators from one Hilbert space H_1 into another H_2. We define $T_1 \succeq T_2$, if $T_1^*T_1 \geq T_2^*T_2$, i.e. $< T_1^*T_1 f, f > \geq < T_2^*T_2 f, f >$ for all $f \in H_1$; and we define $T_1 \succ T_2$, if $T_1^*T_1 > T_2^*T_2$, i.e. $< T_1^*T_1 f, f > \geq < T_2^*T_2 f, f >$ for all $f \in H_1$ and there exists $f_0 \in H_1$ such that $< T_1^*T_1 f_0, f_0 > > < T_2^*T_2 f_0, f_0 >$.*

Remark. By the min-max principle (see Reed and Simon [12] Theorem XIII.1), we see that if $T_1 \succeq T_2$, then $\|T_1\|_{S_p} \geq \|T_2\|_{S_p}, 0 < p \leq \infty$, and if $T_1 \succeq T_2, T_1$ is compact, then T_2 is also compact.

Now we consider the operators from $A^{\alpha,2}(D)$ into $L^{\alpha,2}(D)$.

Let $\widetilde{H}_b' f = \widetilde{H}_b f - \int_D \bar{b} f d\mu_\alpha$, then $\widetilde{H}_b - \widetilde{H}_b'$ has rank (at most) one. Consequently, $\widetilde{H}_b \in S_p$ (or is compact) iff $\widetilde{H}_b' \in S_p$ (or is compact).

It is obvious that

$$\begin{aligned}< H_b^* H_b f, f > &= \|H_b f\|^2 = \|\bar{b} f\|^2 - \|P^\alpha \bar{b} f\|^2 \\ &\geq \|\overline{P}^\alpha \bar{b} f\|^2 - |\int_D \bar{b} f d\mu_\alpha|^2 = \|\widetilde{H}_b' f\|^2 =< \widetilde{H}_b'^* \widetilde{H}_b' f, f >\end{aligned}$$

and there exists f_0 such that the exact inequality holds, so

$$H_b \succ \widetilde{H}_b'. \tag{1.3}$$

This is the reason that H_b and $\widetilde{H}_b$ have became known as big and small Hankel operators.

Now we construct our middle Hankel opeartor.

First we define a subspace $B_t^{\alpha,2}(D)$ of $L^{\alpha,2}(D)$, which is in some way a shift of the space $A^{\alpha,2}(D)$. Let $e_n = e_n(z) = r^n e^{in\theta}/\gamma_{n,\alpha}, z = re^{i\theta}$, where $\gamma_{n,\alpha}^2 = \int_D |z^n|^2 d\mu_\alpha = \frac{\Gamma(n+1)\Gamma(\alpha+2)}{\Gamma(n+\alpha+2)}$, and let, for $t > -\frac{1}{2}$ and $t \neq 0, g_n^t = g_n^t(z) = r^{n+2t} e^{in\theta}$. Then $< e_n, g_m^t >= 0$ if $n \neq m$. and $< e_n, g_n^t > \neq 0$. Let

$$\begin{aligned}f_n &= g_n^t - < g_n^t, e_n > e_n \\ &= (r^{2t} - \gamma_{n,\alpha}^{-2} \frac{\Gamma(n+t+1)\Gamma(\alpha+2)}{\Gamma(n+t+\alpha+2)}) r^n e^{in\theta} \\ &= (r^{2t} - \frac{\gamma_{n+t,\alpha}^2}{\gamma_{n,\alpha}^2}) r^n e^{in\theta},\end{aligned}$$

where the $\gamma_{s,\alpha}$ has been generalized to the case $s > -1$. Then $\|f_n\|_\alpha^2 = \gamma_{n+2t,\alpha}^2 - \frac{\gamma_{n+t,\alpha}^4}{\gamma_{n,\alpha}^2}$. Now let $f_n^t = f_n/\|f_n\|_\alpha$. Thus $\{f_n^t\}$ is orthonormal. We know that $A^{\alpha,2}(D) = span\{e_n, n \geq 0\}$, we let $B_t^{\alpha,2}(D) = span\{f_n^t, n \geq 0\}$, and P_t^α the corresponding projection from $L^{\alpha,2}(D)$ onto $B_t^{\alpha,2}(D)$. So $B_t^{\alpha,2}(D) \perp A^{\alpha,2}(D)$.

DEFINITION 1.2. *For b an analytic function on D, we define a middle Hankel operator H_b^t by*

$$H_b^t f = (I - P^\alpha - P_t^\alpha)\bar{b} f. \tag{1.4}$$

Remark. As in (1.1) and (1.2), one can study H_b^t as either an operator from $A^{\alpha,2}(D)$ into $L^{\alpha,2}(D)$ or an operator from $A^{\beta,2}(D)$ into $L^{\alpha,2}(D)$.

For every $f \in A^{\alpha,2}(D)$ (or $f \in A^{\beta,2}(D)$), we have

$$\begin{aligned}< H_b^* H_b f, f > &= \|H_b f\|_\alpha^2 = \|\bar{b} f\|_\alpha^2 - \|P^\alpha \bar{b} f\|_\alpha^2 \\ &\geq \|\bar{b} f\|_\alpha^2 - \|P^\alpha \bar{b} f\|_\alpha^2 - \|P_t^\alpha \bar{b} f\|_\alpha^2 \\ &=< H_b^{t*} H_b^t f, f > \\ &\geq \|\overline{P}^\alpha \bar{b} f\|_\alpha^2 - |\int_D \bar{b} f d\mu_\alpha|^2 \\ &=< \widetilde{H}_b'^* \widetilde{H}_b' f, f >\end{aligned}$$

and there exist functions f such that the exact inequalities hold. So, by Definition1.1, we have

(1.5) $$H_b \succ H_b^t \succ \widetilde{H}_b'.$$

For technical reasons we study H_b^t only in the case $\alpha = 0$. For the general case $\alpha > -1$, the formulas in the arguments involve the gamma functions so that the estimates become very complicated, we will not pursue that direction.

The main result of this paper is the following

THEOREM. *Let $\alpha = 0, t > -1/2, t \neq 0$ and $0 < p \leq \infty$.*
(i) If $p > 1/2$, then $H_b^t \in S_p$ iff $b \in B_p^{\frac{1}{p}}$.
(ii) If $p \leq 1/2$, then $H_b^t \in S_p$ only if $H_b^t = 0$, and thus b is constant.
(iii) H_b^t is compact iff $b \in b_\infty^0$.

The part (ii) of Theorem shows that H_b^t has the cut-off at $p_0 = 1/2$. So we say that H_b^t is the strict middle Hankel operator.

If we start with $h_n = h_n(z) = r^{n^s} e^{in\theta} (s > 1)$ instead of g_n, we get $H_b^s = (I - P^\alpha - P_s^\alpha) M_{\overline{b}}$. It is easy to show that it is a middle Hankel operator, which is close the big Hankel operator, i.e. it has the cut-off at $p_0 = 1$ as the same for H_b. We omit the details.

By virtue of Theorem A and B, (1.5) and the remark after Definition 1.1, we need only to prove

(i) If $\frac{1}{2} < p \leq 1$ and $b \in B_p^{\frac{1}{p}}$, then $H_b^t \in S_p$.
(ii)If $0 < p \leq \frac{1}{2}$ and $H_b^t \in S_p$, then $H_b^t = 0$.

In §2 we study the cut-off phenomenon, a more general result is proved. In §3 we give a method to estimate direct S_p-results for general operators T_b for $0 < p \leq 1$. In §4, the S_p-results for the middle Hankel operators are proved. In fact we give the results for H_b^t from $A^{\beta,2}(D)$ to $L^{0,2}(D)$. In §5 we propose some open problems.

§2. The cut-off

First let us calculate the reproducing kernel $G^t(z, w)$ of the subspace $B_t^{0,2}(D)$ of $L^{0,2}(D)$. If $\alpha = 0$, $\gamma_s = \gamma_{s,0} = \frac{1}{\sqrt{s+1}}$, we have

$$f_n^t(z) = \frac{1}{|t|} \sqrt{n + 2t + 1}(n + t + 1)(|z|^{2t} - \frac{n+1}{n+t+1}) z^n,$$

and

(2.1)
$$G^t(z, w) = \sum_{n=0}^{\infty} \overline{f_n^t(w)} f_n^t(z)$$
$$= \sum_{n=0}^{\infty} \frac{(n + 2t + 1)(n + t + 1)^2}{t^2} (|w|^{2t} - \frac{n+1}{n+t+1})(|z|^{2t} - \frac{n+1}{n+t+1})(\overline{w}z)^n.$$

For $t = 1$, we have

$$\text{(2.2)} \quad \begin{aligned} G^t(z,w) = {} & |w|^2|z|^2 \frac{-2(\overline{w}z)^3 + 8(\overline{w}z)^2 - 12(\overline{w}z) + 12}{(1-\overline{w}z)^4} \\ & - 6(|z|^2 + |w|^2)\frac{1}{(1-\overline{w}z)^4} + \frac{-9(\overline{w}z)^2 + 4(\overline{w}z) + 3}{(1-\overline{w}z)^4}. \end{aligned}$$

The middle Hankel operator H_b^t has the form

$$\text{(2.3)} \qquad H_b^t f(z) = \int_D (\overline{b(z)}K(z,w) - \overline{b(w)}K(z,w) - \overline{b(w)}G^t(z,w))f(w)dm(w),$$

where $K(z,w) = \frac{1}{(1-\overline{w}z)^2}$ is the reproducing kernel of $A^{0,2}(D)$.

From (2.1) and (2.3), we know that H_b^t has no invariance under the whole Moebius group, which makes the proof of Theorem rather complicated. Meanwhile we see that H_b^t is invariant under the subgroup of rotations, which will be used later on.

Now let $\beta > -1$. We define a operator from $A^{\beta,2}(D)$ to $L^{\alpha,2}(D)$ by the formula (2.3) and denote it also by H_b^t. It is more convenient when using interpolation. We show that $H_{\overline{z}^j}^t$ has the cut-off at $p_0 = \frac{2}{4-\beta}$. Note that for $n \geq j$, we have

(2.4)

$$\begin{aligned} H_{\overline{z}^j}^t\left(\frac{z^n}{\gamma_{n,\beta}}\right) &= \frac{\overline{z}^j z^n}{\gamma_{n,\beta}} - \left\langle \frac{\overline{z}^j z^n}{\gamma_{n,\beta}}, \frac{z^{n-j}}{\gamma_{n-j}} \right\rangle \frac{z^{n-j}}{\gamma_{n-j}} - \left\langle \frac{\overline{z}^j z^n}{\gamma_{n,\beta}}, f_{n-j}^t \right\rangle f_{n-j}^t \\ &= \frac{1}{\gamma_{n,\beta}}\left[r^{2j} - \frac{\gamma_n^2}{\gamma_{n-j}^2} - \frac{\gamma_{n+t}^2\gamma_{n-j}^2 - \gamma_{n-j+t}^2\gamma_n^2}{\gamma_{n-j}^2\gamma_{n-j+2t}^2 - \gamma_{n-j+t}^4}\left(r^{2t} - \frac{\gamma_{n-j+t}^2}{\gamma_{n-j}^2}\right)\right] r^{n-j} e^{i(n-j)\theta}, \end{aligned}$$

and for $n < j$, we have

$$\text{(2.5)} \qquad H_{\overline{z}^j}^t\left(\frac{z^n}{\gamma_{n,\beta}}\right) = \frac{1}{\gamma_{n,\beta}} r^{n+j} e^{i(n-j)\theta}.$$

It is easy seen that $H_{\overline{z}^j}^t(\frac{z^n}{\gamma_{n,\beta}})$ are orthogonal, and since $\{z^n/\gamma_{n,\beta}\}$ is an ON-basis in $A^{\beta,2}(D)$, it follows that the singular numbers of $H_{\overline{z}^j}^t$ are givin by $\{\|H_{\overline{z}^j}^t(z^n/\gamma_{n,\beta})\|\}_0^\infty$, rearranged in decreasing order. For $n \geq j$, we have

$$\begin{aligned}
\|H^t_{\bar z^j}(\frac{z^n}{\gamma_{n,\beta}})\|^2 &= \frac{1}{\gamma^2_{n,\beta}}\{\|\bar z^j z^n\|^2 - |<\bar z^j z^n, \frac{z^{n-j}}{\gamma_{n-j}}>|^2 - |<\bar z^j z^n, f^t_{n-j}>|^2\} \\
&= \frac{1}{\gamma^2_{n,\beta}}\{\gamma^2_{n+j} - \frac{\gamma^4_n}{\gamma^2_{n-j}} - \frac{(\gamma^2_{n+t}\gamma^2_{n-j} - \gamma^2_{n-j+t}\gamma^2_n)^2}{\gamma^2_{n-j}(\gamma^2_{n-j}\gamma^2_{n-j+2t} - \gamma^4_{n-j+t})}\} \\
&= \frac{1}{\gamma^2_{n,\beta}}\{\frac{1}{n+j+1} - \frac{n-j+1}{(n+1)^2} - \frac{j^2(n-j+2t+1)}{(n+1)^2(n+t+1)^2}\} \\
&= \frac{1}{\gamma^2_{n,\beta}}\frac{j^2(t-j)^2}{(n+1)^2(n+t+1)^2(n+j+1)} \\
&\asymp \frac{j^2(t-j)^2}{n^{4-\beta}}.
\end{aligned}$$

For $n < j$, we have

$$\|H^t_{\bar z^j}(\frac{z^n}{\gamma_{n,\beta}})\|^2 = \frac{1}{\gamma^2_{n,\beta}}\frac{1}{n+j+1}.$$

Therefore we get that
$\|H^t_{\bar z^j}\|_{S^{\beta o}_p} < \infty$ iff $\frac{1}{p} < 2 - \frac{\beta}{2}$;
$\|H^t_{\bar z^j}\|_{S^{\beta o}_\infty} < \infty$ iff $\beta \le 4$;
$H^t_{\bar z^j}$ is compact iff $\beta < 4$.
Using this fact, we can prove the following

PROPOSITION 2.1. *Let $t > -1/2, t \neq 0, \beta > -1$ and $0 < p \le \infty$.*
(i) If $p < \infty$, $\frac{1}{p} \ge 2 - \beta/2$ and $H^t_b \in S^{\beta o}_p$, then $H^t_b = 0$, and thus $b = c$.
(ii) If $\beta \ge 4$ and H^t_b is compact, then $H^t_b = 0$, and thus $b = c$.

PROOF: We give the proof only for part (i), the argument for part (ii) is the same.
Define, for $|\xi| \le 1$,

$$b^{(j)}_\xi(z) = \begin{cases} \dfrac{b(\xi z) - \sum_{k=0}^{j-1}\frac{b^{(k)}(0)}{k!}(\xi z)^k}{\xi^j}, & \text{if } \xi \neq 0, \\ \dfrac{b^{(j)}(0)}{j!} z^j, & \text{if } \xi = 0. \end{cases}$$

Then $\xi \to b^{(j)}_\xi$ is analytic and $\xi \to H^t_{b^{(j)}_\xi}$ is anti-analytic in the unit disc D. If $|\xi| = 1$, then $\|H^t_{b^{(1)}_\xi}\|_{S^{\beta o}_p} = \|H^t_b\|_{S^{\beta o}_p} < \infty$ by the invariance of H^t_b under the subgroup of rotations. Since the maximum modulus principle holds in S_p, it follows that $H^t_{b^{(1)}_\xi} \in S^{\beta o}_p$ for every $\xi \in D$. In particular, we may take $\xi = 0$, it follows that $H^t_{b'(0)z} = 0$.

Assume that $H^t_{\sum_{k=0}^{j-1}\frac{b^{(k)}(0)}{k!}z^k} = 0$, if $|\xi| = 1$, then $\|H^t_{b^{(j)}_\xi}\|_{S^{\beta o}_p} = \|H^t_b\|_{S^{\beta o}_p} < \infty$ again by the invariance of H^t_b under the subgroup of rotations. It follows that $H^t_{b^{(j)}_\xi} \in S^{\beta o}_p$

for every $\xi \in D$. In particular, we may take $\xi = 0$, it follows that $H^t_{b^{(j)}(0)z^j} = 0$, thus $H^t_{\sum_{k=0}^{j} \frac{b^{(k)}(0)}{k!} z^k} = 0$. By induction, it follows that $H^t_b = 0$, and thus $b = c$.

§3. S_p- estimates for general operators T_b for $0 < p \leq 1$.

We need the following Lemma, the proof can be found in Peller [8].

LEMMA 3.1. *Let n be a positive integer, and let $\xi_j = e^{2\pi(j/4n)i}, 0 \leq j \leq 4n-1$. Then*

$$C_1 \|\phi\|_p^p \leq \frac{1}{4n} \sum_{0 \leq j \leq 4n-1} |\phi(\xi_j)|^p \leq C_2 \|\phi\|_p^p,$$

for every polynomial ϕ of degree $n-1$, where the positive constants C_1 and C_2 do not depend on n.

(Compare with Lemma 6 in Peng [10].)

We start with a general operator T_b, which satisfies the conditions D1: T_b maps $A^{\beta,2}(D)$ into $L^{\alpha,2}(D)$, D2: $T_b(f)$ is linear in f and anti-linear in b, and D3: $T_{b \circ e^{i\theta}} = e^{i\theta} \circ T_b \circ e^{i\theta}$.

It is obvious that $H_b, \widetilde{H}_b$ and H^t_b satisfy D1, D2 and D3.

Now we define a partition of unity. We take a function $\psi \in S(R)$ such that $supp\hat{\psi} \subset [1/2, 4]$ and such that, putting

$$\psi_n(z) = \sum_{k \in Z} \hat{\psi}(\frac{k}{2^n}) z^k, \quad n \geq 1$$

$$\psi_n = \overline{\psi}_n, \quad n < 0, \quad \psi_0 = 1 + z + \overline{z},$$

holds

$$\sum_{n \in Z} \hat{\psi}_n(k) = 1$$

for every k. And we take another function $\psi_1 \in S(R)$ such that $supp\hat{\psi}_1 \subset [\frac{1}{3}, 5]$ and $\hat{\psi}_1(\xi) \equiv 1$ on $[\frac{1}{2}, 4]$, $\hat{\psi}_{1,n} = \hat{\psi}_1(\frac{\cdot}{2^n})$, $\psi_{1,0} = \psi_0$.

Let $b_l = \psi_l * b$. Then for $0 < p \leq 1$ it holds

$$\|T_b\|^p_{S_p^{\beta o}} \leq \sum_{l=0}^{\infty} \|T_{b_l}\|^p_{S_p^{\beta o}} = \sum_{l=0}^{\infty} \| \sum_{k \in Z^+} \overline{\hat{b}_l(k)} \hat{\psi}_{1,l}(k) T_{z^k} \|^p_{S_p^{\beta o}}.$$

Since

$$\hat{\phi}(k) = \frac{1}{2m} \sum_{0 \leq j \leq 2m-1} \phi(\xi_j) \overline{\xi}_j^k$$

holds for any polynomial of degree $< m$, where $\xi_j = e^{2\pi i(j/2m)}$, we have, for $l \geq 0$,

$$\begin{aligned} \| \sum_{k \in Z^+} \overline{\hat{b}_l(k)} \hat{\psi}_{1,l}(k) T_{z^k} \|^p_{S_p^{\beta o}} &\leq \frac{1}{2^{(l+2)p}} \sum_{o \leq j \leq 2^{l+2}-1} |b_l(\xi_j)|^p \| \sum_{k \leq 2^{l+2}} \xi_j^k \hat{\psi}_{1,l}(k) T_{z^k} \|^p_{S_p^{\beta o}} \\ &\leq C 2^{l(1-p)} \|b_l\|_p^p \| \sum_{k \leq 2^{l+2}} \hat{\psi}_{1,l}(k) T_{z^k} \|^p_{S_p^{\beta o}}, \end{aligned}$$

by Lemma 3.1. Thus we get the following lemma.

LEMMA 3.2. *If T_b satisfies D1, D2 and D3, and* $\|\sum_{k\le 2^{l+2}}\hat{\psi}_{1,l}(k)T_{z^k}\|^p_{S^{\beta 0}_p} \le C2^{l(p+\frac{\beta p}{2})}$, *then*

$$\|T_b\|^p_{S^{\beta 0}_p} \le C\|b\|^p_{B_p^{\frac{1}{p}+\frac{\beta}{2}}}.$$

Let $I^l_h = span\{z^n/\gamma_{n,\beta}: \quad h2^{l+2} \le n < (h+1)2^{l+2}\}$, then $\sum_{h=0}^{\infty}\oplus I^l_h = A^{\beta,2}(D)$. Since $\|T\|^p_{S_p} = \|T^*T\|^{\frac{p}{2}}_{S_{\frac{p}{2}}}$ (see Zhu [15] , p. 34) we have

$$\begin{aligned}\|\sum_{k\le 2^{l+2}}\hat{\psi}_{1,l}(k)T_{z^k}\|^p_{S^{\beta 0}_p} &\le \sum_{h=0}^{\infty}\|\sum_{k\le 2^{l+2}}\hat{\psi}_{1,l}(k)T_{z^k}\|^p_{S_p(L^{0,2}(D),I^l_h)} \\ &= \sum_{h=0}^{\infty}\|(\sum_{k\le 2^{l+2}}\hat{\psi}_{1,l}(k)T_{z^k})^*(\sum_{k\le 2^{l+2}}\hat{\psi}_{1,l}(k)T_{z^k})\|^{\frac{p}{2}}_{S_{\frac{p}{2}}(I^l_h,I^l_h)}.\end{aligned}$$

Now the operator $T^l_h = \left(\sum_{k\le 2^{l+2}}\hat{\psi}_{1,l}(k)T_{z^k}\right)^*\left(\sum_{k\le 2^{l+2}}\hat{\psi}_{1,l}(k)T_{z^k}\right)|_{I^l_h\times I^l_h}$ maps I^l_h into itsself. It is given by the matrix $\{< T^l_h(e_n), e_m >\}$. And for $e_n = z^n/\gamma_{n,\beta}$ and $e_m = z^m/\gamma_{n,\beta} \in I^l_h$, we have

$$\begin{aligned}< T^l_h(e_n), e_m > &=< \sum_{k_2\le 2^{l+2}}\hat{\psi}_{1,l}(k_1)T_{z^{k_1}}(e_n), \sum_{k_2\le 2^{l+2}}\hat{\psi}_{1,l}(k_2)T_{z^{k_2}}(e_m) > \\ &= \sum_{k^1\le 2^{l+2}}\sum_{k_2\le 2^{l+2}}\hat{\psi}_{1,l}(k_1)\hat{\psi}_{1,l}(k_2) < T_{z^{k_1}}(e_n), T_{z^{k_2}}(e_m) > .\end{aligned}$$

Thus we get the following lemma by Lemma 3.1.

LEMMA 3.3. *If T_b satisfies D1, D2 and D3, and*

$$\|\{< T^l_h(e_n), e_m >\}\|^{\frac{p}{2}}_{S_{\frac{p}{2}}} \le C2^{l(p+\frac{\beta p}{2})}h^{-sp}$$

for $h \ge 1$ and $\|\{< T^l_0(e_n), e_m >\}\|^{\frac{p}{2}}_{S_{\frac{p}{2}}} \le C2^{l(p+\frac{\beta p}{2})}$, then, for $\frac{1}{p} < s$,

$$\|\sum_{k\le 2^{l+2}}T_{z^k}\|^p_{S^{\beta 0}_p} \le C2^{l(p+\frac{\beta p}{2})}.$$

To get the estimates in Lemma 3.3, we need the following Lemma very often.

LEMMA 3.4. *Let $\psi \in \mathcal{S}(R)$ such that supp $\hat{\psi} \subset [\frac{1}{3},5]$, $\hat{\psi}(\xi) \equiv 1$ on $[\frac{1}{2},4]$, and let $(\hat{\psi})_e(\xi)$ denote the periodic extension of $\hat{\psi}$ with the period 6π. Then*

$$\|(\hat{\psi})_e\left(\frac{m-n+k_1}{2^l}\right)K(m,n)\|^p_{S_p(I^l_h\times I^l_h)} \le C\|K(m,n)\|^p_{S_p(I^l_h\times I^l_h)}$$

holds for any $0 \leq k_1 \leq 2^{l+2}$, *and* $0 < p \leq 1$.

PROOF: We expand the function $(\hat{\psi})_e$ into the Fourier series

$$(\hat{\psi})_e(\xi) = \sum_{j \in Z} b(j) e^{ij\xi/3}.$$

Thus

$$(\hat{\psi})_e \left(\frac{m-n+k_1}{2^l} \right) = \sum_{j \in Z} b(j) e^{ij \frac{m-h2^{l+2}}{3 \cdot 2^{l+1}}} e^{ij \frac{h2^{l+2}-n+k_1}{3 \cdot 2^{l+2}}}.$$

Therefore

$$\begin{aligned} \|(\hat{\psi})_e \left(\frac{m-n+k_1}{2^l} \right) K(m,n)\|^p_{s_p(I^l_h \times I^l_h)} &\leq \sum_{j \in Z} |b(j)|^p \|K(m,n)\|^p_{s_p(I^l_h \times I^l_h)} \\ &\leq C \|K(m,n)\|^p_{s_p(I^l_h \times I^l_h)}. \end{aligned}$$

As we said, for example, the big Hankel operator H_b satisfies D1, D2 and D3. If $\beta < 0, 1 \leq \frac{1}{p} < 1 - \frac{\beta}{2}$ and $h \geq 1$, note that $H_{z^{k_1}}(e_n) \perp H_{z^{k_2}}(e_m)$ when $n - k_1 \neq m - k_2$, then we have, for $n - k_1 = m - k_2$,

$$\begin{aligned} < T^l_h(e_n), e_m > &= \sum_{k_1 \leq 2^{l+2}} \hat{\psi}_{1,l}(k_1) \hat{\psi}_{1,l}(k_2) < H_{z^{k_1}}(e_n), H_{z^{m-n-k_1}}(e_m) > \\ &= \sum_{k_1 \leq 2^{l+2}} \frac{\hat{\psi}_{1,l}(k_1)\hat{\psi}_{1,l}(k_2)}{\gamma_{n,\beta}\gamma_{m,\beta}} \frac{k_1(m-n+k_1)}{(n+1)(m+1)(m+k_1+1)} \\ &= \frac{1}{\gamma_{n,\beta}(n+1)} \sum_{k_1 \leq 2^{l+2}} \frac{\hat{\psi}_{1,l}(k_1)\hat{\psi}_{1,l}(k_2) k_1 (m - h2^{l+2} + k_1)}{\gamma_{m,\beta}(m+1)(m+k_1+1)} \\ &\quad + \frac{(h2^{l+2}-n)}{\gamma_{n,\beta}(n+1)} \sum_{k_1 \leq 2^{l+2}} \frac{\hat{\psi}_{1,l}(k_1)\hat{\psi}_{1,l}(k_2) k_1}{\gamma_{m,\beta}(m+1)(m+k_1+1)} \end{aligned}$$

i.e., $\{< T^l_h(e_n), e_m >\}$ become a sum of two matrices, and by Lemma 3.4, each one can be estimated by an operator of rank one, which $S_{\frac{p}{2}}$-norm is dominated by $C\frac{2^{l(2+\beta)}}{h^{2-\beta}}$, so the conditions in Lemma 3.3 were satisfied with $s = 1 - \frac{\beta}{2}$. (The estimate for $h = 0$ is trivial.) Finally we obtain that for $\beta < 0$ and $1 \leq \frac{1}{p} < 1 - \frac{\beta}{2}$,

$$\|H_b\|_{S^{\beta o}_p} \leq C \|b\|_{B_p^{\frac{1}{p}+\frac{\beta}{2}}},$$

i.e., we obtain a new proof of the main part of Theorem A.

§4. S_p-estimates for the middle Hankel operator

We will prove the following proposition which is rather general.

PROPOSITION 4.1. *Let* $-1 < \beta < 2, 1 \le \frac{1}{p} < 2 - \frac{\beta}{2}$ *and* $b \in B_p^{\frac{1}{p}+\frac{\beta}{2}}$, *then* $H_b^t \in S_p^{\beta 0}$ *and*

$$\|H_b^t\|_{S_p^{\beta 0}} \le C\|b\|_{B_p^{\frac{1}{p}+\frac{\beta}{2}}}.$$

By Lemma 3.2, 3.3 and 3.4, it suffices to prove the following

LEMMA 4.1. *If* $n - k_1 = m - k_2$, *then for* $n - k_1 \ge 0$,

$$< H^t_{z^{k_1}}(e_n), H^t_{z^{k_2}}(e_m) > = \frac{1}{\gamma_{n,\beta}\gamma_{m,\beta}} \frac{k_1(m-n+k_1)(t-k_1)(n-m+t-k_1)}{(n+1)(m+1)(n+t+1)(m+t+1)(m+k_1+1)},$$

for $n - k_1 \ge 0$,

$$< H^t_{z^{k_1}}(e_n), H^t_{z^{k_2}}(e_m) >= \frac{1}{\gamma_{n,\beta}\gamma_{m,\beta}} \frac{1}{m+k_1+1}.$$

Assuming Lemma 4.1 holds, we prove our results as follows. Note that $H^t_{z^{k_1}}(e_n) \perp H^t_{z^{k_2}}(e_m)$ when $n - k_1 \ne m - k_2$, we have, for $n - k_1 = m - k_2$ and $h \ge 1$,

$$\begin{aligned}
\langle T_h^l(e_n), e_m\rangle &= \sum_{k_1 \le 2^{l+2}} \hat{\psi}_{1,l}(k_1)\hat{\psi}_{1,l}(k_2) < H^t_{z^{k_1}}(e_n), H^t_{z^{m-n+k_1}}(e_m) > \\
&= \sum_{k_1 \le 2^{l+2}} \frac{\hat{\psi}_{1,l}(k_1)\hat{\psi}_{1,l}(k_2)}{\gamma_{n,\beta}\gamma_{m,\beta}} \frac{k_1(m-n+k_1)(t-k_1)(n-m+t-k_1)}{(n+1)(m+1)(n+t+1)(m+t+1)(m+k_1+1)} \\
&= \frac{n - h2^{l+2}}{\gamma_{n,\beta}(n+1)(n+t+1)} \sum_{k_1 \le 2^{l+2}} \frac{\hat{\psi}_{1,l}(k_1)\hat{\psi}_{1,l}(k_2)k_1(t-k_1)(m-h2^{l+2}+k_1)}{\gamma_{m,\beta}(m+1)(m+t+1)(m+k_1+1)} \\
&- \frac{(n-h2^{l+2})^2}{\gamma_{n,\beta}(n+1)(n+t+1)} \sum_{k_1 \le 2^{l+2}} \frac{\hat{\psi}_{1,l}(k_1)\hat{\psi}_{1,l}(k_2)k_1(t-k_1)}{\gamma_{m,\beta}(m+1)(m+t+1)(m+k_1+1)} \\
&+ \frac{1}{\gamma_{n,\beta}(n+1)(n+t+1)} \cdot \\
&\sum_{k_1 \le 2^{l+2}} \frac{\hat{\psi}_{1,l}(k_1)\hat{\psi}_{1,l}(k_2)k_1(t-k_1)(m-h2^{l+2}+k_1)(h2^{l+2}-m+t-k_1)}{\gamma_{m,\beta}(m+1)(m+t+1)(m+k_1+1)} \\
&+ \frac{h2^{l+2}-1}{\gamma_{n,\beta}(n+1)(n+t+1)} \sum_{k_1 \le 2^{l+2}} \frac{\hat{\psi}_{1,l}(k_1)\hat{\psi}_{1,l}(k_2)k_1(t-k_1)(h2^{l+2}-m+t-k_1)}{\gamma_{m,\beta}(m+1)(m+t+1)(m+k_1+1)},
\end{aligned}$$

i.e.,$\{< T_h^l(e_n), e_m >\}$ become a sum of four matrices, each of which can be estimated by an operator of rank one, which S_p-norm is dominated by $C\frac{2^{2l+\beta}}{h^{4-\beta}}$. So

$$\|\{< T_h^l(e_n), e_m >\}\|^{\frac{p}{2}}_{S_{\frac{p}{2}}} \le C2^{l(p+\frac{\beta p}{2})}h^{-(2-\frac{\beta}{2})p}.$$

For $h = 0$, it is easy to show that $\|\{< T_0^l(e_n), e_m >\}\|_{S_{\frac{p}{2}}}^{\frac{p}{2}} \leq C2^{l(p+\frac{\beta p}{2})}$.

Therefore if $-1 < \beta < 2$, $1 \leq \frac{1}{p} < 2 - \frac{\beta}{2}$ and $b \in B_p^{\frac{1}{p}+\frac{\beta}{2}}$, Lemma 3.2 and 3.3 give us $H_b^t \in S_p^{\beta 0}$ and $\|H_b^t\|_{S_p^{\beta 0}} \leq C\|b\|_{B_p^{\frac{1}{p}+\frac{\beta}{2}}}$.

Now we prove Lemma 4.1 for $n - k_1 \geq 0$ (the case $n - k_1 < 0$ is trivial).

For $m - k_2 = n - k_1$, by (2.4), it follows that

$$
\begin{aligned}
&\langle H_{z^{k_1}}^t(e_n), H_{z^{k_2}}^t(e_m)\rangle \\
&= \frac{1}{\gamma_{n,\beta}\gamma_{m\beta}} \int_0^1 [r^{2k_1} - \frac{\gamma_n^2}{\gamma_{n-k_1}^2} - \frac{\gamma_{n+t}^2\gamma_{n-k_1}^2 - \gamma_{n-k_1+t}^2\gamma_n^2}{\gamma_{n-k_1}^2\gamma_{n-k_1+2t}^2 - \gamma_{n-k_1+t}^4}(r^{2t} - \frac{\gamma_{n-k_1+t}^2}{\gamma_{n-k_1}^2})] \\
&\quad \times [r^{2k_2} - \frac{\gamma_m^2}{\gamma_{m-k_2}} - \frac{\gamma_{m+t}^2\gamma_{m-k_2}^2 - \gamma_{m-k_2+t}^2\gamma_m^2}{\gamma_{m-k_2}^2\gamma_{m-k_2+2t}^2 - \gamma_{m-k_2+t}^4}(r^{2t} - \frac{\gamma_{m-k_2+t}^2}{\gamma_{m-k_2}^2})]r^{n+m-k_1-k_2}dr^2 \\
&= \frac{1}{\gamma_{n,\beta}\gamma_{m,\beta}}[\gamma_{m+k}^2 - \frac{\gamma_m^2\gamma_m^2}{\gamma_{n-k_1}^2} - \frac{(\gamma_{n+t}^2\gamma_{n-k_1}^2 - \gamma_{n-k_1+t}^2\gamma_n^2)(\gamma_{m+t}^2\gamma_{m-k_2}^2 - \gamma_{m-k_1+t}^2\gamma_m^2)}{\cdot\gamma_{n-k_1}^2(\gamma_{n-k_1+2t}^2\gamma_{n-k_1}^2 - \gamma_{n-k_1+t}^4)}] \\
&= \frac{1}{\gamma_{n,\beta}\gamma_{m,\beta}} \frac{k_1(m-n+k_1)(t-k_1)(n-m+t-k_1)}{(n+1)(m+1)(n+t+1)(m+t+1)(m+k_1+1)}.
\end{aligned}
$$

So Lemma 4.1 is proved.

§5. Open problems

1. How to generalize the above results to the general case $\alpha > -1$? In the proof in §4 for the case $\alpha = 0$, we have used the fact that each entry in the matrix $\{< T_h^l(e_n), e_m >\}$ is a sum of four terms, and in each term the variables (n, m) are separated. In the general case $\alpha > -1$, the formula for $< T_h^l(e_n), e_m >$ involves the gamma function, and we do not know if this fact still holds.

2. For any $p_0 < 1$, how to construct a "middle" Hankel operator H_b' such that $\widetilde{H}_b \prec H_b' \prec H_b$ and H_b' has the cut-off at p_0? (In particular, $p_0 < \frac{1}{2}$). let us give an example. Let $B_t^{0,2}(D)$ be the subspace of $L^{0,2}(D)$ generated by $\{|z|^{2t}z^n, n \geq 0\}$, and $A^{0,2}(D)$ the Bergman space. Let $A_t^{0,2}(D) = B_t^{0,2}(D) \ominus A^{0,2}(D)$. Denote P and P_0 the corresponding projections from $L^{0,2}(D)$ onto $A^{0,2}(D)$ and $A_t^{0,2}(D)$, and H_b^t the middle Hankel operator defined as before. Let us take another subspace $B_{2t}^{0,2}(D)$ generated by $\{|z|^{4t}z^n, n \geq 0\}$ and $A_{2t}^{0,2}(D) = B^{0,2}(D) \ominus A^{0,2}(D) \ominus A_t^{0,2}(D)$, and denote the correponding projection by P_{2t}, then we define

$$H_b^{2t} = (I - P - P_t - P_{2t})M_{\overline{b}}P.$$

It is obvious that $\widetilde{H}_b \prec H_b^{2t} \prec H_b^t \prec H_b$, and we guess that H_b^{2t} has the cut-off at $p_0 = \frac{1}{3}$.

3. In Arazy et al [1] it is showed that the big Hankel operators can be viewed as a vector -valued paracommutators, in fact the same is true for the above middle Hankel operators. Other examples of vector-valued paracommutators are the big Hankel operators of higher-weights (see Boman et al [3]). How to settle the S_p estimates for the vector-valued paracommutators? We will study this in a subsequent publication.

Acknowledgements. The authors would like to thank Richard Rochberg for the problem and Svante Janson for helpful discussion. Special thanks go to Jaak Peetre for cordial hospitality and advice.

References

[1] J. Arazy, S. Fisher and J. Peetre, *Hankel operators in Bergman spaces* Amer. J. Math.**110** (1988), 989–1053.

[2] S. Axler, *The Bergman kernel, the Bloch space, and commutators of multiplication operators*, Duke Math. J. **53** (1986), 315–332.

[3] J. Boman, S. Janson and J. Peetre, *Big Hankel operators of higher weights*, Rend. Circ. Mat. Palermo **38** (1989), 65–78.

[4] S. Janson, *Hankel operators between weighted Bergman spaces*, Ark. Mat., (1988), 205–219.

[5] S. Janson and R. Rochberg, *personal communication.*

[6] J. Peetre, *Hankel operators, rational approximation and allied questions of analysis*, CMS Conference Procedings, **3** (1985), 287–332.

[7] V. V. Peller, *Hankel operators of class γ_p and applications (rational approximation, Gaussian processes, the majorant problem for operators)*, Math. USSR Sbornik **41** (1982), 443–479.

[8] V. V. Peller, *A description of Hankel operators of class γ_p for $p > 0$, and investigation of the rate of rational approximation, and other applications*, Math. USSR Sbornik **50** (1985), 465–494.

[9] V. V. Peller, *Vectorial Hankel operators, commutators and related operators of Schatten - von Neumann class γ_p*, Integral Equations and Operator Theory **5** (1982), 244–272.

[10] L. Peng, *Paracommutators of Schatten-von Neumann class S_p*, $0 < p < 1$, Math. Scand. **61** (1987), 68–92.

[11] L. Peng, *Ha-plitz operators on Bergman space.* In preparation.

[12] M. Reed and B. Simon, "Methods of modern mathematical physics, IV," Academic Press, New York, San Francisco, London, 1978.

[13] R. Rochberg, *Trace ideal criteria for Hankel operators and commutators*, Indiana Univ. Math. J. **31** (1982), 913–925.

[14] S. Semmes, *Trace ideal criterion for Hankel operators* , $0 < p < 1$, Integral Equation and Operator Theory **7** (1984), 241–281.

[15] K. Zhu, "Operator Theory in Function Spaces." To appear.

The Bishop-Phelps Theorem in Complex Spaces: An Open Problem

R. R. PHELPS, Department of Mathematics GN-50, University of Washington, Seattle, WA 98195

Recall, first, the statement of the Bishop-Phelps theorem [BP] for a real Banach space E: If C is a nonempty closed convex subset of E, if $f \in E^*$ is bounded above on C and if $\varepsilon > 0$, then there exists $g \in E^*$, $g \neq 0$, which attains its supremum on C at some point x of C and which satisfies $\|f - g\| < \varepsilon$. (We say that g is a *support functional* of C and that x is a *support point* of C.) Moreover, for any closed convex C the set of support points is dense in the boundary of C.

Suppose, now, that E is a *complex* Banach space and that C is a nonempty closed convex subset of E. By working with the underlying real space E_r, one can use the results cited above to obtain the same kinds of theorems, *provided one replaces the functionals involved by their real parts.* If, however, one wants an intrinsically complex version of these density theorems, then it is necessary to replace the functionals by their absolute values, which leads to the following two problems:

Suppose that C is a nonempty closed convex subset of the complex Banach space E, that $f \in E^*$ and that $|f|$ is bounded on C. Given $\varepsilon > 0$, does their exist $g \in E^*$, $g \neq 0$, such that $\|f - g\| < \varepsilon$ and such that $|g(x)| = \sup |g|(C)$ for some $x \in C$? Furthermore, are such "modulus support points" x dense in the boundary of C?

The second question is trivially answered in the negative (an example is given below), but the first question – which has been around for some years as a "folk problem" – remains open. We will reformulate it in terms of real linear functionals on the underlying real Banach space E_r, exhibit some situations where it is trivially true and show that the analogue for real Banach spaces is valid. None of this is difficult; our motivation is to call wider attention to the problem.

At the outset, we will simplify the problem by assuming that the set C is *bounded*, so that the hypothesis that $|f|$ be bounded on C is automatically satisfied. Boundedness of

C also guarantees that the set

$$\Delta C = \{\alpha x : |\alpha| \leq 1,\ x \in C\}$$

is closed, an easily proved but useful fact. Note that the supremum of $|f|$ on C is the same as its supremum on ΔC. The first problem can now be stated as follows:

1. Problem. *Given a nonempty bounded closed convex subset C of the complex Banach space E, $f \in E^*$ and $\varepsilon > 0$, do there exist $g \in E^*$, $g \neq 0$, and $x \in C$ such that $\|f - g\| < \varepsilon$ and $|g(x)| = \sup\{|g(y)| : y \in C\}$?*

The key to reformulating Problem 1 is the following elementary fact.

2. Fact. *If $g \in E^*$, $g \neq 0$, then $\sup |g|(C) = \sup(\operatorname{Re} g)(\Delta C)$. Moreover, g attains its maximum modulus at a point of C if and only if $\operatorname{Re} g$ attains its supremum on the closed set ΔC.*

(The proof of this is standard and will be left to the reader.) All the assertions in the above Fact are valid for the smaller set

$$\{\alpha x : |\alpha| = 1 \text{ and } x \in C\}$$

in place of ΔC.

3. Proposition. *Problem 1 has an affirmative solution if and only if the (real) linear functionals in $(E_r)^* \equiv E_r^*$ which support the closed set ΔC are dense in E_r^*.*

Proof. We make use of the standard fact that if $h \in E_r^*$, then there exists a unique $f \in E^*$ such that $h = \operatorname{Re} f$ and $\|h\| = \|f\|$, that is, the map from E^* to E_r^* which sends a functional to its real part is a surjective linear isometry. Suppose, now, that Problem 1 has an affirmative answer and that $h \in E_r^*$ and $\varepsilon > 0$. Write $h = \operatorname{Re} f$ for f in E^* and choose g in E^* such that $\|f - g\| < \varepsilon$ and $|g|$ attains its supremum on C. By the Fact proved above, $\operatorname{Re} g$ attains its supremum on ΔC and $\|\operatorname{Re} f - \operatorname{Re} g\| = \|f - g\| < \varepsilon$. On the other hand, suppose that the support functionals for ΔC are dense in E_r^*. Given $f \in E^*$ and $\varepsilon > 0$, choose $g \in E^*$ such that $\operatorname{Re} g$ supports ΔC and $\|\operatorname{Re} g - \operatorname{Re} f\| < \varepsilon$. Then $|g|$ supports C and $\|f - g\| < \varepsilon$.

From this proposition it is clear that if ΔC is *convex*, then, by the original Bishop-Phelps theorem, Problem 1 has an affirmative answer. Unfortunately, every space of (complex) dimension greater than 1 contains a closed bounded convex set C such that ΔC is not convex, as shown by the following example.

4. Example. *Choose $x_0 \in E$ such that $\|x_0\| = 1$ and choose $f \in E^*$ such that $\|f\| = 1$ and $f(x_0) = 1$. Choose $y_0 \in E$ such that $f(y_0) = 0$ and $\|y_0\| = 3$. (This requires $\dim E > 1$.) Let $C = y_0 + B$, where B is the closed unit ball of E. Then ΔC is not convex.*

Proof. It is clear that $y_0 \pm x_0 \in C$, and hence $y_0 + x_0$ and $-(y_0 - x_0)$ are in ΔC. Obviously, $x_0 = (1/2)[(y_0 + x_0) + (-y_0 + x_0)]$ is in the convex hull of ΔC. It is not in ΔC, however.

If it were, we could write $x_0 = \alpha(y_0 + x)$, where $|\alpha| \leq 1$ and $\|x\| \leq 1$. Consequently, $1 = \|x_0\| \geq |\alpha|(\|y_0\| - \|x\|) \geq 2|\alpha|$. On the other hand, $1 = |f(x_0)| = |\alpha f(y_0 + x)| \leq |\alpha|\,\|x\| \leq |\alpha|$, a contradiction.

One might ask whether this nonconvex ΔC provides a counterexample for Problem 1, but the following observation shows that it does not, since C is a translate of the unit ball.

5. Observation. *If a bounded closed convex set D is circled, that is $\alpha D \subset D$ for every $|\alpha| = 1$, then Problem 1 has an affirmative answer for any translate C of D.*

Proof. Since D is circled and convex, we have $\Delta D = D$, so by our earlier remarks, it suffices to show that if $|g|$ supports D then it supports any translate $C = D + y$. By hypothesis, there exists $x \in D$ such that $|g(x)| = \sup |g|(D)$. Since D is circled, we can assume that $g(x) = |g(x)|$. Write $g(y) = \alpha|g(y)|$, for some $|\alpha| = 1$. Then $\alpha x + y \in C$ and

$$|g(\alpha y + x)| = |\alpha\{g(x) + |g(y)|\}| = |g(x) + |g(y)|\,| = g(x) + |g(y)|,$$

while $\sup |g|(C) \leq \sup |g|(D) + |g(y)| = g(x) + |g(y)|$.

Here is an example which shows that the modulus support points of C need not be dense in the boundary of C.

6. Example. *In the complex plane, let C be the convex hull of the closed unit disk and the real number 2. The latter is the unique modulus support point of C.*

The following easy proposition shows that, in *real* Banach spaces, the modulus support functionals will be dense in E^*.

7. Proposition. *Suppose that E is a real Banach space and that C is a nonempty, closed convex and bounded subset of E. Given $f \in E^*$, $\|f\| = 1$ and $\varepsilon > 0$ there exists $g \in E^*$, $\|g\| = 1$ such that $\sup |g(C)| = |g(x)|$ for some $x \in C$ and $\|f - g\| < \varepsilon$.*

Proof. Let M be such that $\|x\| \leq M$ for all $x \in C$. By the Bishop-Phelps theorem there exists $h \in E^*$, $\|h\| = 1$, such that $\|f - h\| < \varepsilon/2$ and h attains its supremum on C. If $\sup h(-C) \leq \sup h(C)$, then $\sup h(C) = \sup |h(C)|$ and we can let $g = h$. Suppose, then, that $\sup h(-C) > \sup h(C)$, say

$$\sup h(-C) = \sup h(C) + \delta, \qquad \text{for some } \delta > 0.$$

Choose $g \in E^*$, $\|g\| = 1$, such that $\|h - g\| < \min\{\varepsilon/2,\ \delta/2M\}$ and g attains its supremum on $-C$. If $y \in C$, then

$$g(y) \leq \|h - g\| \cdot \|y\| + h(y) \leq \|h - g\| \cdot M + \sup h(C),$$

so

$$\sup g(C) \leq \|h - g\| \cdot M + \sup h(C).$$

Similarly,

$$\sup h(-C) \leq \|h - g\| \cdot M + \sup g(-C),$$

hence

$$\sup g(C) \leq 2\|h - g\| \cdot M - \delta + \sup g(-C) < \sup g(-C),$$

which implies that $\sup |g(C)| = g(-x)$ for some $x \in C$ (therefore $g(-x) = |g(x)|$). By the triangle inequality, $\|g - f\| < \varepsilon$.

It is clear that a counterexample to the validity of Problem 1 for bounded convex sets will not be found in a reflexive space. More generally, *Problem 1 has an affirmative solution for any bounded closed nonempty convex subset C of a Banach space E with the Radon-Nikodym property.* (By definition, this is the same as saying that the underlying real space E_r has the RNP.) To see this, let D denote the closed convex hull of ΔC. Since D is bounded closed convex and nonempty, the set of functionals in E_r^* which strongly expose a point of D form a dense G_δ subset of E_r^* [B, p.55]. By the standard isometry $f \to \operatorname{Re} f$ between E^* and E_r^*, they are the real parts of a dense G_δ subset G of E^*. It is easily checked that any strongly exposed point of D is actually contained in ΔC, so by the Fact (above), each functional in G actually attains it maximum modulus on C.

References

[BP] E. Bishop and R. R. Phelps, The support functionals of a convex set, Proc. Sympos. Pure Math. VII, Convexity, Amer. Math. Soc., Providence, R. I. (1963).

[B] Richard D. Bourgin, "Geometric aspects of convex sets with the Radon-Nikodym Property", Lecture Notes in Math. 993, Springer-Verlag, Berlin-Heidelberg-New York-Tokyo (1983).

On Hyperconvex Hulls of Some Normed Spaces

N. V. Rao. Department of Mathematics, University of Toledo, Toledo, Ohio 43606

1 Introduction

Herrlich [3] has recently shown that $\mathbf{R}^n$ with the sum metric $\|x\| = |x_1| + |x_2| + \ldots + |x_n|$ has its hyperconvex hull isometric to $\mathbf{R}^{2^{n-1}}$ In this note, we describe the hyperconvex hull of any normed space X, the unit ball of which has only finitely many faces i.e. the set of extreme points of the unit ball of the dual X^* of X is finite. This restriction on X makes it finite dimensional and its unit ball a polytope and our result will include the above mentioned result of Herrlich.

Lacey and Cohen [6] have more general results than these but our proof of these results is more direct and simple.

We say that (X, ρ) *is hyperconvex* if given any family of closed balls $B(x_\alpha, r_\alpha), \alpha \in I$ such that $|x_\alpha - x_\beta| \leq r_\alpha + r_\beta$ for all pairs $\alpha, \beta \in I$, then there exists a point x_0 in X such that $|x_0 - x_\alpha| \leq r_\alpha$ for all $\alpha \in I$. For a detailed discussion of various levels of hyperconvexity we refer the reader to [1]. For some other aspects of this problem, one may see Dress [2]. The hyperconvexity (binary intersection property) arose in connection with a generalisation of the Hahn-Banach theorem. Nachbin [7] has investigated the problem of extending the Hahn-Banach theorem in the following manner:

> Let Y be a Banach space and $X \subset Z$ be any two normed spaces X a subspace of Z. If for any pair $X \subset Z$ and any continuous linear functional $f : X \to Y$, there exits an extension $F : Z \to Y$ such that $\|F\| = \|f\|$, $F(x) = f(x)\ \forall\, x \in X$; then we say that Y has the Hahn-Banach extension property.

Nachbin has characterized all such spaces as hyperconvex and also as spaces of continuous functions on compact Hausdorff extremally disconnected spaces with the assumption that the closed unit ball in Y has at least one extreme point. Later on Kelley [5] showed that the existence of an extreme point is unnecessary.

We say that a metric space Y is a **hyperconvex hull** of a metric space X if there exists an isometry

$$e : X \to Y$$

such that is no proper subspace Z of Y containing $e(X)$, which is hyperconvex. The mapping e is essential for uniqueness. The existence and uniqueness of hyperconvex hulls for arbitrary metric spaces was established by Isbell [4].

2

THEOREM: *Let X be any normed space and suppose that X^* is its dual space, and U is the closed unit ball of X^*. Assume that U has only finitely many extreme points and let E be the set of all extreme points of the closed unit ball. (This implies that X is finite dimensional but obviously the converse is false). Let W be a subset of E such that $W \cap (-W) = \emptyset$ and $W \cup (-W) = E$. Then the hyperconvex hull of X is canonically isometric to $l^\infty(W)$.*

PROOF. Since U is the closed convex hull of E and by the Hahn-Banach theorem, we get that for any $x \in E$

$$\|x\| = \sup_U |\hat{x}| = \sup_E |\hat{x}|$$

Thus the mapping from $X \to l^\infty(E)$ given by

$$x \hookrightarrow \hat{x}/E$$

is an isometry. Further we notice that

$$\hat{x}(w) = -\hat{x}(-w) \text{ for all } w \in X^*$$

and so $\|\hat{x}\|_E = \|\hat{x}\|_W = \|x\|$. This gives us that

$$x \hookrightarrow \hat{x}/W$$

also is an isometry of X into $l^\infty(W)$

It is well-known and easy to show that $l^\infty(G)$, where G is any set, is hyperconvex (See [7]).

Now we shall show that any function $\xi \in l^\infty(W)$ is the **exact intersection** of a certain family $\mathcal{F}_\xi$ of balls with centers coming from X. This means that $l^\infty(W)$ is a hyperconvex hull of X. For if E is hyperconvex and $X \subseteq E \subseteq l^\infty(W)$, and if $\xi \in l^\infty(W)$, the above family $\mathcal{F}_\xi$ has a non-void intersection in E. Thus $\xi \in E$. Now we need a

LEMMA: *Given any $\xi \in l^\infty(W)$ and any vector $e \in W$, there exists an $x \in X$ such that*

$$\hat{x}(e) - \xi(e) \geq \sup_{w \in W} |\hat{x}(w) - \hat{\xi}(w)| = \|\hat{x} - \xi\|$$

PROOF. Suppose not. Then

$$\hat{x}(e) - \xi(e) < \sup_{w \in W \setminus \{e\}} |\hat{x}(w) - \hat{\xi}(w)| \text{ for all } x \in X$$

(Let W_e denote $W \setminus \{e\}$.) Hence for any $t > 0$,

$$t\hat{x}(e) - \xi(e) < \sup_{W_e} |t\hat{x}(w) - \hat{\xi}(w)|$$

Divide by t and take limits as $t \to \infty$. Then we have

$$\hat{x}(e) \leq \sup_{W_e} |\hat{x}(w)|.$$

Consider the set L of all linear combinations:

$$\sum_{W_e} \lambda(w)w, \quad \sum_{w \in W_e} |\lambda(w)| \leq 1.$$

This is a compact convex set and $e \notin L$ because if $e \in L$, e will not be an extreme point of the closed unit ball, i.e. because if it were (say)

$$e = \sum_{w \in W_E} \lambda(w)w, \quad \sum |\lambda(w)| \leq 1$$

We can rewrite $\lambda(w)w = \lambda(w')w'$ where $\lambda(w') = |\lambda(w)|$, $w' = \frac{\lambda(w)}{|\lambda(w)|}w$ if $\lambda(w) \neq 0$ and otherwise simply w. Then $e = \sum \lambda(w')w'$ is in the interior of the unit ball if $\sum|\lambda(w)| < 1$ and that is impossible. So e is a convex combination of w' which is impossible unless $e = w'$ for some w and $\lambda(w') = 0$ except perhaps for one w. But then it means either e or $-e \in W \setminus \{e\}$ which is a contradiction. Certainly $e \notin W_e$ and if $-e \in W$, then $W \cap (-W)$ contains e at least. But we assumed that $W \cap (-W) = \emptyset$.

Since $e \notin L$ and L is a compact convex set, by the Hahn-Banach theorem, there exists an element $y \in X^{**}$ which is X since we are considering X^* with weak* topology such that

$$\hat{y}(e) > \sup_{u \in L} \hat{y}(u)$$

since L is symmetric with respect to origin, $\hat{y}(e) > \sup_{u \in L} |\hat{y}(u)|$. From the definition of L, $W_e \subset L$ and so $\hat{y}(e) > \sup_{W_e} |\hat{y}(w)|$, a contradiction. Thus we have proved the lemma.

By applying the lemma to $-\xi$, we can restate the lemma as follows:

Given any $\xi \in l^\infty(W)$ and any $e \in W$, there exists an $x \in X$ such that

$$\|\hat{x} - \xi\| = \hat{x}(e) - \xi(e) \tag{1}$$

and a y in X such that

$$\|\hat{y} - \xi\| = \xi(e) - \hat{y}(e). \tag{2}$$

Now for each fixed $e \in W$, let us choose this pair of elements x, y from X satisfying (1) and (2). Take any $\eta \in l^\infty(W)$ such that

$$\|\hat{x} - \eta\| \leq \hat{x}(e) - \xi(e) = \|\hat{x} - \xi\|$$

$$\|\hat{y} - \eta\| \leq \xi(e) - \hat{y}(e) = \|\hat{y} - \xi\|$$

then

$$\hat{x}(e) - \eta(e) \leq \hat{x}(e) - \xi(e)$$

$$\eta(e) - \hat{y}(e) \leq \xi(e) - \hat{y}(e)$$

and so $\eta(e) = \xi(e)$ for all η in the intersection of the balls $B(\hat{x}, \hat{x}(e) - \xi(e))$ and $B(\hat{y}, \xi(e) - \hat{y}(e))$.

Thus for each $e \in W$, we get points x_e, y_e from X such that whenever

$$\|\hat{x}_e - \eta\| \leq \|\hat{x}_e - \xi\|, \quad \|\hat{y}_e - \eta\| \leq \|\hat{y}_e - \xi\|, \qquad \xi(e) = \eta(e). \tag{3}$$

By the triangle inequality, the set of balls $B(\hat{x}_e\ ,\ \|\hat{x}_e - \xi\|)\ ,\ B(\hat{y}_e\ ,\ \|\hat{y}_e - \xi\|)$ have the binary intersection property and, by (3), their only common point is exactly ξ. This shows that $l^\infty(W)$ is a minimal hyperconvex extension of X.

Herrlich [3] has also considered $X = C_0$, the space of all sequences $\{x_n\}$ such that $x_n \to 0$ as $n \to \infty$ and the norm is $\sup |x_n| = \|x\|$. We know that $X^* = l^1$ and the extreme points of the closed unit ball of l^1 is the set of sequences e_n where $e_n(k) = 0$ if $n \neq k$ and $e_n(n) = \pm 1$. We can take W to be the set of all sequences (a_n) such that $a_n = 0$ except for one integer n and there it is equal to 1. Let us call this sequence e_k, i.e. $e_k(n) = 0$ for $k \neq n$, $e_k(k) = 1$. Clearly $W \cap (-W) = \emptyset$, $W \cup (-W) = E$ is the set of all extreme points.

We claim that the natural mapping $i : X \to l^\infty(W)$ is the injective hull of X. All we need to check is whether given any sequence $\xi \in l^\infty(W)$ and any positive integer k there exists an $x \in C_0$ such that

$$\hat{x}(e_k) - \xi_k \geq \sup_{n \neq k} |\hat{x}(e_n) - \xi_n|.$$

Since $\{\xi_n\}$ is bounded, let us say $|\xi_n| \leq C$. Then let us define $\hat{x}(e_k) = x_k = \xi_k + C$, $\hat{x}(e_n) = 0$ for $n \neq k$. For this, it is easy to verify that $\hat{x}(e_k) - \xi_k = C$, $\sup_{n \neq k} |\hat{x}(e_n) - \xi_n| = \sup_{n \neq k} |\xi_n| \leq C$. Thus we can apply our lemma and obtain the theorem.

References

[1] Aronszajn and Panitchpakdi, *Extension of Uniformly continuous transformations and hyperconvex metric spaces,* Pac. J. Math. 6(1956), 405-439.

[2] A. W. M. Dress, *Trees, tight extension of metric spaces, and the cohomological dimension of certain groups: a note on combinatorial properties of metric spaces,* Advances in Math. 53(1984), 321-402.

[3] Horst Herrlich, *On Hyperconvex Hulls of Metric Spaces,* Manuscript, 1989.

[4] J. R. Isbell, *Six theorems about injective metric spaces,* Comment. Math. Helv. 39(1964/1965), 65-74.

[5] J. L. Kelley, *Banach spaces with the extension property,* Trans. Amer. Math. Soc. 72(1952), 323-326.

[6] H. E. Lacey and H. B. Cohen,*On Injective Envelopes of Banach Spaces,* J. Functional Analysis, 4(1969), 11-30.

[7] Leopoldo Nachbin, *A Theorem of the Hahn-Banach type for linear transformations,* Trans. Amer. Math. Soc. 68(1950), 28-46.

Eigenvalue Estimates for Calderon-Toeplitz Operators

RICHARD ROCHBERG Department of Mathematics, Washington University, St. Louis, Missouri

Introduction

We give eigenvalue estimates for certain compact (generalized) Toeplitz operators. One of our results is that for certain Calderon–Toeplitz operators there is a cut off in the Schatten–von Neumann ideal behavior at $p = 1/2$. By a Calderon–Toeplitz operator we mean a Toeplitz–like operator built using the Calderon reproducing formula. By a "cut off at 1/2" we mean that (for appropriate classes of symbols) there are operators in S_p for $p > 1/2$ but no non–trivial operators is S_p for $p = 1/2$.

Background and definitions are collected in Section II. In Section III we give Berezin–Lieb style estimates for the eigenvalues. In Section IV we use the results from Section III to show that certain operators are in S_p for $p > 1/2$. For the same operators we also show that the N–th eigenvalue is at least cN^{-2} and thus we have a cut off. In the final section we discuss variations including the applicability of these results to more familiar operators such as Toeplitz operators on Bergman spaces.

Although we will give a bit of background and motivation as we go, [Z], [R] and [C] are suggested for a look at the bigger picture.

II. Definitions

1. Toeplitz operators on the Bergman Space

To help with orientation we recall the definition of the Bergman space of the upper half plane and the associated Toeplitz operators. Let U be the upper half plane and $L^2 = L^2(U, dxdy)$. The associated Bergman space, A^2, is the Hilbert space of functions in L^2

Supported in part by NSF grant DSM 8701271.

which are holomorphic. For each $\zeta = x + iy$ in U there is the Bergman kernel function $k_\zeta(z) = -\pi^{-1}(z - \overline{\zeta})^{-2}$. k_ζ is a reproducing kernel for ζ; that is, for f in A^2, $f(\zeta) = \langle f, k_\zeta \rangle$. Applying the reproducing formula a second time we find that for any z in U,

$$f(z) = \int_U \langle f, k_\zeta \rangle\, k_\zeta(z)\, dx\, dy.$$

This is the Bergman reproducing formula. The projection P of L^2 onto A^2 is given by a similar formula: for any g in L^2,

$$Pg(z) = \int \langle g, k_\zeta \rangle\, k_\zeta(z)\, dx\, dy.$$

We will often find it more convenient to work with unit vectors; let $\tilde{k}_\zeta = \|k_\zeta\|^{-1} k_\zeta = k_\zeta(\zeta)^{-1/2} k_\zeta$. Set $d\mu(\zeta) = k_\zeta(\zeta)\, dx\, dy = c\, y^{-2}\, dx\, dy$. The integral formula for the projection can now be rewritten as

(2.1) $$Pg(z) = \int \langle g, \tilde{k}_\zeta \rangle\, \tilde{k}_\zeta(z)\, d\mu(\zeta)$$

Given a function $b(\zeta)$ defined on U we define T_b, the Toeplitz operator with symbol b, to be the map from A^2 to A^2 given by $T_b(f) = P(bf)$. Using (2.1) we obtain an integral representation for T_b;

(2.2) $$T_b f(z) = \int b(\zeta)\, \langle f, \tilde{k}_\zeta \rangle\, \tilde{k}_\zeta(z)\, d\mu(\zeta).$$

Although there are interesting things to be said if b is unbounded we will suppose b is bounded. In that case it is immediate that T_b is bounded and $\|T_b\| \le \|b\|_\infty$. We will generally assume b is real and non–negative. In that case T_b is self–adjoint and is a non–negative operator. If T_b is compact then it will have a sequence of eigenvalues which tends to zero. We denote the sequence $\{\lambda_i\}_{i=1}^\infty$ and assume it is ordered to be non–increasing.

Associated to T_b is a transformed version of b which we will call Berezin symbol [B], $\tilde{b}$, given by

(2.3) $$\tilde{b}(z) = \langle T_b \tilde{k}_z, \tilde{k}_z \rangle.$$

Using (2.2) we find

(2.4) $$\tilde{b}(z) = \int b(\zeta)\, |\langle \tilde{k}_z, \tilde{k}_\zeta \rangle|^2 d\mu(\zeta).$$

For fixed z, $|\langle \tilde{k}_z, \tilde{k}_\zeta \rangle|^2 d\mu(\zeta)$ is a probability measure. In this case, and also for the operators we will introduce in the next section, this measure is concentrated near $\zeta = z$. Thus $\tilde{b}$ is a smoothed or smeared version of b.

The pair $(b,\tilde{b})$ goes by various names; a very suggestive one is the "upper" and "lower" symbol. The idea behind the names is that in some contexts the two functions give upper and lower estimates for the operator T_b. An elementary example is

$$\|\tilde{b}\|_\infty \leq \|T_b\| \leq \|b\|_\infty. \tag{2.5}$$

We will see a subtler example in the next section.

2. The Calderon Reproducing Formula

Suppose ψ is a function in $L^2(\mathbb{R})$ and $\hat{\psi}$ is its Fourier transform. Suppose that

$$\int_{-\infty}^{0} |\hat{\psi}(t)|^2\, t^{-1}\, dt = \int_{0}^{\infty} |\hat{\psi}(t)|^2\, t^{-1}\, dt = 1. \tag{2.6}$$

(The fact that both integrals are finite is a mild regularity condition on ψ, that fact that they both equal one is a normalization.) It is often convenient to think of ψ as smooth, compactly supported and having mean zero. Later we will work with ψ which is C on $(-A,0)$, $-C$ on $(0, A)$ and zero elsewhere. (C and A are chosen so that (2.6) holds and so that ψ is a unit vector. More generally, if the two integrals in (2.6) are equal then by replacing $\psi(x)$ with $a\psi(bx)$ for appropriate constants a and b we can insure that (2.6) is satisfied and $\|\psi\| = 1$)

Fix ψ for now. For $\zeta = x + i\,y$ in U define the function $\tilde{k}_\zeta$ by

$$\tilde{k}_\zeta(s) = y^{-1/2}\, \psi((s-x)/y).$$

Thus $\tilde{k}_\zeta$ is the function ψ translated to be centered at x, scaled to a width of y, and renormalized to be a unit vector. We then have the following reproducing formula due to Calderon: for any f in $L^2(\mathbb{R})$, x in $\mathbb{R}$

$$f(x) = \int_U \langle f,\tilde{k}_\zeta\rangle\, \tilde{k}_\zeta(x)\, d\mu(\zeta). \tag{2.7}$$

(Recall that $d\mu = cy^{-2}dxdy$.) Furthermore the map from functions f in $L^2(\mathbb{R})$ to their Calderon transforms, Cf, defined by

$$Cf(\zeta) = \langle f,k_\zeta\rangle$$

is an isometry from $L^2(\mathbb{R})$ onto a closed subspace H of $L^2(U,d\mu)$. Also,

$$\left\| \int_U h(\zeta)\, \tilde{k}_\zeta(x)\, d\mu(\zeta) \right\|_{L^2(dx)} \leq \|h(\zeta)\|_{L^2(d\zeta)} \tag{2.8}$$

All these facts are discussed in detail in [C] and [N].

Given a bounded function b defined on U we define T_b, the Calderon Toeplitz operator with symbol b, to be the map of $L^2(\mathbb{R})$ to $\dot{L}^2(\mathbb{R})$ obtained by multiplying Cf by b before using the reconstruction formula (2.7). That is, for f in $L^2(\mathbb{R})$

(2.9) $$T_b f(x) = \int_U b(\zeta) \langle f, \tilde{k}_\zeta \rangle \tilde{k}_\zeta(x)\, d\mu(\zeta).$$

Certainly there is a formal analogy between the operator defined by (2.2) and the one defined by (2.9). (The analogy is even stronger between the operator acting on the Bergman space of the upper halfplane defined by (2.2) and the operator $\tilde{T}_b$ which maps H = $C(L^2(\mathbb{R}))$ to itself which is given by multiplication by b followed by projection back to H. It is straightforward to check that $\tilde{T}_b$ is unitarily equivalent to T_b given by (2.9). We choose to work with T_b on $L^2(\mathbb{R})$, and slightly weaken the analogy, because the operators T_b are of intrinsic interest on $L^2(\mathbb{R})$.)

The operators T_b given by (2.9) are introduced in [R] and studied systematically in [N]. This class of operators includes as special cases many interesting class of Fourier multiplier operators, singular integral operators, and natural bilinear (in b and f) operators (such as paracommutators and paraproducts). Related operators show up in physics when working with "coherent states", see [KS] and [Pa].

We again define $\tilde{b}$ by (2.3). (2.4) still holds and using (2.8) we see that (2.5) is correct. Also, the interpretation of $\tilde{b}$ as a smeared version of b is still valid.

III. Berezin–Lieb inequalities

We now work with the operators defined by (2.9); however, as we discuss further in the final section, similar considerations hold for other classes of operators.

Suppose b is non–negative and bounded. In this case $T = T_b$ is a bounded positive operator and we can use spectral theory. Suppose further T is compact. (It is straightforward to see this will happen if b is continuous and has compact support, for a fuller analysis see [N].) We are denoting the eigenvalues of T by $\{\lambda_i\}$. Let $\{\phi_i\}$ be the corresponding normalized eigenvectors. $\{\phi_i\}$ is an orthonormal set and by including eigenvectors corresponding to the eigenvalue 0 we can insure that it is a complete orthonormal set. Any f in $L^2(\mathbb{R})$ can be described using (2.7) or by

(3.1) $$f = \Sigma \langle f, \phi_i \rangle \phi_i.$$

Similarly, the action of T can be described by (2.9) or

(3.2) $$Tf = \Sigma\, \lambda_i \langle f, \phi_i \rangle \phi_i.$$

By playing these representations against each other and using Jensen's inequality we can estimate the λ's using the upper and lower symbol. Before doing that we need to recall that if S is any positive trace class operator then the trace of S, denoted Tr(S), can be evaluated by

(3.3) $$\mathrm{Tr}(S) = \Sigma <S\phi_i,\phi_i>$$

of by

(3.4) $$\mathrm{Tr}(S) = \int_U <S\tilde{k}_\zeta,\tilde{k}_\zeta> d\mu(\zeta).$$

The first representation is well known, the second follows easily. (The proof of (3.4) in [Z] (Prop. 6.3.2) for the Bergman space carries over immediately.)

Suppose now that F(t) is a convex function of t for $t \geq 0$. Then F(T) is given by

(3.5) $$F(T)f = \Sigma\, F(\lambda_i)<f,\phi_i>\phi_i.$$

For any unit vector f, direct computation using (3.1) and (3.5) gives

$$<F(T)f,f> = \Sigma|<f,\phi_i>|^2F(\lambda_i).$$

Because f is a unit vector $\Sigma|<f,\phi_i>|^2 = 1$. Hence we can continue with Jensen's inequality and get

$$<F(T)f,f> \geq F(\Sigma|<f,\phi_i>|^2\lambda_i) = F(<Tf,f>).$$

Now let $f = \tilde{k}_\zeta$, integrate with respect to $d\mu(\zeta)$ and take into account (3.4) and the definition of $\tilde{b}$. This gives

(3.6) $$\int_U F(\tilde{b}(\zeta))d\mu(\zeta) \leq \mathrm{Tr}(F(T)).$$

In the other direction

$$<F(T)\phi_i,\phi_i> = F(<T\phi_i,\phi_i>) = F(\int b|<\phi_i,\tilde{k}_\zeta>|^2d\mu).$$

Because ϕ_i is a unit vector

$$\int |<\phi_i,\tilde{k}_\zeta>|^2d\mu = 1.$$

We now use Jensen's inequality and find

$$<F(T)\phi_i,\phi_i> \leq \int_U F(b)|<\phi_i,\tilde{k}_\zeta>|^2d\mu.$$

Summing on i and taking note of (3.3) gives

(3.7) $$\mathrm{Tr}(F(T)) \leq \int_U F(b(\zeta))d\mu(\zeta).$$

Thus we have

Theorem 3.1: If F is convex and T is a compact positive operator then

$$\int_U F(\tilde{b})d\mu \leq \mathrm{Tr}(F(T)) \leq \int_U F(b)d\mu.$$

If F is concave then

$$\int_U F(\tilde{b})\,d\mu \geq \mathrm{Tr}(F(T)) \geq \int_U F(b)\,d\mu.$$

We have already proved the first statement. Jensen's inequality goes in the opposite direction for concave functions; that gives the second statement.

For $\lambda > 0$, let $M(\lambda) = \mu(\{\lambda: \tilde{b} > \lambda\})$, let $m(\lambda) = \mu(\{\lambda: b > \lambda\})$, and let $N(\lambda) = |\{i: \lambda_i > \lambda\}|$.

Corollary 3.2: If $a > 0$ then

$$\int_a^\infty M(x)\,dx \leq \int_a^\infty N(x)\,dx \leq \int_a^\infty m(x)\,dx.$$

Proof: Apply the theorem to the convex function $F(x) = (x-a)_+$ and integrate by parts. (In fact, having the corollary for all positive a is equivalent to the first part of the theorem.)

Results of this general sort (but often concentrating on unbounded operators as opposed to compact ones) were used by Berezin and Lieb in the 1970's. Of particular interest is the case in which b is very smooth. In that case the estimates from above and from below will be close to each other and fairly refined conclusions are possible. (See [B] and the references there.) Although such estimates are referred to in [S] as "Berezin–Lieb" inequalities they certainly go back much further (see, e.g. [Pei]). The proof here is adapted from [KS].

IV. Computation of the cut off

1. The Haar Wavelet.

In this section we work with ψ which is C on $(-A,0)$, $-C$ on $(0, A)$ and zero elsewhere, C and A are chosen so that (2.6) holds and that ψ is a unit vector. We want to show that the associated class of Toeplitz operators with positive symbols is cut off at 1/2.

Theorem 4.1:

A. If b is bounded, non–negative and has compact support then T_b is in S_p for all $p > 1/2$.

B. If b is non–negative and not identically zero then there is a positive constant c such that the eigenvalues of T_b satisfy

$$\forall n \quad \lambda_n > cn^{-2}$$

Proof: For convenience we do the estimates with the unnormalized function ψ which is -1 on $(-1,0)$ and 1 on $(0,1)$.

By the second part of Theorem 3.1 we will be done if we can show that $\tilde{b}^p$ is integrable $d\mu$ for $p > 1/2$. Let S be the support of b and K an upper bound for b. We have

(4.1) $$|\tilde{b}(\zeta')| \le K \int_S |<\tilde{k}_{\zeta'},\tilde{k}_{\zeta}>|^2 d\mu(\zeta).$$

Because S is a compact subset of U and ψ has compact support it is possible to find in U a cone, Γ, and a rectangle, R, with the following properties. The vertex of Γ is on the real axis, one side of R is on the axis, S is contained in $\Gamma \cap R$; and if ζ is in S and ζ' is outside of $\Gamma \cup R$ then $<\tilde{k}_{\zeta'},\tilde{k}_{\zeta}> = 0$. Hence, by (4.1), if ζ' is outside of $\Gamma \cup R$ then $\tilde{b}(\zeta') = 0$.

We now show that $\tilde{b}^p$ is integrable in both $\Gamma \setminus (\Gamma \cap R)$ and in R. Let $\zeta = x + i\,y$ be a point of S and $\zeta' = x' + i\,y'$ a point of $\Gamma \setminus (\Gamma \cap R)$. Then $y' > y$ and hence

(4.2) $$|<\tilde{k}_{\zeta'},\tilde{k}_{\zeta}>| \le c\,(y/y')^{1/2}.$$

Combining this with (4.1) we find $|\tilde{b}(\zeta')| \le c\,y'^{-1}$ in $\Gamma \setminus (\Gamma \cap R)$. Using polar coordinates with origin at the vertex of Γ we see that $\tilde{b}^p$ is integrable in $\Gamma \setminus (\Gamma \cap R)$ for any $p > 0$.

Now let $\zeta = x + i\,y$ be a point of S and $\zeta' = x' + i\,y'$ a point of R. Direct estimates give

(4.3) $$|<\tilde{k}_{\zeta'},\tilde{k}_{\zeta}>| \le c\,(y'/y)^{1/2}.$$

We need an additional observation before using (4.1). Let Q be a thin rectangle along the bottom edge of R. $\tilde{b}$ is bounded and hence integrable on $R \setminus Q$ so we only need verify the required integrability on Q. If ζ' is in Q and ζ is in S then it will often be true that $\tilde{k}_{\zeta}$ is constant on the support of $\tilde{k}_{\zeta'}$. If that happens then $<\tilde{k}_{\zeta'},\tilde{k}_{\zeta}> = 0$. For that to fail to happen, one of the three points of discontinuity of $\tilde{k}_{\zeta}$ must meet the support of $\tilde{k}_{\zeta'}$. For fixed ζ' this restricts attention to (at most) three vertical strips in $R \setminus Q$ with total $d\mu$ measure that is $O(y')$. When we combine this fact with (4.3) and use (4.1) we find that $|\tilde{b}(\zeta')| \le c\,y'^2$. Thus we must look at the integrability over Q of $(y')^{2p}\,y'^{-2}\,dx\,dy'$. This is integrable if $p > 1/2$. This completes the proof of A.

We first prove B in the special case in which b(z) is the characteristic function of the set $S = \{x + iy: 0 \le x \le .1,\ 1 \le y \le 1.1\}$. Pick and fix a large integer N. For $j = 1,2,\dots,N$ let $\zeta_j = (90\,j + i)/1000N$ and let $\tilde{k}_j(x) = \tilde{k}_{\zeta_j}(x)$. Let $X = X_N$ be the span of $\{\tilde{k}_j\}_1^N$. The $\tilde{k}_j$ have disjoint support. Hence the dimension of X is N and (up to a constant factor which we ignore), $\{\tilde{k}_j\}_1^N$ is an orthonormal basis of X. We will show that there is a constant c so that for all unit vectors v in X_N

(4.4) $$<T_b v,v> \ge c\,N^{-2}.$$

By the minimax principle this insures that $\lambda_N \geq c\,N^{-2}$ which is, at least for this choice of b, the desired conclusion. We start with $v = \Sigma\, a_i \tilde{k}_i$ with $\Sigma\, |a_i|^2 = 1$. Using (2.9) we find

$$<T_b v,v> = \sum_{i,j} a_i \bar{a}_j \int_S <\tilde{k}_i, \tilde{k}_\zeta> <\tilde{k}_\zeta, \tilde{k}_j> d\mu(\zeta). \tag{4.5}$$

Now notice, as we did in part A, that for ζ in S the support $\tilde{k}_\zeta$ and the support of $\tilde{k}_i$ may overlap but the general pattern is that $\tilde{k}_\zeta$ is constant across the support of $\tilde{k}_i$. In such a case $<\tilde{k}_i, \tilde{k}_\zeta> = 0$. The possible exceptions to this are for ζ in the strip

$$S_i = S \cap \{ x + i\,y : 8i < 100x < 10i \}.$$

For different i these strips are disjoint. Hence the integral in (4.5) is 0 unless $i = j$. Thus

$$<T_b v,v> = \sum_i |a_i|^2 \int_{S_i} |<\tilde{k}_i, \tilde{k}_\zeta>|^2 d\mu(\zeta). \tag{4.6}$$

Now consider center (1/1000)–th of S_i. By direct computation we find that for some small positive c

$$|<\tilde{k}_i, \tilde{k}_\zeta>| \geq c\,N^{-1/2}.$$

Also, the μ measure of that fraction of S_i is at least k/N for some small k. Combining these these two estimates with (4.6) and with the fact that $\Sigma\, |a_i|^2 = 1$ gives (4.4).

We now consider general positive bounded b. Let S be a rectangle in U of the form $S = \{ x + iy\colon a \leq x \leq a + .1d,\ d \leq y \leq 1.1\, d\}$ on which the integral of b is positive. (If there is no such S then b vanishes identically and we are done.) Let $b' = b\chi_{S'}$. Because $T_b > T_{b'}$ it suffices to obtain the estimate for the eigenvalues of $T_{b'}$. For notational convenience we just suppose that $T = T_{b'}$. We now show how to extend the argument we used in the special case.

Let $I = (a, a + .1d)$. Let B(x) be the function on I given by

$$B(x) = \int_0^\infty b'(x + i\,y)\, y^{-2} dy.$$

Because b' is bounded and has compact support B is bounded by, say, M. Let $A = \int_I B(x)dx$. Pick and fix a large integer N. Divide I into N subintervals of equal length. We claim that on a substantial proportion of these subintervals the integral of B is large. More precisely, we can find small positive α and β so that on a special subcollection of αN of the N subintervals the integral of B is at least β/N. To see this we look at the estimate we would be able to obtain for A otherwise. If we can't find the subintervals then there are at least $N - \alpha N$ intervals on which the integral is less than β/N. This gives a contribution of $(1-\alpha)\beta$ to A. On each of the αN subintervals on which the integral is large it can be at most M times the length of the subinterval, $|I|/N$. Combining these estimates gives

$$A \le (1-\alpha)\beta + \alpha N \times M \times |I|/N = (1-\alpha)\beta + \alpha M |I|.$$

This can't hold for small α and β and hence we have our subintervals. (It is of no concern that the choices α and β depend on M and $|I|$, as long as the choices are independent of N.) By replacing α with $\alpha/2$ we can also insure that no two adjacent subintervals are in our special subcollection. We assume that has been done but don't change notation. Let $\mathcal{I}$ be the collection of special subintervals. Fix K, a large positive integer and for each I in $\mathcal{I}$ break I into K pieces of equal length. Of these K, select one, I^*, on which the integral of B is a maximum, at least β/NK. For $j = 1,2,\ldots,\alpha N$ and I_j in $\mathcal{I}$, let x_j be the center point of the interval I^*, let S_j be the strip in S above I_j, and let S_j^* be the strip in S above I_j^*. Let $\zeta_j = x_j + i\,|I|/100N$, $\tilde{k}_j(x) = \tilde{k}_{\zeta_j}(x)$, and let $X = X_n$ be the span of $\{\tilde{k}_j\}$. The $\tilde{k}_j$ have disjoint support and hence the dimension of X is αN and (again up to a constant factor) $\{\tilde{k}_j\}$ is an orthonormal basis of X. Again we want to show that for any unit vector v in X we have (4.4). As before, if $\tilde{k}_\zeta$ is constant across the support of $\tilde{k}_i$ then $\langle \tilde{k}_i, \tilde{k}_\zeta\rangle = 0$ and hence an expression similar to (4.5) (but containing $b(\zeta)$) can be replaced by one similar to (4.6).

If K is large then, for ζ in S_i^*

$$|\langle \tilde{k}_i, \tilde{k}_\zeta\rangle| \ge c\, N^{-1/2} \tag{4.7}$$

for some positive constant c. This is true for ζ on the center line of S_i^* by direct computation. It remains true as long as the distance from the center line is at most a small fraction of the width of the support of $\tilde{k}_i$. The choice of large K (which is independent of N) insures that we stay sufficiently close to the center.

We make the integrals in the evaluation of $\langle T_b v, v\rangle$ smaller by replacing S_i by S_i^*. Doing that and using (4.7) we obtain

$$\langle T_b v, v\rangle \ge c \sum_i |a_i|^2 \int_{S_i^*} b(\zeta)\, N^{-1} d\mu(\zeta).$$

We now do the y integration and get

$$<T_b v,v> \geq c\, N^{-1} \sum_i |a_i|^2 \int_{I_i^*} B(x)\, dx.$$

By construction each integral is at least β/NK and we can complete the proof as before.

V. Comments

A. The crucial facts about ψ in the previous proof were that it had compact support, was piecewise constant (and hence, among other things, bounded), and was orthogonal to constants. If we change some of these properties then we can obtain different results of the same general sort. For instance, starting with a ψ with compact support which is piecewise linear (and hence, among other things, has bounded derivative) and which is orthogonal to polynomials of degree 1 we obtain a class of operators which is cut off at $p = 1/4$. The proof of the analog of the first part of the theorem proceeds as before. We obtain the same estimate as before for $\tilde{b}$ in Γ. In the region $R \setminus Q$ we get the estimate $|\tilde{b}(\zeta')| < c\, y'^4$ which leads to a critical value of $p = 1/4$. To prove an analog of part B one exhibits a subspace on which the operators are large. The pattern of construction is the same but instead of having the subspace be spanned by translates and dilates of a step function that is orthogonal to the constants, we use a step function which is orthogonal to all linear functions.

Although it is possible to go further, the general pattern is not clear. If the base function ψ is very smooth and has lots of cancellation (for example if its Fourier transform has compact support which omits the origin) T_b can have eigenvalues which decay faster than any power of n. (This can be checked by using the estimates in [FJ] to estimate $\tilde{b}$ and then using the theorem in Section III.)

It would be interesting to know more about the relationship between the properties of ψ and the cut off for the operator.

B. The results in the previous sections apply equally well to Toeplitz operators on the Bergman space. In that context, if b has compact support then the eigenvalues will decay geometrically. Thus, for Toeplitz operators on the Bergman space there is no cut off. On the other hand, there has been interest recently in the cut off behavior of Hankel operators and related operators on the Bergman space and other spaces. ([J] [Pe] [PZ] [PRW] [JR].) It would be interesting to know if there is a common explanation for the various observations about cut offs.

In fact the Berezin–Lieb style estimates apply to any Toeplitz operator on a Hilbert space with reproducing kernel.

C. Other estimated concerning the eigenvalues of operators of the type we have been considering can be found in [D], [DP],[S] [N] [R2].

REFERENCES

[B] Berezin, General concept of quantization, Comm. Math. Phys. 40, 153–174.

[C] Combes et al (eds.), Wavelets, Time–Frequency Methods and Phase space, Springer Verlag, New York, 1989.

[D] I. Daubechies, Time–frequency localization operators: a geometric phase space approach, IEEE Transactions on Information Theory, 34 (1988) 605–612

[DP] I. Daubechies and T. Paul, Time–frequency localization operators—a geometric phase space approach: II. The use of dilations, Inverse problems 4 (1988) 661–680.

[FJ] M. Frazier and B. Jawerth, A Discrete Transform and Decompositions of Distribution spaces, J. Functional Anal., to appear.

[FR] M. Feldman and R. Rochberg, Singular values estimates for commutators and Hankel operators on the unit ball and the Heisenberg Group, Analysis and Partial Differential Equations, C. Sadosky, Ed., M. Decker Inc.,1990, 121–160.

[J] S. Janson, Hankel operators between weighted Bergman spaces, Ark. Math, 26 (1988), 205–219.

[KS] J. Klauder and B. Skagerstam, Coherent States, Applications in Physics and Mathematical Physics, World Scientific, Singapore 1985

[N] K. Nowak, Ph. D. Dissertation, Washington University, in preparation, 1990.

[Pa] T. Paul, Wavelets and Path Integrals, [C], 204–209.

[Pe] L. Peng, Ha–plitz operators on Bergman space, preprint 1990.

[PRW] L. Peng, R. Rochberg, and Z. Wu, Orthogonal polynomials and middle Hankel operators on Bergman spaces, preprint 1990.

[PZ] L. Peng, G. Zhang, Middle Hankel operators on Bergman space, preprint 1990.

[Pei] Peierls, On a Minimization Property of the free energy, Phys. Rev. 54 (1938), p. 918.

[P] A. Perelomov, Generalized Coherent States and Their Applications, Springer–Verlag New York, 1986.

[R] R. Rochberg, Toeplitz and Hankel operators, Wavelets, NWO sequences, and almost diagonalization of operators, to appear, Proc. Symp. Pure Math.

[R2] R. Rochberg, A correspondence principle for Toeplitz and Calderon Toeplitz operators, Preprint 1990.

[RS] R. Rochberg and S. Semmes, End point results for estimates of singular values of singular integral operators, in Contributions to Operator Theory and its Applications, OT35, Gohberg et al eds., Birkhauser, 1988, 217–231.

[S] K. Seip, Reproducing formulas and double orthogonality in Bargmann and Bergman spaces, preprint 1990.

[Si] B. Simon, Classical limit of quantum partition function, Comm. Math. Phys 71 (1980).

[Z] K. Zhu, Operator Theory in Function Spaces, manuscript, 1990.

Stable Rank in Banach Algebras

RUDOLF RUPP Universität Karlsruhe, Math. Institut I, Englerstr. 2, D-7500 Karlsruhe, Germany

Most of the following results are known – at least to the workers in the field – but the usual proofs make use of deep algebraic-topological arguments. We present easier proofs, based on the Arens–Royden Theorem ([5], Theorem 7.2, p 89).

In this paper we only consider complex, commutative Banach algebras A with unit element being denoted by 1. Especially we are interested in such algebras A whose maximal ideal space is homeomorphic to a compact set K of the complex plane $\mathbb{C}$. A standard example is the algebra $C(K)$ of all complex-valued continuous functions on a given compact set $K \subset \mathbb{C}$ (under the usual pointwise operations and the supremum norm on K). Also well known is the algebra $A(K)$ of functions $f \in C(K)$ which are analytic in the interior of K, see [1]. Finally, we look at the set $C(K, S^1)$ of all continuous functions on K with values in $S^1 := \{z \in \mathbb{C} : |z| = 1\}$. Recall the following definitions, see [3], [4].

Let $U_2(A)$ denote the set of all unimodular elements, i.e.,

$$U_2(A) := \{(f,g) \in A^2 : Af + Ag = A\}.$$

The unimodular element (f,g) is called *reducible* (in A), if there exists $h \in A$ such that $f + hg$ is invertible in A. We say that A is *1-stable* at g iff every unimodular element (f,g) is reducible in A. It is said that A has *stable rank one* iff A is 1-stable at every $g \in A$, i.e., every $(f,g) \in U_2(A)$ is reducible in A.

A function $g \in C(K)$ fulfills the *boundary principle* iff the following holds:

(B_1) If G is open in $\mathbb{C}$, $G \subset K$ such that g vanishes identically on the boundary ∂G, then g vanishes identically on G.

This is clearly equivalent to the following condition, where the *zero set* of g is denoted by Z_g, i.e.,

$$Z_g = \{z \in K : g(z) = 0\}.$$

(B_2) If G is open in $\mathbb{C}$, $G \subset K \setminus Z_g$ then there exists $z_0 \in \partial G$ such that $g(z_0) \neq 0$.

The condition (B_2) is necessary and sufficient that every function $f \in C(Z_g, S^1)$ can be extended to $\tilde{f} \in C(K, S^1)$, see [6, Theorem VI.12].

The condition (B_1) was introduced by the author in his thesis [9].

§1. We begin by showing that at least some special unimodular elements are reducible.

Definition 1.1: Suppose $(f, g) \in A^2$ and $\lambda, \mu \in \mathbb{C}$. The numbers λ, μ are called equivalent, for short $\lambda \sim \mu$, if there exists elements $h, k \in A$ such that

$$f - \lambda + hg = (f - \mu)\exp(k)$$

It is not hard to see that for fixed (f, g) the relation $\sim$ is indeed an equivalence relation, the set of numbers equivalent to μ being denoted by $[\mu]$.

Proposition 1.2: Suppose that $(f - \mu, g) \in U_2(A)$. Then the set of equivalences $[\mu]$ is open in $\mathbb{C}$.

Proof: Take $\lambda \in [\mu]$. Since we also have $(f - \lambda, g) \in U_2(A)$, there exists $\alpha, \beta \in A$ such that

$$\alpha(f - \lambda) + \beta g = 1. \tag{1}$$

If we chose the number M large enough, there exists $h \in A$ such that

$$f - M = \exp(h). \tag{2}$$

Multiplying equation (1) with a complex number $\varrho \neq -1$ (to be specified later) and adding $(f - M)^{-1}(f - \lambda)$ on both sides yields

$$[(f - M)^{-1} + \varrho\alpha](f - \lambda) + \varrho\beta g = (\varrho + 1)(f - M)^{-1}\left[f - \frac{\lambda + \varrho M}{1 + \varrho}\right] \tag{3}$$

Remember that as $(f - M)^{-1} = \exp(-h)$ the coefficient of $(f - \lambda)$ is also an exponential, if ϱ is sufficiently small. But then every number τ in an appropriate neighborhood U of λ is obtained, choosing the number ϱ from the equation

$$\frac{\lambda + \varrho M}{1 + \varrho} = \tau.$$

Looking at equation (3) we are done. ■

The following Proposition settles the question of reducibility in a special case, which will be needed later.

Proposition 1.3: Let $(f, g) \in U_2(A)$ and suppose that $\lambda = 0$ belongs to the unbounded connected component of the open set $I(f, g) := \{\lambda \in \mathbb{C} : (f - \lambda, g) \in U_2(A)\}$. Then there exists $h, k \in A$ such that

$$f + hg = \exp(k).$$

Proof: Let U denote the unbounded connected component of the open set $I(f, g)$. Choose the complex number M so large that there exists $l \in A$ such that

$$f - M = \exp(l).$$

Of course we may take $M \in U$. But then Proposition 1.2 implies that $\mu = M$ and $\lambda = 0$ are equivalent. (Otherwise the open connected set U would split into disjoint open sets, namely certain equivalence classes.)

Thus there exist $h, k \in A$ such that

$$f + hg = (f - M)\exp(k) = \exp(k + l).$$

So we are done. ■

§2. The following result is the main ingredient for the discussion of stable rank one in algebras. As I said before it is probably well known among the specialists, although its proof is new. An alternative approach may be found in [11].

Theorem 2.1: If $K \subset \mathbb{C}$ is compact, the following conditions are equivalent:

(i) $C(K)$ is 1-stable at $g \in C(K)$.

(ii) The boundary principle (B_1) holds, i.e., if g vanishes identically on the boundary ∂G of a set $G \subset K, G$ open in $\mathbb{C}$, then g vanishes identically on G.

Proof. As mentioned before, it is enough to show that (i) and (B_2) are equivalent. If g has no zero at all in K, then (i) and (ii) are always fulfilled. So we may assume that Z_g is non-void in the following.

Assume that $C(K)$ is 1-stable at g. We show that every function $f \in C(Z_g, S^1)$ can be extended to $\tilde{f} \in C(K, S^1)$. This gives condition (B_2), see [6, Theorem VI.12].

If h denotes a continuous extension of f to K, then the element (h, g) is unimodular in $C(K)$. Since $C(K)$ is 1-stable at g, there exists a continuous function k such that $h + kg =: u$ is invertible in $C(K)$.

But then the function $\tilde{f} := u/|u|$ is an extension of f and $\tilde{f} \in C(K, S^1)$. So we are done.

For the reverse implication we assume that (B_2) holds, and we have to show that $C(K)$ is 1-stable at g.

To this end suppose that we have $|f(z)| + |g(z)| \geq \delta > 0 \quad (z \in K)$.

Especially $|f(z)| \geq \delta \quad (z \in Z_g)$.

Next consider the following factorisation

$$f(z) = |f(z)| \frac{f(z)}{|f(z)|} \quad (z \in Z_g).$$

For the first factor, Proposition 1.3 implies the existence of $d, k \in C(K)$ such that

$$|f(z)| + d(z)g(z) = \exp(k(z)) \quad (z \in K). \tag{1}$$

(Note that $\{\lambda : \operatorname{Re} \lambda \leq 0\}$ belongs to the unbounded component of $I(|f|, g)$).

Since condition (B_2) holds, there exists a function $u \in C(K, S^1)$ such that

$$\frac{f(z)}{|f(z)|} = u(z) \quad (z \in Z_g) \tag{2}$$

Together with equation (1) this implies

$$f(z) = u(z)\exp(k(z)) \quad (z \in Z_g)$$

Thus we are tempted to introduce the function $w \in C(K)$ by the equation

$$f = uw\exp(k) \tag{3}$$

But then we have

$$w(z) = 1 \quad (z \in Z_g).$$

This shows that $\mathbb{C}\setminus\{1\}$ is contained in the unbounded component of $I(w, g)$.

One more time, Proposition 1.3 implies the existence of $h, l \in C(K)$ such that

$$w + hg = \exp(l)$$

Inserting this in equation (3) gives

$$f + ue^k hg = u\exp(l + k),$$

which was to be shown. ∎

§3. Now we can formulate an "analytic condition" for many algebras to have stable rank one.

Theorem 3.1: Let A denote a complex, commutative Banach algebra with unit element, whose maximal ideal space is homeomorphic to a compact set $K \subset \mathbb{C}$. Then the following conditions are equivalent:

(i) The stable rank of A is one.

(ii) The boundary principle holds in A, i.e., for every Gelfand transform $\hat{g} \in C(K)$ the following is true:
If $\hat{g}$ vanishes identically on the boundary of a set $G \subset K, G$ open in $\mathbb{C}$, then $\hat{g}$ vanishes identically on G.

Proof: We begin by defining certain uniform algebras, which we'll consider closely. Starting from the algebra $\hat{A}$ of Gelfand transforms of A, we take the closure in $C(K)$ under the supremum norm. This gives a uniform algebra $\tilde{A}$ having also K as maximal ideal space ([8], Proposition 3, p 83). Let

A_{Z_g} denote the uniform closure of the restriction algebra $\tilde{A}|_{Z_g}$. It is most important that the maximal ideal space of A_{Z_g} is indeed Z_g (Note that Z_g is – by the very definition – an $\tilde{A}$-convex set, then use [5], Theorem 6.1, p 39).

1) Assume that the stable rank of A is one. We have to show that the boundary principle (B_1), or equivalently (B_2) holds. By ([6], Theorem VI.12) it is enough to show that each function $f \in C(Z_g, S^1)$ can be extended to $\tilde{f} \in C(K, S^1)$.

Since f is an invertible, continuous function on Z_g, the Arens-Royden Theorem ([5], Theorem 7.2, p 89), applied to the algebra A_{Z_g}, gives us $u \in A_{Z_g}$, u invertible in A_{Z_g} and $h \in C(Z_g)$ such that

$$f = u \exp(h) \tag{1}$$

Let H denote any continuous extension of h to K. Observe that u is the uniform limit on Z_g of restricted functions $u_n \in \hat{A}$. Therefore, for all large numbers n, the function u_n also does not vanish on Z_g. This implies that $(u_n, \hat{g})$ is unimodular and hence reducible by assumption. So there exists an invertible function $U_n \in \hat{A}, k \in \hat{A}$ such that

$$u_n + k\hat{g} = U_n.$$

Especially $u_n(z) = U_n(z) \quad (z \in Z_g)$. Remember that u_n was an approximation for u. So for sufficiently large n we have

$$|f(z) - U_n(z) \exp(H(z))| \leq \varepsilon \quad (z \in Z_g). \tag{2}$$

Since $|f(z)| = 1$ on this set, we arrive at

$$|U_n(z) \exp(H(z))| \geq 1 - \varepsilon \quad (z \in Z_g). \tag{3}$$

Abbreviate

$$V := U_n \exp(H).$$

Then

$$\left| \frac{f(z)}{V(z)} - 1 \right| \leq \frac{\varepsilon}{1 - \varepsilon} \quad (z \in Z_g).$$

Thus – for small ε – there exists a continuous logarithm w for $\frac{f}{V}$ on Z_g, which we think of extended continuously to W on K. This implies

$$f(z) = V(z) \exp(W(z)) \quad (z \in Z_g).$$

But then

$$\tilde{f}(z) := \frac{V(z)\exp(W(z))}{|V(z)\exp(W(z))|} \quad (z \in K)$$

is the desired extension of f to $\tilde{f} \in C(K, S^1)$.

2) Assume that the boundary principle (B_2) holds. We must show that the stable rank of A is one, i.e., every unimodular element (f, g) is reducible in A. Of course it's the same to show that $(\hat{f}, \hat{g})$ is reducible in $\hat{A} \subset C(K)$.

By Theorem 2.1 there exists continuous functions $h, u \in C(K)$, u invertible in $C(K)$ such that

$$\hat{f} + h\hat{g} = u. \tag{4}$$

By use of the Arens-Royden Theorem ([5], Theorem 7.2, p 89) – applied to the Banach algebra A itself – we have

$$u = \hat{v}\exp(k).$$

Hereby $\hat{v}$ denotes an invertible function in $\hat{A}$ and k is a continuous function on K. Thus we have

$$\frac{\hat{f}}{\hat{v}} + h\frac{\hat{g}}{\hat{v}} = \exp(k).$$

Of course it is enough to show the assertion is true for the functions $\tilde{f} := \frac{\hat{f}}{\hat{v}}$ and $\tilde{g} = \frac{\hat{g}}{\hat{v}}$.

Especially we arrive at

$$\tilde{f}(z) = \exp(k(z)) \quad (z \in Z_g)$$

Thus the function $\tilde{f} \in A_{Z_g}$ has a continuous logarithm on Z_g and so there exists $\tilde{k} \in A_{Z_g}$ such that

$$\tilde{f}(z) = \exp(\tilde{k}(z)) \quad (z \in Z_g)$$

(see ([5], Corollary 6.2, p 88).

Since A_{Z_g} is the uniform closure of the restriction algebra $\hat{A}|_{Z_g}$, there exist a sequence of function $\hat{k}_n \in \hat{A}$ such that

$$|\tilde{f}(z) - \exp(\hat{k}_n(z))| \le \varepsilon \quad (z \in Z_g) \tag{5}$$

for all sufficiently large numbers n.

Now we use the fact that $(\tilde{f}, \tilde{g})$ is unimodular, i.e., there exists $\delta > 0$ such that

$$|\tilde{f}(z)| + |\tilde{g}(z)| \geq \delta \quad (z \in K),$$

especially

$$|\tilde{f}(z)| \geq \delta \quad (z \in Z_g).$$

Choosing $\varepsilon := \delta/3$ in equation (5) we arrive at

$$|\exp(\hat{k}_n(z))| \geq \frac{2}{3}\delta$$

for all large n. Together with equation (5) this gives

$$|\tilde{f}(z)\exp(-\hat{k}_n(z)) - 1| \leq \frac{3\varepsilon}{2\delta} = \frac{1}{2} \quad (z \in Z_g).$$

Thus there exists $\hat{k} \in \hat{A}$ such that the function

$$\hat{F} := \tilde{f}\exp(-\hat{k})$$

has real part at least 1/2 for all $z \in Z_g$, i. e. $\{: \mathrm{Re}\,\lambda < \frac{1}{2}\}$ is contained in the unbounded component of $I(F, g)$. (Note that we can take $F := \frac{f}{v}\exp(-k)$ – despite the fact that there may exist a radical for the Banach algebra A). By Proposition 1.3 the unimodular element (F, g) is reducible, i.e., there exists $H, U \in A, U$ invertible in A such that

$$F + Hg = \exp(U).$$

But then

$$f + v \cdot H \cdot e^k \cdot g = v \cdot \exp(k + U).$$

So we are done. ∎

Remark: This method also gives information about the so called reducibility to the principal component, as we'll show elsewhere.

Acknowledgement: I want to thank the Karlsruher Hochschulvereinigung for financial support during this conference.

References

1. R. Arens: The maximal ideals of certain function algebras. Pacific J. Math. **8**(1958), 641–648.

2. H. Bass: K-theory and stable algebra. Publ. Math. I.H.E.S. **22**(1964), 5–60.

3. G. Corach, F.D. Suárez: Extension problems and stable rank in commutative Banach algebras. Topology Appl. **21**(1985), 1–8.

4. G. Corach, F.D. Suárez: Stable rank in holomorphic function algebras. Illinois J.Math. **29**(1985), 627–639.

5. T.W. Gamelin: Uniform Algebras. Prentice-Hall, Inc. Englewood Cliffs, N.J., 1969.

6. W. Hurewicz, H. Wallman: Dimension Theory. Princeton Univ. Press, 1941.

7. P.W. Jones, D. Marshall, T. Wolff: Stable rank of the disc algebra. Proc. Amer. Math. Soc. **96**(1986), 603–604.

8. H.L. Royden: Function algebras. B.A.M.S. **69**(1963), 281–298.

9. R. Rupp: Über den Bass-Stable-Rank komplexer Funktionen-Algebren. Dissertation, Karlsruhe 1988.

10. R. Rupp: Stable Rank of Subalgebras of the Disc-Algebra. Proc. Amer. Math. Soc. **108**(1990), 137–142.

11. R. Rupp: Stable Rank and Boundary-Principle. Topology Appl. (to appear)

On Stability Problems of Some Properties in Banach Spaces

ELIAS SAAB, University of Missouri, Columbia, MO 65203

PAULETTE SAAB, University of Missouri, Columbia, MO 65203

If E is a Banach space, $(\Omega, \Sigma, \lambda)$ a probability space and $1 \leq p < \infty$ we denote by $L_p(E)$ the space of p-Bochner integrable functions from Ω to E equipped with the norm $||f||_p = (\int_\Omega ||f(\omega)||^p d\lambda(w))^{\frac{1}{p}}$. If $p = +\infty$, then $L_\infty(E)$ will stand for the space of all measurable and essentially bounded functions from Ω to E equipped with the norm $||f||_\infty = \text{ess sup} \, ||f(\omega)||$.

If K is a compact Hausdorff space, $C(K, E)$ will denote the Banach space of all E-valued continuous functions on K endowed with the uniform norm. In case E is the scalar field then $L_p(E)$ will be denoted by L_p, and $C(K, E)$ will be denoted by $C(K)$.

If F and E are two Banach spaces, we denote by $F \otimes E$ the algebraic tensor product of F and E. The ϵ-norm of an element $u = \sum_{i=1}^{m} x_i \otimes y_i \in F \otimes E$ is defined by

$$||\sum_{i=1}^{m} x_i \otimes y_i|| = \sup\{|\sum_{i=1}^{m} x^*(x_i)y^*(y_i)| \mid ||x^*||, ||y^*|| \leq 1\}.$$

The completion $F \hat{\otimes}_\epsilon E$ of $F \otimes E$ is called the **injective tensor product** of E and F. It is well known that the space $C(K, E)$ is isometrically isomorphic to $C(K) \hat{\otimes}_\epsilon E$.

The π-norm of an element $u \in F \otimes E$ is defined by

$$||u|| = \inf \sum_{i=1}^{m} ||x_i|| \, ||y_i||$$

where the infimum is taken over all the representations $u = \sum_{i=1}^{m} x_i \otimes y_i$. The completion $F \hat{\otimes}_\pi E$ of $F \otimes E$ is called the **projective tensor product** of E and F. It is well known that the space $L_1(E)$ is isometrically isomorphic to $L_1 \hat{\otimes}_\pi E$.

In this paper we will be concerned with the following problem: Given a Banach space E that has a certain property (P) and suppose that L_p or $C(K)$ has this property. We would like to know when does $L_p(E)$ or $C(K, E)$ have the property (P). We will also discuss the stability of some properties under taking injective or projective tensor products.

The properties that we would like to concentrate on are the properties (u), (V) and (V^*) of Pełczynski, the Dunford-Pettis property, the embedding of c_0, ℓ_1 and L_1 in E as subspaces or as complemented subspaces. Some other related properties will also be discussed. For each property considered, we will give some brief background about it by stating well known results and sometimes give some indications of how to prove some of these results. We will supply complete proofs of some new facts and observations. We will also state some of the many problems that are still open.

For a series $\sum_n x_n$ in a Banach space E we say that $\sum_n x_n$ is a **weakly unconditionally Cauchy** series in E if it satisfies one of the following equivalent statements

a) $\sum_n |x^*(x_n)| < \infty$, for every $x^* \in E^*$;

b) $\sup \left\{ ||\sum_{n \in \sigma} x_n|| : \ \sigma \text{ finite subset of } \mathbf{N} \right\} < \infty$

c) $\sup_n \sup_{\epsilon_i = \pm 1} ||\sum_{i=1}^{n} \epsilon_i x_i|| < \infty$.

We say that an operator (bounded linear operator)$T : E \to F$ is **unconditionally converging** if T sends weakly unconditionally Cauchy series into unconditionally convergent series. It is easy to see that T is unconditionally converging if and only if $\lim_{n \to 0} ||T(x_n)|| = 0$ for any weakly unconditionally Cauchy series $\sum_{i=1}^{\infty} x_i$ in E.

We say that an operator $T : E \to F$ fixes a copy of a Banach space G if there is a subspace H (always taken to be closed) of E isomorphic to G such that T restricted to H is an isomorphism. All notion, notations and results used and not defined can be found in [18], [19], [21], or [48].

Proposition 1 **([54])**. *Let E be any Banach space then the following statements are equivalent:*

1. *For any Banach space Y, every bounded linear operator $T : E \to Y$ that is unconditionally converging is weakly compact;*

2. *A set $K \subset E^*$ is relatively weakly compact if and only if for any weakly unconditionally Cauchy series $\sum_{i=1}^{\infty} x_i$ in E one has $\lim_n \sup_{x^* \in K} x^*(x_n) = 0$.*

Definition 1 A Banach space E is said to have the **property (V)** if it satisfies one of the equivalent conditions of the Proposition 1.

It is clear from Definition 1 that any quotient of a space with the property (V) has (V) and any reflexive Banach space has (V).

Theorem 2 **[54])**. *Let S be a compact Hausdorff space, then $C(S)$ has the property (V).*

The proof of the above theorem depends on Grothendieck's characterization of relatively weakly compact subsets of the space $M(S)$ of Radon measures on S [19].

In [39] a more general result is shown:

Theorem 3 **([39])**. *Let E be a Banach space whose dual is isometric to an L_1-space, then E has the property (V).*

The following proposition can be deduced from [54] and [4]:

Proposition 4 *A Banach space E has the property (V) if and only if for every Banach space Y, every bounded linear operator $T : E \to Y$ which is not weakly compact fixes a copy of c_0.*

Bourgain [8] showed that if $E = H^\infty$ and Y is any Banach space then every bounded linear operator $T: E \to Y$ which is not weakly compact fixes a copy of ℓ_∞ and hence by the above proposition H^∞ has the property (V).

Kisliakov [43] (see [52]) showed that the disc algebra has the property property (V).

It follows from Proposition 4 that a non reflexive Banach space that has property (V) must contain a subspace isomorphic to c_0. This shows that ℓ_1 and $L_1[0,1]$ do not have the property (V). In particular a closed subspace of a Banach space with the property (V) does not have in general the property (V).

Definition 2 A Banach space E is said to have the **property** (V^*) if a subset K of E is relatively weakly compact whenever $\lim_n \sup_{x \in K} x(x_n^*) = 0$ for every weakly unconditionally Cauchy series $\sum_{i=1}^{\infty} x_i^*$ in E^*.

It follows from that definition that:

1. any closed subspace of a space having property (V^*) has property (V^*);
2. if E has (V) then E^* has V^*.

Statements 1 and 2 above imply that if E^* has property (V) then E^{**} has property (V^*) and in particular E has property (V^*). Statement 2 shows that any abstract L-space [54] has property (V^*).

The converse of statement 2 above is not true. Bourgain and Delbaen [10] constructed a Banach space E with the Schur property such that E^* is isomorphic to ℓ_1.

The following theorem is due to Pełczynski [54]; its proof is simple enough to deserve presentation.

Theorem 5 ([54]). *If E has property (V^*) then E is weakly sequentially complete.*

Proof: Let $K = \{x_1, x_2, x_3, \cdots, x_n, \cdots\}$ be a weak Cauchy sequence in E. It is enough to show that $\lim_n \sup_{x \in K} x(x_n^*) = 0$ whenever $\sum_{i=1}^{\infty} x_i^*$ is a weakly unconditionally

Cauchy series in E^*. To do that define $T : E \to \ell_1$ by $T(x) = (x_i^*(x))$. It is easy to check that T is continuous, hence $T(K)$ is relatively weakly compact in ℓ_1 and therefore it is relatively compact in ℓ_1. It follows now from the well known characterization of relatively norm compact subsets of ℓ_1 that $\lim_n \sup_{x \in K} x(x_n^*) = 0$.□

So if E has (V) then E^* is weakly sequentially complete.

The following corollary is now evident.

Corollary 6 *Let E be a Banach space, then the following statements are equivalent:*

1. *E has property (V) and property (V^*);*
2. *E and E^* have property (V^*);*
3. *E has property (V^*) and c_0 does not embed in E^*;*
4. *E is reflexive;*
5. *E and E^* have (V).*

The converse of Theorem 5 is not true in general. The Bourgain Delbaen space mentioned above is weakly sequentially complete, is not reflexive and its dual does not contain c_0. This space cannot have property (V^*) by statement 3 of Corollary 6. However if E is a Banach lattice one has the following theorem:

Theorem 7 ([64]) *A Banach lattice E has property (V^*) if and only if E is weakly sequentially complete.*

So a Banach lattice E has property (V^*) if and only if c_0 does not embed in E. In fact more can be said

Theorem 8 ([64], [6]). *A closed subspace E of an order continuous Banach lattice has property (V^*) if and only if c_0 does not embed in E.*

As a consequence of the above theorem and [12] one can obtain:

Theorem 9 ([64]). *A closed subspace E of an order continuous Banach lattice has property (V^*) if and only if $L_1(E)$ has property (V^*).*

The following question is still open for general Banach spaces:

Question 1 Does $L_p(E)(1 \leq p < \infty)$ have the property (V^*) whenever E has the property (V^*)?

The following proposition should be compared to Proposition 4:

Proposition 10 ([31]). *Let E be a Banach space, then the following statements are equivalent:*

1. *E has (V^*);*
2. *Every subset K of E which is not relatively weakly compact contains a sequence (x_n) that generates a complemented copy of E that is isomorphic to ℓ_1;*
3. *If $T: Y \to E$ is not weakly compact then it fixes a complemented copy of ℓ_1.*

This proposition implies that every non reflexive Banach space having the property (V^*) contains a complemented copy of ℓ_1. Godefroy and Talagrand [33] introduced a property of a Banach space that is stronger than the property (V^*).

Definition 3 We say that a Banach space E has the **property (X)** if and only if any element $u \in E^{**}$ such that $u(\text{weak}^* \sum_{i=1}^{\infty} x_i^*) = \sum_{i=1}^{\infty} u(x_i^*)$ for every weakly unconditionally Cauchy series $\sum_{i=1}^{\infty} x_i^*$ in E^* must be in E.

In [33], it was shown that the property (X) is stable by subspaces and if E has the property (X) then E is the unique predual of its dual E^*.

Theorem 11 ([31]). *Any Banach space with the property (X) has the property (V^*).*

Proof: Edgar [24] showed that a Banach space E has the property (X) if and only if $E \prec \ell_1$ where $\prec$ is the ordering of Banach spaces introduced in [24], then he showed that if a Banach space $E \prec \ell_1$, then every bounded sequence in E that is not relatively weakly compact has a subsequence equivalent to the unit vector basis of ℓ_1 whose closed span is complemented in E. This fact coupled with Proposition 10 finishes the proof. □

Theorem 12 ([24]). *Let Γ be a set then $\ell_1(\Gamma)$ has the property (X) if and only if card Γ is not a (real-valued) measurable cardinal.*

This theorem shows that the property (X) is stronger than the property (V^*). It was thought first that the property (X) might be equivalent to the property (V^*) in the separable case. Talagrand [68] showed that this is not the case by constructing a separable Banach space having the property (V^*) but failing the property (X).

In [33], Godefroy and Talagrand showed that the following spaces have the property (X) at least when they are separable:

1. Weakly sequentially complete Banach spaces that are complemented in a Banach lattice.
2. $\mathcal{L}_1$ spaces.
3. Preduals of W^* algebra.
4. L_1/H_1.
5. Spaces with local unconditional structure which do not contain ℓ_n^∞ uniformly.

The following proposition deals with the three space problem for the property (V^*) and property (X).

Proposition 13 ([31]). *Let E be a Banach space and R a reflexive subspace of it, then E has the property (V^*) (resp., (X)) if and only if E/R has the property (V^*) (resp., (X)).*

The three space problem for the properties (V^*) and (X) seems to be open in general.

In view of Question 1 one can ask:

Question 2 Suppose that $L_p(E)$ $(1 \leq p < \infty)$ is separable. Does it follow that $L_p(E)(1 \leq p < \infty)$ has the property (X) or at least property (V^*) whenever E has the property (X)?

Definition 4 (Pełczynski). A Banach space E is said to have **property (u)** if for any weakly Cauchy sequence (e_n) in E there exists a weakly unconditionally Cauchy series $\sum_n x_n$ in E such that the sequence $(e_n - \sum_{i=1}^{n} x_i)$ converges weakly to zero in E.

Every closed subspace of a space with the property (u) has the property (u) [48].

Examples of Banach spaces with the property (u)

1. Any weakly sequentially complete Banach space has property (u). In particular any $L_p, 1 \leq p < \infty$ has the property (u).
2. (Pełczynski) Any Banach space E with unconditional basis or more generally any space with unconditional reflexive decomposition has property (u).
3. (Tzafriri [73]) Any order continuous Banach lattice has property (u).
4. (Godefroy and Li [30]). Any Banach space that is an M-ideal in its bidual has property (u).
5. (Godefroy and P. Saab [32]) Under certain conditions, spaces of compact operators on a Banach space X have property (u).
6. (Knaust-Odell [44]). If a Banach space E has property (S) (Every subspace of E has the Dunford-Pettis property), then E has property (u).

It is clear that a Banach space E that has the property (u) is weakly sequentially complete if and only if E does not contain a copy of c_0.

Theorem 14 ([54]). *A Banach space E that has the property (u) and does not contain a copy of ℓ_1 has the property (V).*

Actually every closed subspace of a space that has property (u) and does not contain a copy of ℓ_1 has (V). It would be interesting to know the answer to the following question:

Question 3 Give a characterization of those Banach spaces whose all closed subspaces have the property (V). For example, do they have the property (u)?

F. Lust Picard [49] showed that the property (u) is not stable under taking the injective or the projective tensor products. She actually showed that $c_0\hat{\oplus}_\pi c_0$ does not have the property (u) and she was able to exhibit a Banach space E_p that has the property (u) such that $\ell_p\hat{\oplus}_\epsilon E_p(1 < p < 2)$ does not have the property (u).

However the following stability result for the property (u) was shown in [40]:

Theorem 15 *Let E be a Banach space having property (u) then $L_p(E)$ enjoys the same property for $1 \leq p < \infty$.*

Of course it is enough to show the above theorem in the case where E is separable. The main tools used to prove Theorem 15 are a result of Talagrand [71] and a selection theorem. To be able to state this selection theorem one needs to introduce the following notation:

Let W be the set

$$W = \{(x_n)_{n\geq 1} \in E^{\mathbf{N}} \quad | \quad (x_n)_{n\geq 1} \text{ is a weakly Cauchy sequence in} E\}$$

Theorem 16 *Let E be a separable Banach space that has the property (u), then there is a constant $C > 0$ and an analytic subset M of $E^{\mathbf{N}}$ that contains W and a sequence $(S_i)_{i\geq 1}$ of universally measurable maps from M to E such that if $x = (x_n)_{n\geq 1}$ is a weakly Cauchy sequence in E we have*

i) $\left(x_n - \sum_{i=1}^{n} S_i(x)\right)$ *converges weakly to zero in E;*

ii) $\sup_{\sigma\in\mathcal{F}} \|\sum_{i\in\sigma} S_i(x)\| \leq C \liminf_{n\to\infty} \|x_n\|$. *(where $\mathcal{F}$ stands for the set of all finite subsets of* $\mathbf{N}$*)*

We will sketch the proof of Theorem 15 for $p = 1$. Let $(h_n)_{n\geq 1}$ be a weakly Cauchy sequence in $L_1(E)$. By [71], the sequence $(h_n)_{n\geq 1}$ can be decomposed as $h_n = g_n + w_n$ for every $n \geq 1$, where for almost all ω, the sequence $(g_n(\omega))_{n\geq 1}$ is a weakly Cauchy sequence in X and w_n converges weakly to zero in $L_1(E)$. We can suppose that for every ω the sequence $g_n(\omega))_{n\geq 1}$ is a weakly Cauchy sequence in X. Let W and M be as in Theorem 16 and consider the map $g : \Omega \to W$ defined by $g(\omega) = (g_n(\omega))_{n\geq 1}$. For each $i \geq 1$ let $S_i : M \to X$ be the universally measurable map obtained by Theorem 16 and let $f_i = S_i \circ g$. Since S_i is universally measurable, it follows that f_i is λ-measurable. The properties of the sequence $(S_i)_{i\geq 1}$ and a uniform integrability argument imply that (for more details see [40])

$$g_n - \sum_{i=1}^{n} f_i \text{ converges weakly to zero in } L_1(E),$$

and that $\sum_{i=1}^{\infty} f_i$ is weakly unconditionally Cauchy in $L_1(E)$. To finish the proof, notice that

$$h_n - \sum_{i=1}^{n} f_i = g_n - \sum_{i=1}^{n} f_i + w_n$$

and the sequence $(w_n)_{n\geq 1}$ converges weakly to zero in $L_1(E)$. □

Before stating another application of Theorem 16, let us recall some definitions.

Let $T: E \to F$ be a bounded linear operator from a Banach space E into a Banach space F. We say that T is **weakly completely continuous** (w.c.c) (also called a Dieudonné operator) if for every weakly Cauchy sequence (x_n) in E, the sequence $(T(x_n))$ converges weakly in F. It is clear that if T is weakly compact, then T is weakly completely continuous which implies that T is unconditionally converging. It is also evident that if a Banach space E has property (u) then every unconditionally converging operator on E is weakly completely continuous. We say that a Banach space E has the **property semi-(V)** if every unconditionally converging operator is weakly completely continuous. It is clear that if E has property (V), then it has property semi-(V), the converse is of course not true (i.e. $E = \ell_1$).

The following theorem extends and strengthens Theorem 3 of [66].

Theorem 17 ([40]). *Let K be a compact Hausdorff space and let E be a Banach space that has the property (u), then $C(K, E)$ has the property semi-(V).*

In [15] the following theorem was shown to be true

Theorem 18 *If E has the property (u) and ℓ_1 does not embed in E then $C(K, E)$ has the property (V).*

The proof of Theorem 18 depends on James's Theorem [36] and the following selection theorem:

Theorem 19 *Let E be a separable Banach space that has the property (u) and does not contain a copy of ℓ_1. Then there is a sequence of maps θ_n : $(E^*, \text{weak}^*) \to E$ so that each θ_n is universally measurable and for each $e^* \in E^*$ on has*

1. $\sum_{i=1}^{\infty} \theta_i(e^*)$ *is weakly unconditionally Cauchy;*

2. $\|\sum_{i=1}^{n} \theta_i(e^*)\| \leq 1 + \frac{1}{n}$, $n \geq 1$;

3. $\sum_{i=1}^{n} \langle \theta_i(e^*), e^* \rangle = \|e^*\|$.

At this stage one can ask the following:

Question 4 Give a characterization of those Banach spaces for which Theorem 19 is valid. Or one can ask for which Banach spaces E, does every element $e^* \in E^*$ attain its norm on a weakly unconditionally Cauchy series $\sum_{i=1}^{\infty} u_i$ in E. (i.e. $\|e^*\| = \sum_{i=1}^{\infty} e^*(u_i)$)

As a corollary of Theorem 18 one obtains [15]

Theorem 20 *Let E be a Banach space isomorphic to a closed subspace of an order continuous Banach lattice. Then E has the property (V) if and only if $C(K, E)$ has property (V).*

Question 5 Does $C(K, E)$ have the property (V) whenever E has the property (V)?

Similarly

Question 6 Does $C(K, E)$ have the property semi-(V) whenever E has the property semi-(V)?

In [5] Bombal observed that if E is a closed subspace of an order continuous Banach lattice, then $L_p(E)$ has the property (V) if $1 < p < \infty$ and E has the property (V). In the case where $p = \infty$, Bombal's result does not hold true as we will show later.

Definition 5 A Banach space E is said to have the **Dieudonné property** if for any Banach space F, every operator $T: E \to F$ that is weakly completely continuous is weakly compact.

If K is any compact Hausdorff space, the space $C(K)$ has the Dieudonné property. It follows from Rosenthal's theorem [57] that any Banach space E that does not contain a copy of ℓ_1 has the Dieudonné property.

In [7], the authors showed that if E^* is separable then $C(K, E)$ has the Dieudonné property and they asked whether the same result is true when replacing the assumption that E^* is separable by the assumption that ℓ_1 does not embed in E. This question was answered in the affirmative in [41]. Namely one has:

Theorem 21 *Let E be a Banach space not containing a copy of ℓ_1 then $C(K, E)$ has the Dieudonné property.*

See [26] for another proof of the above theorem.

Question 7 Does $C(K, E)$ have the Dieudonné property whenever E has the Dieudonné property?

In ([3], Theorem 8.3) Aron, Cole and Gamelin isolated a family of Banach spaces that satisfies the following condition:

A Banach space E belongs to this family if any symmetric operator $T: E \to E^*$ is weakly compact. We recall that an operator T is said to be symmetric if $\langle Tx, y\rangle = \langle x, Ty\rangle$ for every $x, y \in E$. Influenced by that let us give the following definition:

Definition 6 ([62]) A Banach space E is said to have the **property (w)** if and only if every bounded linear operator $T: E \to E^*$ is weakly compact.

In [3] the authors noticed that any $C(K)$ space has the property (w). Actually more can be said:

Proposition 22 ([62]) *Any Banach space E such that E^* has the property (V^*) has the property (w). In particular any Banach space with the property (V) has the property (w).*

The spaces X constructed by Pisier [55] are such that every bounded linear operator from X to X^* is integral and hence is weakly compact so these spaces X have the property (w).

It is easy to see that any complemented subspace of a Banach space with the property (w) has this property. Since ℓ_1 does not have the property (w), then any space E with the property (w) cannot contain a complemented copy of ℓ_1 and therefore E^* cannot contain a copy of c_0 [4]. With the help of the above proposition and Theorem 9 one can deduce the following:

Corollary 23 **([62])** *Let E be a Banach space such that E^* is complemented in a Banach lattice or E^* is isomorphic to a closed subspace of an order continuous Banach lattice. Then the following statements are equivalent:*

1. *E has the property (w);*
2. *E does not contain a complemented copy of ℓ_1;*
3. *E^* has the property (V^*).*

Corollary 24 *Any $\mathcal{L}_\infty$ space has the property (w).*

Proof: Let E be an $\mathcal{L}_\infty$ space, then E^* is a $\mathcal{L}_1$-space and hence E^* has property (V^*). Apply Proposition 22 to conclude. □

The space constructed by Bourgain and Delbaen in [10] is an $\mathcal{L}_\infty$ space so it has property (w) but does not have (V) since it is a Schur space.

If a Banach space E does not contain a copy of ℓ_1 and E^* is weakly sequentially complete, then E has the property (w). To see this notice that any operator $T : E \longrightarrow E^*$ is weakly precompact (the image of the unit ball of E by T does not contain a copy of ℓ_1) and therefore T is weakly compact since E^* is weakly sequentially complete.

We can now offer the following theorem:

Theorem 25 **([62])** *Let F be a Banach space whose dual is isometric to an L_1-space and let E be a Banach space that does not contain a copy of ℓ_1 and assume that E^* is weakly sequentially complete, then $F\hat{\otimes}_\epsilon E$ has the property (w).*

In the special case where $F = C(K)$, the above theorem can be deduced from an unpublished result of C. Fierro [28].

The following question arises naturally now:

Question 8 Does the property (w) pass from a Banach space E to $C(K, E)$?

By Proposition 22 and [61], the answer is yes if E is a Banach lattice. Actually more can be said (see Proposition 47).

Most of the Banach spaces E known to us that have (w) are such that their duals have the property (V^*). However the spaces constructed by Pisier [55] have the property (w) but nothing can be said about their duals according to Pisier (private communication).

Question 9 Suppose that a Banach space E has the property (w), does it follow that E^* has property (V^*) or at least that E^* is weakly sequentially complete?

Let us now turn our attention to the containment of c_0 in a given Banach space. For this let us recall the following basic theorems in this direction:

Theorem 26 ([4]) *A Banach space E does not contain a copy of c_0 if and only if every weakly unconditionally Cauchy series in E is unconditionally converging.*

Theorem 27 ([4]) *A Banach space E contains a complemented copy of ℓ_1 if and only if E^* contains a copy of c_0.*

Theorem 26 has been extended by Elton:

Theorem 28 ([25]) *A Banach space E does not contain a copy of c_0 if and only if every series $\sum_{i=1}^{\infty} x_i$ in E such that $\sum_{i=1}^{\infty} |x^*(x_i)| < +\infty$ for every extreme point x^* in the unit ball of E^* is unconditionally converging.*

The following theorem of R.C. James says that whenever E contains an isomorphic copy of c_0, then it contains very good copies of c_0.

Theorem 29 (James, see [19]) *If E contains an isomorphic copy of c_0, then for any $\epsilon > 0$ there is a sequence $(y_n)_{n\geq 1}$ in the unit ball of E such that*

$$(1-\epsilon)\sup_{i\geq 1}|a_i| \leq ||\sum_{i=1}^{\infty} a_i y_i|| \leq \sup_{i\geq 1}|a_i|$$

for any $(a_i)_{i\geq 1}$ in c_0.

In the context of Banach lattices more can be said:

Theorem 30 ([48]) *A Banach lattice E does not contain a copy of c_0 if and only if E is weakly sequentially complete.*

Notice that there are Banach spaces that do not contain c_0 but they are not weakly sequentially complete. For example the James space JT [21].

Let us now discuss when c_0 can embed as a complemented subspace of a given Banach space E. The first result in this direction is due to Sobczyk [67].

Theorem 31 *c_0 is not complemented in ℓ_∞.*

Actually more can be said:

Theorem 32 ([4]) *A dual Banach space can never contain a complemented copy of c_0.*

This is a direct consequence of Theorem 31. To see this, let $L : c_0 \longrightarrow E^*$ be the embedding of c_0 in E^* and let $Q : E^* \longrightarrow L(c_0)$ be the projection of E^* onto $L(c_0)$. Let $P = L^{-1}QSL^{**}$ where S is the natural projection of E^{***} onto E^*. It is easily verified that P is a projection of ℓ_∞ onto c_0.

When the space containing c_0 is separable the story is different. We say that a Banach space E is **separably injective** if any time E embeds in a separable space F then E is complemented in F.

Theorem 33 ([67]) *The Banach space c_0 is separably injective.*

Zippin [74] proved the converse by showing:

Theorem 34 *Any separably injective Banach space must be isomorphic to* c_0.

In [45] Kwapian showed the following result:

Theorem 35 *A Banach space* E *contains a copy of* c_0 *if and only if* $L_p(E)$ *contains a copy of* c_0 *for* $1 \leq p < \infty$.

Kwapien's result led to the natural question whether weak sequential completeness passes from E to $L_p(E)$ $1 \leq p < \infty$. In [71], Talagrand gave a positive answer to this question:

Theorem 36 ([71]) *If a Banach space* E *is weakly sequentially complete then* $L_p(E) 1 \leq p < \infty$ *is weakly sequentially complete.*

Let us mention the following result of F. Lust Picard:

Theorem 37 ([49]) *Let* E *and* F *be two Banach spaces that are weakly sequentially complete. Suppose in addition that* E *has an unconditional basis, then* $E\hat{\otimes}_\pi F$ *is weakly sequentially complete.*

For the injective tensor product of weakly sequentially complete Banach spaces, D.R. Lewis [47] gave this characterization:

Theorem 38 ([47]) *Let* E *and* F *be two Banach spaces that are weakly sequentially complete and one of them has the approximation property, then* $E\hat{\otimes}_\epsilon F$ *is weakly sequentially complete if and only if every bounded linear operator* $T : E^* \to F$ *that is weak* to weak continuous is compact.*

In [11] Bourgain and Pisier constructed a space X that is weakly sequentially complete but $X\hat{\otimes}_\pi X$ contains a copy of c_0.

The following result due to Cembranos seems surprising:

Theorem 39 *If* K *is an infinite compact Hausdorff space and* E *is an infinite dimensional Banach space, then* $C(K, E)$ *contains a complemented copy of* c_0.

In [63] Cembrano's result was generalized as follows:

Theorem 40 **([63])** *Let E and F be two infinite dimensional Banach spaces. If E contains a copy of c_0 then $E\hat{\otimes}_\epsilon F$ contains a complemented copy of c_0.*

Combining the above theorem with Theorem 32 yields the following:

Theorem 41 *Let E and F be two infinite dimensional Banach spaces. If E contains a copy of c_0 then $E\hat{\otimes}_\epsilon F$ is not isomorphic to a dual space.*

We say that a Banach space E is a **Grothendieck space** if weak* and weak sequential convergence in E^* coincide. It is well known that a Banach space E is a Grothendieck space if and only if every bounded linear operator $T: E \to c_0$ is weakly compact. Hence a Grothendieck space cannot contain a complemented copy of c_0.

The following result generalizes a result of Khurana [42] and is a direct consequence of Theorem 40.

Corollary 42 *Let E be a Banach space that contains an isomorphic copy of c_0 and let F be an infinite dimensional Banach space then, $E\hat{\otimes}_\epsilon F$ is not a Grothendieck space.*

Two other results similar to Theorem 39 and Theorem 40 were discovered by Emmanuelle [27] and Dowling [23].

Theorem 43 **([27])** *If E contains a copy of c_0 then $L_p(E)$ contains a complemented copy c_0 if $1 \leq p < \infty$.*

To be able to state Dowling's result we need to introduce the vector-valued Hardy spaces. Let $\mathbf{T}$ be the unit circle, let m be the normalized Lebesgue measure on $\mathbf{T}$ and let $\mathbf{D}$ denote the open unit disk in the complex plane.

Let E be a complex Banach space and let $1 \leq p < \infty$. The space $H_p(\mathbf{D}, E)$ is the collection of all E-valued analytic functions on $\mathbf{D}$ with $||f||_p < \infty$ where

$$||f||_p = \sup_{0 \leq r < 1} \left\{ \int_0^{2\pi} ||f(re^{i\theta})||^p dm \right\}^{1/p}$$

If $f \in L_p(\mathbf{T}, E)$, then its Fourier coefficients are

$$\hat{f}(n) = \text{Bochner}-\int_0^{2\pi} f(e^{i\theta})e^{-in\theta}dm$$

where $n \in \mathbf{Z}$. Now we define

$$H_p(\mathbf{T}, E) = \left\{f \in L_p(\mathbf{T}, E) \mid \hat{f}(n) = 0 \text{ for all } n < 0\right\}.$$

Theorem 44 [23] *Let E be a complex Banach space that contains a copy of c_0, then $H_p(\mathbf{T}, E)$ contains a complemented copy of c_0.*

The above theorem does not hold for $H_p(\mathbf{D}, E)$ as observed in [23]. For example if $E = \ell_\infty$ then $H_p(\mathbf{D}, E)$ is a dual space [22] which cannot contain any complemented copy of c_0 by Theorem 32. This shows that $H_p(\mathbf{T}, E)$ for $E = \ell_\infty$ is not isomorphic to a complemented subspace of $H_p(\mathbf{D}, E)$ [22].

Theorem 40 shows that the presence of c_0 in E gives birth to the presence of a complemented copy of c_0 in the injective tensor product $E\hat{\otimes}_\epsilon F$ if F is infinite dimensional. Actually if c_0 is not present in any factor it might become present in the injective or projective tensor product, for example if one takes $E = F = \ell_2$, then $E\hat{\otimes}_\epsilon F$ contains c_0. For the projective tensor product see [55] or [11].

The search of the presence of ℓ_1 in a given Banach space has been studied by many people in recent years. The most celebrated theorem in this direction is Rosenthal's result. In [57] Rosenthal showed the following fundamental theorem:

Theorem 45 *A Banach space E contains no subspace isomorphic to ℓ_1 if and only if every bounded sequence in E has a weak Cauchy subsequence.*

As a consequence of Rosenthal's theorem one can conclude that a weakly sequentially complete Banach space is either reflexive or contains an isomorphic copy of ℓ_1.

In [51] Odell and Rosenthal showed that a separable Banach space E does not contain an isomorphic copy of ℓ_1 if and only if every element $x^{**} \in E^{**}$ is of first Baire class on the unit ball of X^* when it is equipped with the weak* topology. Haydon [35] extended this result to the non separable case by showing that a Banach space E does not contain an isomorphic copy of ℓ_1 if and only if every element $x^{**} \in E^{**}$ is universally measurable on the unit Ball of X^* when it is equipped with the weak* topology.

In [60] a point of continuity argument was given to characterize spaces not containing a copy of ℓ_1. Namely a Banach space E does not contain an isomorphic copy of ℓ_1 if and only if for every element $x^{**} \in E^{**}$ and every weak* compact convex subset C of E^*, the intersection $Z \cap \text{Ext(C)}$ of the set Z of the points of continuity of x^{**} restricted to (C, weak*) with the set of extreme points of C is a G_δ subset of (Ext(C), weak*) and C is equal the weak* closed convex hull of $Z \cap \text{Ext(C)}$.

In [53] Pełczynski showed that a Banach space E contains a copy of ℓ_1 if and only if E^* contains a copy of L_1.

Recall that a Banach space F is said to have the **weak Radon Nikodym property (WRNP)** if every operator $T : L_1 \to E$ is Pettis representable [50]. If one considers an operator $T : L_1 \to E^*$ one can always represent this operator by a weak* derivative (See [70]), in case E does not contain a copy of ℓ_1 one can show [37] that this derivative can be taken to be Pettis integrable by using the Haydon's result [35] and hence E^* will have the WRNP. Actually the converse is also true [37], one way of seeing that is to use the fact that a space having the WRNP cannot contain a copy of L_1 [29], hence E^* cannot contain L_1 and therefore E cannot contain ℓ_1. So the WRNP is well understood in dual spaces. In [29] it was shown that if a Banach lattice has the WRNP then it has the RNP.

The study of the stability property of spaces containing ℓ_1 started with the result of Pisier [56] where he showed the following theorem:

Theorem 46 ([56]) *If $L_p(E)$ $(1 < p < \infty)$ contains a copy of ℓ_1 then E contains a copy of ℓ_1.*

In [61] a somewhat similar result was shown, namely if $C(K, E)$ contains a complemented copy of ℓ_1 then E contains a complemented copy of ℓ_1. Actually a slightly more general result is true, namely it was observed in [65] that if F is a Banach space such that F^* is isometric to an L_1 space and if $F\hat{\otimes}_\epsilon E$ contains a complemented copy of ℓ_1 then E contains a complemented copy of ℓ_1.

Proposition 47 *Let F be a Banach space whose dual is isometric to an L_1-space and let E be a Banach lattice having the property (w), then $F\hat{\otimes}_\epsilon E$ has the property (w).*

Proof: The dual of $F\hat{\otimes}_\epsilon E$ is isometric to a subspace of the dual of a $C(K, E)$ space [65]. The space $C(K, E)$ cannot contain a complemented copy of ℓ_1 [61] and therefore its dual which is a Banach lattice has property (V^*). Apply Proposition 22 to conclude. □

Bombal [5] showed that if E is a Banach lattice and if $L_p(E)$ $(1 < p < \infty)$ contains a complemented copy of ℓ_1 then E contains a complemented copy of ℓ_1.

A question can now be asked: What happens if in Pisier's result [56] one replaces ℓ_1 by L_1? More precisely:

Question 10 If $L_p(E)$; $1 < p < \infty$ contains a copy of L_1 does E contain a copy of L_1?

We can offer a positive answer to this question in the case $E = F^*$ is dual space. To do that let us recall that the dual of $L_p(F)\,(1 < p < \infty)$ can be identified with the space $L_q(F^*_\sigma)$ of all map h from the measure space to F^* such that h is weak* scalarly measurable and

$$||h||_q = \left(\int ||h(t)||^q d\lambda\right)^{1/q} < \infty$$

where $\frac{1}{p} + \frac{1}{q} = 1$ (see [72]). It is clear that $L_q(F^*)$ is a closed subspace of $L_q(F^*_\sigma)$ and $L_q(F^*) = L_q(F^*_\sigma)$ if and only if F^* has the Radon Nikodym property. If $L_p(F^*)$ contains a copy of L_1, then $L_p(F^*_\sigma)$ contains a copy of L_1. Apply now Pełczynski's result [53] to deduce that $L_q(F)$ contains a copy of ℓ_1 and hence by Pisier [56] F contains a copy of ℓ_1 which in turns implies that F^* contains a copy of L_1. This shows in particular that if F^* has the WRNP then $L_p(F^*_\sigma)\,(1 < p < \infty)$ has the WRNP. One can now ask:

Question 11 Is it true that $L_p(F^*)$ has the WRNP whenever F^* has the WRNP, and in general does $L_p(E)$ have the WRNP if and only if E has the WRNP for $1 < p < \infty$.

A result of this nature is true in the case of the Radon-Nikodym property (See [21]).

Another question in this direction is:

Question 12 If $C(K, E)$ contains a complemented copy of L_1 must E contain a complemented copy of L_1?

In [1] the authors showed that if F is a Banach space such that F^* has the Radon Nikodym property and contains no subspace isomorphic to ℓ_1, and if G is any Banach space and K a compact Hausdorff space, then an operator $T: C(K, F) \to G$ is unconditionally converging if and only if its adjoint T^* is weakly precompact and they asked whether or not the result is still true if one assumes only that F^* does not contain a subspace isomorphic to ℓ_1. In [65] a positive answer was given to their question. Actually a more general result was proved in [65], namely if E, F and G are Banach spaces such that E^* is isometric to an L_1-space, and F^* contains no subspace isomorphic to ℓ_1, a bounded linear operator $T: E\hat{\otimes}_\epsilon F \to G$ is unconditionally converging if and only if its adjoint T^* is weakly precompact.

Definition 7 We say that an operator $T: E \to F$ between two Banach spaces is a **Dunford-Pettis operator** (also called completely continuous) if T sends weakly convergent sequences into norm convergent sequences.

It is clear by Rosenthal's theorem [57] that if E does not contain an isomorphic copy of ℓ_1 then every Dunford-Pettis operator on E is compact. The converse is also true and is due to Odell [58].

We say that a Banach space E has the **complete continuity property** if every operator $T: L_1 \to E$ is a Dunford-Pettis operator. It is well known that E^* has the WRNP if and only if E^* has the complete continuity property (see [59]).

The natural question that comes to mind is the following:

Question 13 Suppose that $1 < p < \infty$, does $L_p(E)$ have the complete continuity property whenever E has the same property?

By what we did above the answer to this question is yes if E is a dual space.

Definition 8 We say that a Banach space E has the **Dunford-Pettis property** if and only if every weakly compact operator $T: E \to F$ is a Dunford-Pettis operator.

For more information about the Dunford-Pettis property we refer the reader to the excellent survey on the subject by J. Diestel [20]. Here we recall that any $C(K)$ space and any L_1 space have the Dunford-Pettis property. The problem of the stability of the Dunford-Pettis was discussed in [20], namely Bourgain's positive results about when $C(K, E)$ and $L_1(E)$ have the Dunford-Pettis property were stated. In [9] Bourgain showed among others that $C(K, L_1)$ and $L_1(C(K))$ both have the Dunford-Pettis property. In [20] the question was asked whether $C(K, E)$ and $L_1(E)$ have the Dunford-Pettis property whenever E does. Talagrand [69] constructed a counter example to that question.

Let us say that a Banach space is **hereditary Dunford-Pettis** if every closed subspace of it has the Dunford-Pettis property. Such a space is said to have the property (S) (see [44] and [14]). The space c_0 has the hereditary Dunford-Pettis property [20], the James-Hagler space JH also has this property [34]. In [44], it was shown that a Banach space that has (S) has also the property (u). Theorem 1 of [69] and the result of [44] show that the Talagrand space E is hereditary Dunford-Pettis but $C(K, E)$ and $L_1(E^*)$ fail the Dunford-Pettis property. Since the dual E^* of the Talagrand space has the Schur property, then it follows from [2] that $L_1(E)$ has the Dunford-Pettis property.

In [17] it was shown that a C^*-algebra A has the Dunford-Pettis if and only if, given any weakly null sequence $(a_n)_{n\geq 1}$ in A, the sequence $(a_n^* a_n)_{n\geq 1}$ is also weakly null in A, where $a \to a^*$ is the involution in A. It was also shown [17] that A has the Dunford-Pettis property if and only if A^* has the Dunford-Pettis property.

As a nice and easy observation we offer the following consequence of the above result:

Theorem 48 *Let K be a compact Hausdorff space and A be a C^*-algebra. Then A has the Dunford-Pettis property if and only if $C(K, A)$ has the Dunford-Pettis property.*

Proof: It is clear that $C(K, A)$ is a C^*-algebra with the natural operations and pointwise involution. Let $(f_n)_{n\geq 1}$ be a weakly null sequence in $C(K, A)$. This implies that $(f_n)_{n\geq 1}$ is bounded and for every $t \in K$ one has $\lim_{n\to\infty} f_n(t) = 0$ weakly in E. This shows that $\lim_{n\to\infty}(f_n(t))^* f_n(t) = 0$ weakly in A. Therefore the sequence $(f_n^* f_n)_{n\geq 1}$ is weakly null, thus $C(K, A)$ has the Dunford-Pettis property by [17].□

It is well known that the disk Algebra as well as its dual has the Dunford-Pettis property [16]. The following question arises naturally:

Question 14 Suppose A is the disk algebra and K is a compact Hausdorff space, does $C(K, A)$ have the Dunford-Pettis property?

Let us finish by noticing that there are many properties of a Banach space E and L_∞ that do not pass to the space $L_\infty(E)$.

Delbaen noticed that if $E = (\Sigma \oplus \ell_2^n)_{\ell_1}$, then E has the Schur property but $L_\infty(E)$ fails the Dunford-Pettis property. (see [20]).

Now let $E = (\sum \oplus \ell_1^n)_{c_0}$, then E^* has the Radon-Nikodym property and hence E does not contain a copy of ℓ_1, but $E^{**} = (\sum \oplus \ell_1^n)_{\ell_\infty}$ contain a complemented copy of ℓ_1 (see [38]). It is now easily seen that E^{**} is complemented in $\ell_\infty(E)$ which in turns is complemented in $L_\infty(E)$. This shows that $L_\infty(E)$ can contain a complemented copy of ℓ_1 without E containing a copy of ℓ_1. This answers in the negative a question of I. Labuda [46], the question of Labuda was motivated by the result of [61]. This example was kindly shown to us by S. Montgomery-Smith and is due to him.

One can get more out of this space $E = (\sum \oplus \ell_1^n)_{c_0}$. This space does not contain a copy of ℓ_1 and has the property (u) but $L_\infty(E)$ does not have the property (V) even though the space $C(K, E)$ has the property (V) [15]. One can also notice that the space $L_\infty(E)$ does not have the property (w) nor the Dieudonné property despite the fact that $C(K, E)$ has these two properties.

Finally let us notice that a result similar to Cembranos's result [13] does not hold true for $L_\infty(E)$. It is enough to take $E = \ell_1$ and notice that $L_\infty(E)$ is a dual space and therefore cannot contain a complemented copy of c_0 by Theorem 32 [4].

References

[1] C.A. Abbot, E.M. Bator, R.G., Bilyeu, and P.W. Lewis. *Weak precompactness, Strong boundedness and weak complete continuity* . Math. Proc. Camb. Phil. Soc. , **208**:325–335, (1990).

[2] K. Andrews. *Dunford-Pettis set in the space of Bochner integrable functions.* Math. Ann., **241**:35–42, (1979).

[3] R.M. Aron, B.J. Cole, and T.W. Gamelin. *Spectra of algebras of analytic functions on a Banach space.* (1990). To appear.

[4] C. Bessaga and A. Pełczynski. *On bases and unconditional convergence of series in Banach spaces.* Studia Math., **17**:151–164, (1958).

[5] F. Bombal. *On ℓ_1 subspaces of vector-valued function spaces..* Math. Proc. Camb. Phil. Soc., **101**:107–112, (1987).

[6] F. Bombal. *On (V^*) sets and Pełczynski's property (V^*).* Glasgow Math. J., **32**:109–120, (1990).

[7] F. Bombal and P. Cembranos. *The Dieudonné property for $C(K, E)$.* Trans. Amer. Math. Soc., **285**:649–656, (1984).

[8] J. Bourgain. *H^∞ is a Grothendieck space.* Studia Math, **75**:193–216, (1983).

[9] J. Bourgain. *New Classes of $\mathcal{L}^p$-spaces.* Lecture notes in Mathematics, Springer-Verlag, **889**:, (1981).

[10] J. Bourgain and F. Delbaen. *A class of $\mathcal{L}^\infty$-spaces.* Acta Math., **145**:155–170, (1980).

[11] J. Bourgain and G. Pisier. *A construction of $\mathcal{L}^\infty$-spaces and related Banach spaces.* Bol. Soc. Brasil. Mat., **14**:109–123, (1983).

[12] D. Cartwritht. *The order completeness of some spaces of vector-valued functions.* Bull. Austr. Math. Soc., **II**:57–61, (1974).

[13] P. Cembranos. *$C(K, E)$ contains a complemented copy of c_0.* Proc. Amer. Math. Soc., **91**:556–558, (1984).

[14] P. Cembranos. *The hereditary Dunford-Pettis property for $C(K, E)$.* Illinois J. of Math., **31**:365–373, (1987).

[15] P. Cembranos, N.J. Kalton, E. Saab, and P. Saab. *Pełczynski's Property (V) on $C(\Omega, E)$ spaces.* Math. Ann., **271**:91–97, (1985).

[16] J. Chaumat. *Une généralisation d'un théorème de Dunford-Pettis.* Université de Paris XI, Orsay, :, (1974).

[17] C.-H Chu and B. Iochum. *The Dunford-Pettis property in C^*-algebras.* Studia Math., **97**:59–64, (1990).

[18] D.L. Cohn. *Measure Theory.* Birkhäuser, Basel, Stuttgart, (1980).

[19] J. Diestel. *Sequences and Series in Banach Spaces.* Volume **92** of Graduate Text in Mathematics, Springer Verlag, New York, first edition, (1984).

[20] J. Diestel. *The Dunford-Pettis property.* Contemp. Math., **2**:15–60, (1980).

[21] J. Diestel and Jr. J.J. Uhl. *Vector Measures.* Volume **15** of Math Surveys, AMS, Providence, RI, first edition, (1977).

[22] P. Dowling. *Duality in some vector-valued function spaces.* preprint.

[23] P. Dowling. *On complemented copy of c_0 in vector-valued Hardy spaces.* Proc. Amer. Math. Soc., **107**:251–254, (1989).

[24] G. Edgar. *An ordering for the Banach spaces.* Pacific. J. of Math., **108**:83–98, (1983).

[25] J. Elton. *Extremely weakly unconditionally convergent series.* Israel J. of Math., **40**:255–258, (1981).

[26] G. Emmanuelle. *Another proof of a result of N.J. Kalton, E. Saab and P. Saab on the Dieudonné property in $C(K, E)$.* Glasgow Math. J., **31**:137–140, (1989).

[27] G. Emmanuelle. *On complemented copy of c_0 in L^p_X, $1 \leq p < \infty$.* Proc. Amer. Math. Soc., **104**:785–786, (1988).

[28] C. Fierro. *Compacidad débil en espacies de funciones y medidas ectoriales.* PhD thesis, Compultense, Madrid, (1980). Spain.

[29] N. Ghoussoub and E. Saab. *On the weak Radon-Nikodym property.* Proc. Amer. Math. Soc., **81**:81–84, (1981).

[30] G. Godefroy and D. Li. *Banach spaces which are M-ideals in their biduals have property (u).* Annales de l'Institut Fourier, **32**:361–371, (1989).

[31] G. Godefroy and P. Saab. *Quelques espaces de Banach ayant les propriétés (V) ou (V^*) de A. Pełczynski.* C.R. Acad. Sc., **303, Serie I**:503–506, (1986).

[32] G. Godefroy and P. Saab. *Weakly unconditionally convergent series in M-ideals.* Math. Scand., **64**:307–318, (1990).

[33] G. Godefroy and M. Talagrand. *Nouvelles classes d'espaces de Banach a predual unique.* Séminaire d'Analyse Fonctionelle, Ecole Polythechnique, Exposé VI, :, (1980-1981).

[34] J. Hagler. *A counter example to several questions about Banach spaces.* Studia Math., **60**:289–307, (1977).

[35] R. Haydon. *Some more characterizations of Banach spaces containing ℓ_1.* Math. Proc. Cambridge Phil. Soc., **80**:269–276, (1976).

[36] R.C. James. *Weakly compact sets.* Trans. Amer. Math. Soc., **13**:129–140, (1964).

[37] L. Janicka. *Some measure-theoretical characterizations of Banach spaces not containing ℓ_1.* Bull. Acad. Polon. Sci. Sér. Sci., **27**:561–565, (1979).

[38] W. Johnson. *A complementary universal conjugate Banach space and its relation to the approximation problem.* Israel J. of Math., **13**:301–310, (1972).

[39] W.B. Johnson and M. Zippin. *Separable L_1 preduals are quotients of $C(\Delta)$.* Israel J. of Math., **16**:198–202, (1973).

[40] N. Kalton, E. Saab, and P. Saab. *$L^p(X)$, $1 \leq p < \infty$ has the property (u) whenever X does.* To appear in Bull. Sci. Math., Paris.

[41] N. Kalton, E. Saab, and P. Saab. *On the Dieudonné property for $C(\Omega, E)$.* Proc. Amer. Math. Soc., **96**:50–52, (1986).

[42] S.S. Khurana. *Grothendieck spaces.* Illinois J. of Math., **22**:79–80, (1978).

[43] S.V. Kisliakov. *Uncomplemented uniform algebras.* Mat. Zametki, **18**:91–96, (1975).

[44] H. Knaust and E. Odell. *On c_0 sequences in Banach spaces.* Israel J. of Math., **67**:153–169, (1989).

[45] S. Kwapien. *On Banach spaces containing c_0.* Studia Math., **52**:187–188, (1974).

[46] I. Labuda. *Private communication.* (1984).

[47] D. R. Lewis. *Conditional weak compactness in certain inductive tensor products.* Math. Ann., **201**:201–209, (1973).

[48] J. Lindenstraus and L. Tzafriri. *Classical Banach spaces.* Volume **92** of Modern Survey In Mathematics , Springer-Verlag, Berlin-Heidelberg-New York, first edition, (1977).

[49] F. Lust Picard. *Produits tensoriels projectifs d'espace de Banach faiblement sequentiellment complets.* Coll. Math., **36**:255–267, (1976).

[50] K. Musial. *The weak Radon-Nikodym property.* Studia Math., **64**:151–174, (1978).

[51] T. Odell and H.P. Rosenthal. *A double dual characterization of separable Banach spaces containing* ℓ_1. Israel J. of Math., **20**:375–384, (1975).

[52] A. Pełczynski. *Banach spaces of analytic functions and absolutely summing operators.* Volume **30** of CBMS conference series in mathematics, Amer. Math. Soc., (1977).

[53] A. Pełczynski. *On Banach spaces containing* $L^1(\mu)$. Studia Math., **30**:231–246, (1968).

[54] A. Pełczynski. *On Banach spaces on which every unconditionally converging operator is weakly compact.* Bull. Acad. Polon. Sci., **10**:641–648, (1962).

[55] G. Pisier. *Factorization of linear operators and geometry of Banach spaces.* Volume **60** of CBMS conference series in mathematics, Amer. Math. Soc., first edition, (1986).

[56] G. Pisier. *Une propriéte de satbilité de la classe des espaces ne contenant pas* ℓ_1. C. R. Acad. Sci. Paris Sér A, **86**:747–749, (1978).

[57] H.P. Rosenthal. *A characterization of Banach spaces containing* ℓ_1. Proc. Nat. Acad. Sci. USA, **71**:2411–2413, (1974).

[58] H.P. Rosenthal. *Pointwise compact subsets of the first Baire class.* Amer. J. of Math., **99**:362–378, (1977).

[59] E. Saab. *Some characterizations of weak Radon-Nikodym sets.* Proc. Amer. Math. Soc., **86**:307–311, (1982).

[60] E. Saab and P. Saab. *A dual geometric characterization of Banach spaces not containing* ℓ_1. Pacific J. of Math., **105**:415–425, (1983).

[61] E. Saab and P. Saab. *A stability property of a class of Banach spaces not containing a complemented copy of ℓ_1*. Proc. Amer. Math. Soc., **84**:44–46, (1982).

[62] E. Saab and P. Saab. *Extensions of some classes of Operators and applications.* Submitted.

[63] E. Saab and P. Saab. *On complemented copies of c_0 in the inductive tensor products.* Contemp. Math, **52**:131–135, (1986).

[64] E. Saab and P. Saab. *On Pełczynski's properties (V) and* (V^*). Pacific Journal of Mathematics, **125**:205–210, (1986).

[65] E. Saab and P. Saab. *On unconditionally converging and weakly precompact operators.* To appear in the Illinois J. of Math.

[66] P. Saab and B. Smith. *Spaces on which unconditionally converging operators are weakly completely continuous.* To appear in Rocky Mountain J. of Math.

[67] A Sobczyk. *Projection of the space m and c_0*. Bull. Amer. Math. Soc., **47**:938–947, (1941).

[68] M. Talagrand. *A new type of affine Borel function.* Math. Scand., **54**:183–188, (1984).

[69] M. Talagrand. *La Propriété de Dunford-Pettis dans $C(K, E)$ et $L_1(E)$*. Israel J. of Math., **44**:317–321, (1983).

[70] M. Talagrand. *Pettis integral and measure theory.* Mem. Amer. Math. Soc., **51**:307, (1984).

[71] M. Talagrand. *Weak Cauchy sequences in $L^1(E)$*. Amer. J. Math., **106**:703–724, (1984).

[72] A. Tulecea and C. Ionescu. *Topics in the theory of lifting.* Volume **48** of Ergebnisse der Mathematik und Ihrer Grenzgebiete, Springer-Verlag, Berlin and New York, first edition, (1969).

[73] L. Tzafriri. *Reflexivity in Banach lattices and their subspaces.* J. Funct. Anal., **10**:1–18, (1972).

[74] M. Zippin. *The separable extension problem.* Israel J. of Math., **26**:312–387, (1977).

Aspects of Ultraseparability in Banach Function Spaces

S.J. SIDNEY Department of Mathematics, The University of Connecticut, Storrs, Connecticut 06269

1. **Preliminaries.** By a *Banach function space* on a compact Hausdorff space X we mean a linear subspace E of $C(X)$ which contains the constant functions and is a Banach space in a norm N which dominates the uniform norm -- that is, $N(f) \geq \|f\|_X \equiv \sup\{|f(x)|: x \in X\}$ -- and is normalized so that $N(1) = 1$. Here $C(X)$ denotes the Banach algebra of continuous scalar-valued functions on X with pointwise algebraic operations and the supremum norm; the scalars may be the real numbers or the complex numbers, and when the choice is important, devices such as subscripts $\mathbb{C}$ or $\mathbb{R}$ will be used to specify it.

$\ell^\infty(\mathbb{N},E)$ denotes the Banach space of sequences $\langle f_n\rangle_{n\in\mathbb{N}}$ from E which are N-bounded, in the norm $\tilde{N}(\langle f_n\rangle) = \sup\{N(f_n): n \in \mathbb{N}\}$. Each such sequence may be regarded as a bounded continuous function on $\mathbb{N} \times X$, namely, $(n, x) \to f_n(x)$; as such it has an extension to a continuous function $\tilde{f}$ on $\tilde{X} \equiv \beta(\mathbb{N} \times X)$, the Stone-Cech compactification of $\mathbb{N} \times X$. We identify $\langle f_n\rangle$ with $\tilde{f}$, and the set of all such $\tilde{f}$ is a Banach function space $\tilde{E}$ on $\tilde{X}$ in the norm $\tilde{N}(\tilde{f}) = \tilde{N}(\langle f_n\rangle)$; evidently $C(X)^\sim = C(\tilde{X})$. $\tilde{E}$ is a Banach algebra with pointwise multiplication if and only if E is. E is said to be *ultraseparating* (on X) if $\tilde{E}$ separates the points of $\tilde{X}$.

This property was introduced by A. Bernard ([**Be**]; see also [**Bu**]) as a tool in the study of the symbolic calculus of $E = \mathrm{Re}A$, where A is a uniform algebra on X. This study has been continued by the author [**S**] and O.Hatori ([H1], [H2], [H4], [H5], [H6]). At this point the property of being ultraseparating is of interest in its own right. In particular, the following two broad problems suggest themselves:

Problem 1. Characterize ultraseparability. In particular, find ways to decide whether E is ultraseparating without passing from X to $\tilde{X}$.

Problem 2. Investigate the hereditary properties of ultraseparability.

Problem 1 has a long pedigree. Bernard showed [**Be**] that if A is a *Dirichlet algebra* on X -- that is, a uniform algebra such that ReA is uniformly dense in $C_R(X)$ -- then A and ReA are ultraseparating on X. Then Batikyan and Gorin [**B-G**] gave a characterization in the "Dirichlet spirit" of when a Banach function algebra is ultraseparating. In Hatori's papers ([**H3**], [**H4**], [**H6**]) variations of this characterization are extended to more general (sometimes real) Banach function spaces. In section 2 we shall expose some of these characterizations.

Except for essentially trivial observations, little seems to have been done about hereditary properties. Section 3, while recording some of these observations, is mainly devoted to discussing whether ultraseparability is inherited by tensor products. This area is in a very early stage of development, and may prove to be quite interesting.

2. **Characterizations.** Probably the main characterization result is the following, essentially due to Hatori ([**H6**], prop. 1):

THEOREM 1. Let E be a Banach function space on X, and fix a number r, $0 < r < 1$. The following three assertions about E are equivalent:

(1.1) E is ultraseparating on X.

(1.2) There is a positive integer J such that whenever $u \in \text{ball}C_R(X)$, there are functions $g_1,\dots g_J$, $h_1,\dots,h_J$ in ball(E) such that
$$\|u - \sum_{j=1}^{J} (|g_j| - |h_j|)\|_X \leq r.$$

(1.3) There is a positive integer J such that whenever K and L are disjoint closed subsets of X, there are functions $g_1,\dots,$

g_J, $h_1,\ldots, h_J$ in ball(E) such that

$$\sum_{j=1}^{J} (|g_j| - |h_j|) \geq r \text{ on } K, \text{ and}$$

$$\sum_{j=1}^{J} (|g_j| - |h_j|) \leq -r \text{ on } L.$$

The implication (1.2) ⇒ (1.3) is trivial: given r in (1.3), take the J that works for $1-r$ in (1.2). The implication (1.3) ⇒ (1.1) is obtained by taking disjoint closed neighborhoods K, L in $\tilde{X}$ of distinct points φ, ψ of $\tilde{X}$, producing for each $n \in \mathbb{N}$ disjoint closed subsets K_n, L_n of X by the rule $K \cap (\{n\} \times X) = \{n\} \times K_n$, $L \cap (\{n\} \times X) = \{n\} \times L_n$, and using (1.3) to produce $g_1^{(n)},\ldots,g_J^{(n)}, h_1^{(n)},\ldots,h_J^{(n)}$ in ball(E) satisfying the appropriate inequalities. Then $\tilde{g}_j = < g_j^{(n)} >$ and $\tilde{h}_j = < h_j^{(n)} >$ in $\tilde{E}$ satisfy $\sum_{j=1}^{J} (|\tilde{g}_j| - |\tilde{h}_j|)(\varphi) \geq r$, $\sum_{j=1}^{J} (|\tilde{g}_j| - |\tilde{h}_j|)(\psi) \leq -r$.

The implication (1.1) ⇒ (1.2) is more difficult. The crucial observation here is

LEMMA 1.4. Let F be a Banach function space on Y, and suppose F separates the points of Y. Then the real linear span of $|F| \equiv \{|f|: f \in F\}$ is uniformly dense in $C_R(Y)$.

Assuming that (1.4) has been proven, we can prove (1.1) ⇒ (1.2). Indeed, suppose (1.1) is true and (1.2) is false. For each $n \in \mathbb{N}$ there is $u_n \in$ ball $C_R(X)$ such that whenever $g_1,\ldots,g_n, h_1,\ldots,h_n$ belong to ball(E) it follows that

$$(1.5) \quad \|u_n - \sum_{j=1}^{n} (|g_j| - |h_j|)\|_X > r.$$

$\tilde{u} = < u_n > \in C_R(\tilde{X})$, so by (1.4) applied to $Y = \tilde{X}$ and $F = \tilde{E}$ there is a function $\tilde{v}$ in the real linear span of $|\tilde{E}|$ such that $\|\tilde{u} - \tilde{v}\|_X \leq r$. It is easy to write $\tilde{v}$ in the form

$$\tilde{v} = \sum_{j=1}^{J} (|\tilde{g}_j| - |\tilde{h}_j|)$$

where $\tilde{g}_j$ and $\tilde{h}_j$ belong to ball($\tilde{E}$). Thus $\tilde{g}_j = < g_j^{(n)} >_{n \in \mathbb{N}}$ and $\tilde{h}_j = < h_j^{(n)} >_{n \in \mathbb{N}}$ where $g_j^{(n)}$ and $h_j^{(n)}$ belong to ball(E) and $\|u_n - \sum_{j=1}^{J} (|g_j^{(n)}| - |h_j^{(n)}|)\|_X \leq r$. For $n = J$ this contradicts (1.5).

It remains to prove (1.4). In case F is closed under multiplication (which corresponds to E being closed under multiplication in the theorem), (1.4) follows immediately from the Stone-Weierstrass Theorem. In the general case we need

LEMMA 1.6. The real linear span of the functions $t \to |t - \lambda|$ (λ a scalar) is dense in the space of all continuous real-valued function on the scalar field, where the topology is that of uniform convergence on compact sets.

This implies (1.4) as follows: If H is the closed real linear span of $|F|$, by (1.6) H contains all functions u^k where $u \in W \equiv \mathrm{Re}F$ (=F if the scalars are real) and $k \in \mathbb{N}$. It now follows by an algebraic argument that H contains all functions of the form $u_1^{k_1} \ldots u_m^{k_m}$ with all $u_j \in W$ and $k_j \in \mathbb{N}$, and so $H = C_R(Y)$ by the Stone-Weierstrass Theorem.

If the scalars are real, (1.6) is easy to prove, since the span of the $|t - \lambda|$ consists precisely of the piecewise linear functions whose graphs have finitely many corners and whose initial and terminal slopes sum to zero.

The complex case of (1.6) is more difficult; we outline a proof. Let M denote the closed $\mathbb{C}$ - linear span of the functions $t \to |t - \lambda|$ ($\lambda \in \mathbb{C}$) in $C_{\mathbb{C}}(\mathbb{C})$. $1 \in M$, and if $u \in M$ and $\alpha, \beta \in \mathbb{C}$ then $t \to u(\alpha t + \beta) \in M$.

Manipulating binomial series, if $\lambda > 0$ then M contains

$t \to \lambda^{-1}|t - \lambda| = (|1 - \frac{t}{\lambda}|^2)^{\frac{1}{2}} = \sum_{m=0}^{\infty} \lambda^{-m} P_m(t)$ where the summation is valid for (at least) $|t| < (\sqrt{2} - 1)\lambda$.

Here $P_m(t) = \sum_{k=0}^{m} a_{m,k} t^{m-k} \bar{t}^k$ where, if $m \geq 3$,

$$a_{m,k} = \sum_{j=\max\{k,m-k\}}^{m} (-1)^{j-m-1} \frac{(2j-3)(2j-5)\ldots(3)(1)}{(j-k)!(j-m+k)!(m-j)!2^j}$$

Letting λ tend to infinity and using easy manipulations, we find that $P_m(t) \in M$ for $m = 0,1,2,\ldots$.

It turns out that $a_{m,k} \neq 0$. For $m \leq 2$ this is checked by direct computation, while if $m \geq 3$ it holds because the denominator of the summand in $a_{m,k}$ corresponding to $j = m$ is divisible by a higher power of 2 than is any other denominator.

For given m choose complex numbers $\zeta_0,...,\zeta_m$ of modulus 1 such that $\zeta_0^2,...,\zeta_m^2$ are distinct. Then for $\ell = 0,...m$, M contains $t \to P_m(\zeta_\ell t) = \sum_{k=0}^{m} \zeta_\ell^{m-2k}\, a_{m,k}\, t^{m-k}\, \bar{t}^k$. The matrix $(\zeta_\ell^{m-k})_{k,\ell}$ is nonsingular, so finally $t \to t^{m-k}\, \bar{t}^k$ belongs to M for $k = 0,...,m$. Since M contains all monomials $t^j\bar{t}^k$, $M = C_{\mathbb{C}}(\mathbb{C})$; taking real parts gives (1.6). Thus the theorem is proved.

Substantially the same argument allows us (in both lemmas and the theorem) to replace $|\cdot|$ by $|\cdot|^{\alpha}$ where α is a positive *rational* number which is not an even integer. With further modifications the results can be obtained with α a positive *irrational* number; in this case some $a_{m,k}$ may be zero, but enough will not be zero to guarantee that "sufficiently many" $t^j\bar{t}^k$ belong to M.

Hatori, who proved the theorem in the real case, was aware that it holds in the complex case, and offered a more concise proof of (1.6) with $|\cdot|^{\alpha}$, where α is positive and not an even integer. Let Δ_t denote the usual Laplace operator with respect to t in $\mathbb{C} = \mathbb{R}^2$. Then if $c \in \mathbb{C}\setminus\{0\}$ and $w \in \mathbb{C}$, for $|t| < |c| / |w|$ we have

$$\Delta_t (|c + tw|^{\alpha}) = \alpha^2 |w|^2\, |c + tw|^{\alpha-2}.$$

Approximation by difference quotients shows that, for each $k \in \mathbb{N}$, each k-tuple of functions $f_1,...,f_k$ from E, and all small enough complex numbers $t_1,...,t_k$,

$$\Delta_{t_1}...\Delta_{t_k} (|c + t_1f_1 + ... + t_kf_k|^{\alpha}) =$$
$$= \alpha^2(\alpha - 2)^2...(\alpha - 2k + 2)^2\, |f_1|^2...|f_k|^2\, |c + t_1f_1 + ...+ t_kf_k|^{\alpha-2k}$$

belongs to M provided $c \neq 0$; for $t_1 =...= t_k = 0$ this gives $|f_1|^2...|f_k|^2 \in M$, showing that M contains a real point-separating subalgebra of $C_R(X)$. This argument applies not only to $|\cdot|^{\alpha}$ but in fact to any C^{∞} function ψ defined on an open subset U of $\mathbb{C}$, provided ψ is not "preharmonic:" There is no $k \in \mathbb{N}$ such that $\Delta^k\psi \equiv 0$ in U. This leads to other extensions of (1.2), and a (much simpler) version with real scalars already appears in [H3].

We close this section with a condition easily shown to be equivalent to (1.2) holding for all r, $0 < r < 1$:

(1.7) There is a summable series $\sum_{j=1}^{\infty} c_j$ of positive numbers c_j such that whenever $u \in \text{ball } C_R(X)$, there are functions g_j and h_j in ball(E) which satisfy

$$\sum_{j=1}^{\infty} c_j(|g_j| - |h_j|) = u.$$

3. **Tensor products.** We list a few constructions under which ultraseparability is preserved. Proofs are straightforward, and will not be given.

* If E and F are Banach function spaces on X and $E \subset F$, by the closed graph theorem the inclusion mapping of E into F is continuous. Thus $\tilde{E} \subset \tilde{F}$, so if E is ultraseparating on X, then so is F.

* If E is a Banach function space on X and K is a nonempty closed subset of X, then $E|_K$ with the norm

$$N_{E|_K}(g) \equiv \inf\{N_E(f): f \in E, f|_K = g\}$$

is a Banach function space on X. $\tilde{K}$ is naturally contained in $\tilde{X}$ as a closed subset. Then $(E|_K)^{\sim} = \tilde{E}|_{\tilde{K}}$ (isometrically!), so if E is ultraseparating on X then $E|_K$ is ultraseparating on K.

* Conversely, if E is point-separating Banach function space on $X = K_1 \cup ... \cup K_n$ where the K_j are closed subsets of X such that each $E|_{K_j}$ is ultraseparating on K_j, then E is ultraseparating on X.

* If E is a complex Banach function space on X then ReE with the norm

$$N_{ReE}(u) \equiv \inf\{N_E(f): f \in E, \text{Re}f = u\}$$

is a real Banach function space on X. Then $(ReE)^{\sim} = Re(\tilde{E})$ (isometrically!), so ReE is ultraseparating on X if and only if E is.

We shall be concerned in the rest of this paper with the question of whether the tensor product of E and F is ultraseparating on $X \times Y$ if E is ultraseparating on X and F is ultraseparating on Y. To even frame the question properly, we must first decide *which* tensor product we mean.

Recall that the *algebraic* tensor product $E \otimes F$ is the linear span of the "simple tensors" $f \otimes g$ ($f \in E$, $g \in F$), where

$$(f \otimes g)(x,y) = f(x)g(y),\ (x,y) \in X \times Y.$$

The *projective* tensor product $E \hat{\otimes} F$ is the completion of $E \otimes F$ in the norm

$$N_{E \hat{\otimes} F}(h) = \inf\{ \sum_{j=1}^{n} N_E(f_j)N_F(g_j)\}$$

where the infimum is taken over all finite families of simple tensors $f_j \otimes g_j$ ($f_j \in E$, $g_j \in F$) that sum to h. $E \hat{\otimes} F$ is a Banach function space on $X \times Y$. It has an alternative description as the set of all functions h on $X \times Y$ which have a (necessarily uniformly convergent) series representation $h = \sum_{j=1}^{\infty} f_j \otimes g_j$ with $\sum_{j=1}^{\infty} N_E(f_j)N_F(g_j) < \infty$; the norm is the infimum of these sums over all such series representations of h.

It turns out that $E \hat{\otimes} F$ is essentially never ultraseparating! Precisely:

PROPOSITION 2. If X and Y are infinite compact Hansdorff spaces, and if E and F are Banach function spaces on X and Y respectively, then $E \hat{\otimes} F$ *cannot* be ultraseparating on $X \times Y$.

To see this, first recall (see ideas in [G2], Thm. 2.4) that $C(X) \hat{\otimes} C(Y) \subsetneq C(X \times Y)$. As was pointed out to me by S. Hwang, it follows that $C(X) \hat{\otimes} C(Y)$, being self-adjoint, *cannot* be ultraseparating ([**Bu**], Lemma 4.12). But $E \hat{\otimes} F \subset C(X) \hat{\otimes} C(Y)$, so $E \hat{\otimes} F$ in turn cannot be ultraseparating.

In view of the proposition, we turn our attention to the *spatial* tensor product $E \bar{\otimes} F$. This is the uniform closure of $E \otimes F$ in $C(X \times Y)$, and the norm is the uniform norm. $E \bar{\otimes} F$ does not change if E and F are replaced by their uniform closures, and if E is ultraseparating on X then so is its uniform closure. The little we know in this situation is summarized in

THEOREM 3. Let E and F be Banach function spaces on X and Y respectively. Then $E \bar{\otimes} F$ is ultraseparating on $X \times Y$ in each of the following cases:

(3.1) E is a Dirichlet algebra on X and $F = C(Y)$.

(3.2) E is ultraseparating on X, $F = C(Y)$, and Y is totally disconnected.

(3.3) $E \bar{\otimes} C(Y)$ is ultraseparating on $X \times Y$, and F is closed under multiplication and contains enough unimodular functions to separate the points of Y.

Thus we do not know, in general, the answer in such cases as logmodular $\bar{\otimes}$ C(Y) and Dirichlet $\bar{\otimes}$ Dirichlet.

We now indicate the proofs of (3.1) - (3.3).

In case (3.1), the standard DeBranges - Glicksberg argument [G1] shows that $E \bar{\otimes} F$ is Dirichlet on $X \times Y$: Any extreme point of the closed unit ball of the space of real regular Borel measures on $X \times Y$ that annihilate $E \bar{\otimes} F$ is supported on a single slice $X \times \{y_0\}$, so must reduce to 0.

In case (3.2), let $J \in \mathbb{N}$ work for $r = 1/4$ and E in characterization (1.2) of Theorem 1. Let $u \in \text{ball } C_R(X \times Y)$ be given. Cover Y by disjoint open-and-closed sets $Y_1,\ldots,Y_n$ such that $|u(x,y) - u(x,y')| < 1/4$ whenever $x \in X$ and y and y′ belong to the same Y_k. Select $y_k \in Y_k$ for $k = 1,\ldots,n$. Take $g_1^{(k)},\ldots,g_J^{(k)},h_1^{(k)},\ldots,h_J^{(k)}$ in ball(E) such that, for all $x \in X$ and $k = 1,\ldots, n$,

$$\left| \sum_{j=1}^{J} \left(|g_j^{(k)}(x)| - |h_j^{(k)}(x)|\right) - u(x, y_k)\right| \le 1/4.$$

Let $g_j = \sum_{k=1}^{n} g_j^{(k)} \otimes \chi_{Y_k}$ and $h_j = \sum_{k=1}^{n} h_j^{(k)} \otimes \chi_{Y_k}$. Then g_j and h_j belong to ball($E \bar{\otimes} F$), and $\left|\sum_{j=1}^{J} (|g_j| - |h_j|) - u\right| < 3/4$ on $X \times Y$. Thus $E \bar{\otimes} F$ satisfies (1.2) for $r = 3/4$.

In case (3.3), let $J \in \mathbb{N}$ work for $E \bar{\otimes} C(Y)$ and some r in characterization (1.2). Given $u \in \text{ball } C_R(X \times Y)$, take $g_1,\ldots,g_J,h_1,\ldots h_J$ in ball($E \bar{\otimes} C(Y)$) such that

$|\sum_{j=1}^{J}(|g_j| - |h_j|) - u| < r$ on $X \times Y$. Without loss of generality, each g_j and h_j is a finite sum of simple tensors $\varphi \otimes \psi$; further, the Stone-Weierstrass theorem shows that we may suppose each ψ has the form η/τ where η and τ belong to F and τ is unimodular. Multiply each g_j and h_j by the product of all these denominators τ to get g_j' and h_j' in ball($E \bar{\otimes} F$) such that $|g_j'| = |g_j|$ and $|h_j'| = |h_j|$.

References

B-G. B.T Batikyan and E.A. Gorin, *Ultraseparating algebras of continuous functions*. Vestnik Moskov. Univ. Ser. I, Mat. Meh. **31** (1976), 15-20; English transl. Moscow Univ. Math. Bull. 31 (1976), 71-75.

Be. A. Bernard, *Espaces des parties réelles des éléments d'une algèbre de Banach de fonctions*. J. Funct. Anal. **10** (1972), 387-409.

Bu. R. Burckel, *Characterizations of C(X) Among Its Subalgebras*. Marcel Dekker, New York, 1972.

G1. I. Glicksberg, *Measures orthogonal to algebras and sets of antisymmetry*. Trans. Amer. Math. Soc. 105 (1962), 415-435.

G2. __________, *Recent Results on Function Algebras*. CBMS Regional Conference Series in Mathematics, no. 11. American Mathematical Society, Providence, 1971.

H1. O. Hatori, *Functions which operate on the real part of a function algebra*. Proc. Amer. Math. Soc. 83 (1981), 565-568.

H2. __________, *Functions which operate by composition on the real part of a Banach function algebra*. Tokyo J. Math. 6 (1983), 423-429.

H3. __________, *A remark on a theorem of B.T. Batikyan and E.A. Gorin*. Tokyo J. Math. 7 (1984), 157-160.

H4. __________, *Functional calculus for certain Banach function algebras*. J. Math. Soc. Japan **38** (1986), 103-112.

H5. __________, *Range transformations on a Banach function algebra*. Trans. Amer. Math. Soc. 297 (1986), 629-643.

H6. __________, *Range transformations on a Banach function algebra II*. Pacific J. Math. **138** (1989), 89-118.

S. S.J. Sidney, *Functions which operate on the real part of a uniform algebra*. Pacific J. Math. **80** (1979), 265-272.

Multi-Tuple Boundaries of Shilov Type for Function Spaces

TOMA V. TONEV Mathematics Department, The University of Toledo, Toledo, Ohio.

1. INTRODUCTION.

The most utilised boundaries of function spaces probably are their Shilov boundaries. By definition the *Shilov boundary* ∂B of a space of functions B over a compact Hausdorff space X is the smallest B-maximizing subset of X. Recall that a closed subset E of X is B-*maximizing* (or *boundary* in an other terminology) if

$$\max_{x \in X} |f(x)| = \max_{x \in E} |f(x)|$$

for every function f in B. The importance of Shilov boundary stems from the fact that it plays a basic rôle in various maximum modulus principles and integral representation formulas for the functions in B.

Basener [1] and Sibony [5] have introduced the *multi-tuple Shilov boundaries* $\partial^{(n)} A$ of a function algebra A, which have proved to be very essential in the investigation of multi-dimensional analytic structures in algebra spectra (e.g. Basener [1] , Kramm [4] , Kumagai [5] , Sibony [6], Tonev [9]). As shown in [7], $\partial^{(n)} A$ is the smallest A_*^n-minimizing subset of spA. Here A_*^n stands for the set of all *regular* (i.e. without common zeros) n-tuples $F = (f_1, \dots, f_n)$ of functions in A and a closed set $E \subset X$ is called A_*^n-*minimizing* (or *n-tuple boundary* in an other terminology) if

$$\min_{x\in X} \|F(x)\| = \min_{x\in X} \|F(x)\|,$$

for any $F \in A_*^n$, where $\|\mathbf{z}\|$ is any norm in $\mathbf{C}^n$.

In this paper we introduce B_*^n-minimizing sets and n-tuple Shilov boundaries for linear function spaces B and investigate some of their properties.

2. Multi-tuple Shilov boundaries.

Recall that a *function space* is called any linear subspace of the space $C(X)$ (over $\mathbf{R}^1$ or $\mathbf{C}^1$), where X is a compact Hausdorff space, which is closed under the *uniform norm* $\|f(x)\| = \max_{x\in X} |f(x)|$, contains the constants and separates the points of X. In what follows, if not specified, $\mathbf{K}^1$ will denote either $\mathbf{R}^1$ or $\mathbf{C}^1$ and B_* will denote the set of all nonvanishing on X functions in B.

Definition 1. *A closed subset E of X is said to be B_*-minimizing if*

$$\min_{x\in X} |f(x)| = \min_{x\in E} |f(x)|$$

for every nonvanishing on X function f in B.

It is easy the check that B_*-minimizing sets are B-maximizing. Moreover the classes of both sets coinside in the case when B is a function algebra (over $\mathbf{C}^1$) and also if B is a function space over R^1. But in general B-maximizing sets are not B_*-minimizing. For instance let $B = \{\lambda f + \mu : \lambda, \mu \in \mathbf{R}^1\}$, where $f \in C[-1/2, 3\pi/2]$ is the function $f(x) = x + 1$ for $-1/2 \le x \le 0$ and $f(x) = e^{ix}$ for $0 \le x \le \pi$. It is not hard to see that $[0, \pi]$ is a B-maximizing set which is not B_*-minimizing. In fact every B_*-minimizing set contains the point $x = -1/2$, which is not obligatory for B-maximizing sets. These observations motivate the following

Lemma 1. *Let X be a compact Hausdorff space and let B be a function space over X. Then there exists a B_*-minimizing set for B which is contained in every B_*-minimizing set.*

PROOF: We use Bear's idea for proving the existence of smallest B-maximizing sets for function spaces B from [**2**]. We show first that there exists a smallest B_*-minimizing set. Consider the set of all B_*-minimizing sets ordered under inclusion. If $\mathcal{E} = \{E_\alpha\}_{\alpha\in I}$ is a linearly ordered system of B_*-minimizing sets and if $F = \bigcap_{\alpha\in I}\{E_\alpha\}$, then F is a closed and nonempty subset of X. For a fixed $f \in B$ the compact set $M(f) = \{x \in X : f(x) = \min_{x\in X}|f(x)|\}$ meets each set E_α in $\mathcal{E}$, so it meets any finite number of sets in $\mathcal{E}$. By the finite intersection property $M(f)$ meets also F and therefore F is a B_*-minimizing set. According to Zorn's lemma there exists a maximal element $\mathcal{S}$ (with respect to the inclusion) of the set of all B_*-minimizing sets and $\mathcal{S}$ clearly is a B_*-minimizing set as well.

Fix an arbitrary B_*-minimizing set $E \subset X$. Suppose that $\mathcal{S} \setminus E \neq \emptyset$ and let $x_o \in \mathcal{S} \setminus E$. Let

$$V = \{x \in X : |g_j(x_o)| < \varepsilon,\ j = 1, \ldots, n,\ \varepsilon > 0\}$$

be a basis neighborhood in X such that $x_o \in V \subset X \setminus E$. Since $\mathcal{S}$ is the maximal B_*-minimizing set for B, $\mathcal{S} \setminus V$ is not a B_*-minimizing set. Hence there exists an $f \in B$ which does not vanish on X and such that

$$\min_{x\in\mathcal{S}\setminus V} |f(x)| > \min_{x\in X} |f(x)| > 0.$$

Let $g = kf$ for some k big enough so that

$$\min_{x\in\mathcal{S}\setminus V} |g(x)| - \big(\max_{x\in\mathcal{S}\setminus V} |g_1(x)| + \cdots + \max_{x\in\mathcal{S}\setminus V} |g_n(x)|\big) > \min_{x\in X} |g(x)|.$$

Clearly $\min_{x\in X}|g(x)| > 0$ because together with f the function $g = kf$ does not vanish on X. We can choose even bigger k in order to assure $\min_{x\in X}|g(x)| > \varepsilon$ for ε from above. Let α be a complex number with $|\alpha| = 1$ and let j be an integer, $1 \le j \le n$. For every $x \in V$ we have

$$\begin{aligned}|g(x) + \alpha g_j(x)| &\ge |g(x)| - |\alpha g_j(x)| \\ &> \min_{x\in X} |g(x)| - \max_{x\in V} |g_j(x)| = \min_{x\in X} |g(x)| - \varepsilon.\end{aligned}$$

If $x \in \mathcal{S} \setminus V$ then

$$\begin{aligned}|g(x) + \alpha g_j(x)| &\ge \min_{x\in\mathcal{S}\setminus V} |g(x) + \alpha g_j(x)| \\ &\ge \min_{x\in\mathcal{S}\setminus V} |g(x)| - \max_{x\in\mathcal{S}\setminus V} |g_j(x)| \\ &\ge \min_{x\in\mathcal{S}\setminus V} |g(x)| - \big(\max_{x\in\mathcal{S}\setminus V} |g_1(x)| + \cdots + \max_{x\in\mathcal{S}\setminus V} |g_n(x)|\big) \\ &> \min_{x\in X} |g(x)| > \min_{x\in X} |g(x)| - \varepsilon.\end{aligned}$$

Hence $|g(x)+\alpha g_j(x)| > \min_{x\in X}|g(x)|-\varepsilon$ for every $x\in X$ and for each $j=1,\dots,n$. Let now x_1 be some point in X such that

$$|g(x_1)| = \min_{x\in X}|g(x)|.$$

For every $j=1,\dots,n$ we can choose complex numbers α_j, $|\alpha_j|=1$ such that

$$|g(x_1)+\alpha_j g_j(x_1)| = |g(x_1)|-|g_j(x_1)| = \min_{x\in X}|g(x)|-|g_j(x_1)|.$$

Consequently

$$\min_{x\in X}|g(x)|-|g_j(x_1)| = |g(x_1)+\alpha_j g_j(x_1)| > \min_{x\in X}|g(x)|-\varepsilon,$$

wherefrom $|g_j(x_1)|<\varepsilon$ for any $j=1,\dots,n$. We conclude that $g(x)$ does not assume the minimum of its modulus outside $V\subset X\setminus E$ in contradiction with the choice of E. The lemma is proved.

The smallest B_*-minimizing set of B will be denoted by $Sh_1(B)$. Below we define multi-tuple Shilov boundaries for function spaces B. The definition is quite similar to that of the multi-tuple Shilov boundaries for a function algebra, in the form given in [**1**], but instead of function algebra elements we use elements of a function space B.

DEFINITION 2. *The n-tuple Shilov boundary of a function space B we call the set*

$$(1)\qquad Sh_n(B) = \left[\bigcup Sh_1(B|_{V(F)}) : F=(f_1,\dots,f_{n-1})\in B^{n-1}\right]$$

where $V(F)=(f_1,\dots,f_{n-1})^{-1}(\mathbf{0})$.

Next theorem establishes some global characterizations of the n-tuple Shilov boundaries $Sh_n(B)$. Let $\|\ .\ \|$ be an arbitrary norm in $\mathbf{K}^n$ which is equivalent to the Euclidean norm.

THEOREM 1. *The n-tuple Shilov boundary $Sh_n(B)$ of the space B coincides with the smallest among all closed sets E in X such that:*

(1) $bF(X)\subset F(E)$ *for every n-tuple* $F\in B^n$;
(2) *The set* $F(X)\setminus F(E)$ *is open in* $\mathbf{K}^n$ *for any* $F\in B^n$.
(3) $\min_{x\in E}\|F(x)\|\le\min\{\|\mathbf{z}\|:\mathbf{z}\in bF(X)\}$ *for each* $F\in B^n$;
(4) *The open ball* $B_n(\mathbf{0},\min_{x\in E}\|F(x)\|)$ *is either entirely inside, or entirely outside* $F(X)$ *for each* $F\in B^n$;
(5) $\min_{x\in E}\|F(x)\|=\min_{x\in X}\|F(x)\|$ *for every regular* (i.e. *nonvanishing on* X) *n-tuple F over B*;
(6) *Every* $F\in B^n$ *with* $\mathbf{0}\in bF(X)$ *vanishes within* E;
(7) $\overline{B}_n(\mathbf{0},\min_{x\in E}\|F(x)\|)\subset F(X)$ *for every irregular* $F\in B^n$.

PROOF: Since the space B^n is invariant under orthogonal transforms in $\mathbf{K}^n$, the proofs of these results for the case of function algebras from [7] apply for function spaces as well. Here we give an other proof for the basic inclusion (1) which holds for more general spaces (see the remark below).

Let $F = (f_1, \ldots, f_n)$ be an n-tuple over B and suppose that (1) is false. Let $\mathbf{z}_o = F(x_o) \in bF(X) \setminus F(Sh_n(B))$ for some $x_o \in X$. We shall make use of a simple geometrical fact, namely that for any compact set K in $\mathbf{K}^n$

$$bK = \left[\bigcup \{b(K \cap (z_{i_1}, \ldots, z_{i_{n-1}})^{-1}(\mathbf{a})) : \mathbf{a} \in \mathbf{K}^n,\ i_s \neq i_t \text{ as } s \neq t\}\right],$$

where z_{i_k} are the i_k-th coordinate functions in $\mathbf{K}^n$. Consequently without loss of generality we can assume that $\mathbf{z}_o \in b(K \cap (z_2, \ldots, z_n)^{-1}(\mathbf{z}_o))$. Hence

$$\mathbf{z}_o = F(x_o) \in b(K \cap (z_2, \ldots, z_n)^{-1}(\mathbf{z}_o)) =$$

$$b\{F(F) \cap F \circ (f_2, \ldots, f_n)^{-1}(F(x_o))F(X)\} = b\{F \circ (f_2, \ldots, f_n)^{-1}(F(x_o))\}$$

$$= b\{f_1(V(f_2 - f_2(x_o), \ldots, f_n - f_n(x_o))), f_2(x_o), \ldots, f_n(x_o)\}.$$

We conclude that $(\mathbf{z}_o)_1 = f_1(x_o) \in bf_1(V)$ where $V = V(f_2 - f_2(x_o), \ldots, f_n - f_n(x_o))$.

On the other hand $\mathbf{z}_o \notin F(Sh_n(B))$ implies that $(\mathbf{z}_o)_1 \in bf_1(V) \setminus f_1(V \cap Sh_n(B)) \subset bf_1(V) \setminus f_1(Sh_1(B|_{X \cap V}))$. Without loss of generality we can assume (by adding some constant, if necessary) that $f_1 \neq 0$ on $X \cap V$ and the last inclusion shows that we can find a constant $c \in \mathbf{K}^1 \setminus f_1(X \cap V)$ such that the function $g = f_1 - c$ attains the minimum of its modulus out of $Sh_1(B|_{X \cap V})$, in contradiction with the defining property of $Sh_1(B|_{X \cap V})$. We conclude that (1) holds for $E = Sh_n(B)$.

Let now E be a closed subset of X for which (1) holds for each $F \in B^n$. Take an arbitrary $(n-1)$-tuple $G = (g_1, \ldots, g_{n-1})$ of elements in B and suppose that $bf(V(G)) \not\subset f(E \cap V(G))$ for some $f \in B$. There exists an $x_o \in X \cap G^{-1}(\mathbf{0})$ such that $f(x_o) \in b(f(V(G)) \setminus f(E \cap V(G))$. Since $bf(V(G)) \subset b(f, g_1, \ldots, g_{n-1})(X)$ we obtain that $(f(x_o), 0, \ldots, 0) \in b(f, g_1, \ldots, g_{n-1})(X) \setminus (f, g_1, \ldots, g_{n-1})(E)$ in contradiction with (1). Hence $bf(X \cap V(G)) \subset f(E \cap V(G))$ for each $f \in B$. Consequently for each $f \in B$ which does not vanish on $V(G)$ the function $F|_{V(G)}$ attains the minimum of its modulus within $E \cap V(G)$. We conclude that $E \cap V(G) \supset Sh_1(B|_{V(G)})$ for every $(n-1)$-tuple G of functions in B. Consequently

$$Sh_n(B) \supset \bigcup_{G \in B^{n-1}} (V(G) \cap Sh_n(B)) \supset \bigcup_{G \in B^{n-1}} Sh_1(B|_{V(G)})$$

and by taking the closures in both sides we get finally

$$Sh_n(B) \supset \Big[\bigcup_{G \in B^{n-1}} Sh_1(B|_{V(G)}) \Big].$$

From Definition 2 (or from Theorem 1) one can observe that

$$(2) \qquad Sh_1(B) \subset Sh_2(B) \subset \cdots \subset Sh_n(B) \subset \cdots \subset X.$$

Applying an analogical geometrical fact, namely that for every compact subset $K \subset \mathbf{K}^n$ and for every fixed k, $1 \le k \le n$,

$$bK = \Big[\bigcup \{ b(K \cap (z_{i_1}, \ldots, z_{i_k})^{-1}(\mathbf{a})) : \mathbf{a} \in \mathbf{K}^n,\ i_s \ne i_t \text{ as } s \ne t \} \Big],$$

we can obtain in a similar way the following

THEOREM 2. *Let n and k be two integers and let $1 \le k \le n$. Then*

$$(3) \qquad Sh_n(B) = \Big[\bigcup_{F \in B^{n-k}} Sh_k(B|_{V(F)}) \Big].$$

Observe that (3) is trivial for $k = n$ and it restricts to (1) if $k = 1$.

EXAMPLE 1. *Function algebras.*

The algebra $\widehat{A}$ of Gelfand extensions of functions in a function algebra A is a function space over algebra spectrum spA. Now the classes of A-maximizing and A_*-minimizing sets coincide and therefore $Sh_1(B|_K)$ coincides with the Shilov boundary ∂A_K of restricted algebra on $K \subset spA$. $Sh_n(\widehat{A})$ coincides with the boundary $\partial^{(n)} A$ introduced by Basener [1] and Sibony [6] .

EXAMPLE 2. *Minimal n-tuple affine boundaries.*

Let M be a compact connected set in a locally convex linear topological space V and denote by $A(M)$ the set of restrictions on M of all real affine continuous functions in V, i.e. $f \in A(M)$ iff f is continuous and $f(tx + (1-t)y) = tf(x) + (1-t)f(y)$ for every $t \in \mathbf{R}^1$. Clearly $A(M)$ is a function space over M. Remind that a subset N of M is called an *end subset* of M iff it consists of points z which satisfy the following condition: z can not be represented as $z = \lambda x + \mu y$ with $\lambda > 0, \mu > 0, \lambda + \mu = 1$, unless x and y belong to N. The *extreme points* of M are the points which are end subsets of M. Let $E(M)$ stand for the closure of extreme points of M. It is well known that $E(M)$ is the smallest closed subset of M on which every positive affine function in $A(M)$ attains its maximum, i.e. $E(M)$ is the Shilov boundary of $A(M)$. Now again the classes of $A(M)$-maximizing and $A(M)_*$-minimizing sets coincide and $Sh_1(A(K)) = E(K)$ for every $K \subset M$. If M is convex, then $Sh_n(A(M))$ coincides with the *n-tuple minimal affine boundary* $E_n(M)$ of M from [**9**], which coinsides with the closure of all end subsets in M, which are contained in $(n-1)$-dimensional affine subspaces of V.

EXAMPLE 3.

As it follows from Example 2 the smallest among all closed subsets E of a compact convex subset M of a locally convex space V such that the equality

$$\min_E |f(x)| = \min_M |f(x)|$$

holds for every *complex valued affine functional* $f \in A^{\mathbf{C}}(M)$ which does not vanish on M is neither $E(M)$ nor M, but $E_2(M)$. More generally, the smallest among the closed subsets E of M such that

$$\min_E \Big(\sum_{j=1}^n |f_j(x)|^2\Big)^{\frac{1}{2}} = \min_K \Big(\sum_{j=1}^n |f_j(x)|^2\Big)^{\frac{1}{2}}$$

for every non-vanishing on M n-tuple $(f_1, \dots, f_n)$ of complex affine functionals $f_j \in A^{\mathbf{C}}(M)$ is the set $E_{2n}(M)$. Indeed, by Definition 2 we have:

$$\begin{aligned} Sh_n(A^{\mathbf{C}}(M)) &= \Big[\bigcup Sh_1(B|_{Z(g_1,\dots,g_{n-1})})\Big] \\ &= \Big[\bigcup E_2(M \cap Z(g_1, \dots, g_{n-1}))\Big] \\ &= \Big[\bigcup E_2(M \cap Z(\operatorname{Re} g_1, \dots, \operatorname{Re} g_{n-1}, \operatorname{Im} g_1, \dots, \operatorname{Im} g_{n-1}))\Big] \\ &= E_{2n}(M). \end{aligned}$$

EXAMPLE 4.

Let A be a function algebra. Then spA is a weak *-compact subset of A^*, the dual of A. Let A' be the natural image of A into A^{**} (i.e. $A \to A' \subset A^{**} : a \longmapsto \Phi_a : \Phi_a(\varphi) = \varphi(a)$ for every $\varphi \in spA$). The function space $B = A'|_{spA} \equiv \widehat{A}$ is a subspace of the space of *complex-affine* functionals on $spA \subset A^*$, $Sh_1(A'|_K) \subset E_2(K)$ and $Sh_n(A'|_K) \subset E_{2n}(K)$ for every convex compact subset K of spA.

3. Remarks.

Note that a thorough examination shows that most of the results obtained hold for function spaces B which satisfy the following two conditions:

(1) B is a uniformly closed in $C(X)$ point separating space of $\mathbf{K}^1$-valued continuous functions on a compact Hausdorff space X which contains all translations with scalars in $\mathbf{K}^1$ of its elements;
(2) For every subset K of X of type $V(f_1, \ldots, f_n) = (f_1, \ldots f_n)^{-1}(\mathbf{0})$, where $f_j \in B$, there exists a smallest among all closed subsets of K on which all nonvanishing functions in B assume the minimums of their modulus.

For other specific properties of multi-tuple Shilov boundaries $\partial^{(n)}A$ of function algebras and of minimal n-affine boundaries $E_n(M)$ of compact convex subsets of locally convex spaces see **[1]** , **[7]** , **[8]** and **[9]**.

References

1. R. Basener, *A generalized Shilov boundary and analytic structure*, Proc. Amer. Math. Soc. **47** (1975), 98–104.
2. H. Bear, *The Shilov boundary for a linear space of continuous functions*, Amer. Math. Monthly **68** (1961), 483–485.
3. T. Gamelin, *Uniform Algebras*, Prentice-Hall Inc., Englewood Cliffs, New Jersey (1969).
4. B. Kramm, *Nuclearity (resp. schwarzity) helps to embed holomorphic structure into spectra (a survey)*, Banach Algebras and Several Complex Variables, Contemporary Math., **32** (1983), 143–162.
5. D. Kumagai, *Subharmonic functions and uniform algebras*, Proc. Amer. Math. Soc. **78** (1980), 23–29.
6. N. Sibony, *Multi-dimensional analytic structure in the spectrum of a uniform algebra*, Springer Lect. Notes in Math. **512** (1976), p. 139–175.
7. T. Tonev, *New relations between Sibony-Basener boundaries*, Springer Lect. Notes in Math. **1277** (1987), p. 256-262.
8. T. Tonev, *General Complex-Analytic Structures in Uniform Algebra Spectra*, preprint (*in Bulgarian*), Sofia (1987), 300.
9. T. Tonev, *Minimal affine boundaries of convex sets*, Tokyo J. Math. **11** (1989), p. 233-239.
10. T. Tonev, *Multi-dimensional analytic structures and uniform algebras*, Houston J. Math. (to appear) (1991).
11. J. Wermer, *Banach Algebras and Several Complex Variables*, Grad. Texts in Math., Springer Verlag (1976).

Bourgain Algebras

KEITH YALE Department of Mathematical Sciences, University of Montana, Missoula, Montana 59812

1. INTRODUCTION

Let C(X) be the sup norm algebra of continuous complex-valued functions on a compact Hausdorff space X and let $A \subseteq C(X)$ be a uniform algebra. Two algebras can be associated with A and each provides important information about A. In this survey our main interest is in the determination of these algebras for a variety of standard examples. We begin with a brief account of the basic ideas.

J. Cima and R. Timoney [4] introduced the following definition based on ideas of J. Bourgain [1]. The <u>Bourgain algebra</u> A_b consists of those g in C(X) for which

$$\lim_{n \to \infty} \operatorname{dist} \{ \psi_n g , A \} = 0 \qquad (I)$$

or, equivalently,

$$\lim_{n\to\infty} \sup \left\{ \left| \int \psi_n \, g \, d\mu \right| \right\} = 0 \qquad (\text{II})$$

for every weakly null sequence $\{ \psi_n \}$ in A. The sup is taken over the measures μ in the unit ball of $A^{\perp}$. The distance, dist $\{ \psi_n g, A \}$, between $\psi_n g$ and A is the quotient norm of the coset $\psi_n g + A$ in the space $C(X)/A$. The dual $C(X)^*$ is regarded as the space of Radon measures and $A^{\perp} \subseteq C(X)^*$ is the annihilator of A.

It is relatively straightforward to show that A_b is a closed algebra and that $A \subseteq A_b \subseteq C(X)$. From [1] and [4] we have the following

<u>THEOREM</u>: If $A_b = C(X)$ then A has the Dunford-Pettis property.

The Banach space Y has the <u>Dunford-Pettis property</u> (DPP) if whenever $y_n \to 0$ weakly in Y and $y_n^* \to 0$ weakly in Y^* then the pairing $< y_n, y_n^* >$ converges to 0. There are several equivalent definitions. The Dunford-Pettis property has a rich history and the survey articles by J. Diestel [7] and A. Pelczynski [10] are excellent sources of information.

The idea of a tight uniform algebra was introduced by B. Cole and T. Gamelin [6] in connection with (among other things) the solvability of a certain abstract $\bar{\partial}$ - problem.

For each $g \in C(X)$ define an operator $S_g : A \to C(X)/A$ by

$$S_g f = g f + A, \quad f \in A$$

and let

$$A_{wc} = \{ g \in C(X) \mid S_g \text{ is weakly compact} \}.$$

A duality argument ([6], Lemma 4.2) shows that A_{wc} is a closed algebra and clearly we have $A \subseteq A_{wc} \subseteq C(X)$. If $A_{wc} = C(X)$ then A is said to be <u>tight</u>.

We recall that a completely continuous operator is one which takes weakly convergent sequences into norm convergent sequences. We see that in terms of the operators S_g the Bourgain algebra is

$$A_b = \{ g \in C(X) \mid S_g \text{ is completely continuous } \}.$$

and a simple connection with A_{wc} is given by the following

<u>PROPOSITION</u>: If A has the Dunford-Pettis property then $A_{wc} \subseteq A_b$.

This proposition is just the well known fact that weakly compact operators are completely continuous in any space which has the Dunford-Pettis property. If we know *two* of the three items: DPP , A_b , A_{wc} about a uniform algebra A we can often use the theorem and the proposition to determine the third item. A simple example will serve to fix our ideas.

<u>Example</u> Let $A(\mathbb{T})$ be the usual disc algebra where $\mathbb{T} = \partial\mathbb{D}$ is the circle and $\mathbb{D} = \{ z \mid |z| < 1 \}$ is the open unit disc in the plane. We claim that $A_b = A_{wc} = C(\mathbb{T})$. Let $g(z) = \bar{z}$; it suffices to show that g belongs to both A_b and A_{wc} . The observation that $S_g f = \bar{z} f + A = \bar{z} f(0) + A$ tells us that S_g has one dimensional range and, in particular, is weakly compact. We also see that if ψ_n converges to 0 weakly in A then $S_g \psi_n$ converges to 0 in the norm of $C(\mathbb{T})/A$ which tells us that S_g is completely continuous.

The strong F. and M. Riesz Theorem, $A^{\perp} = H_0^1(\mathbb{T})$, and version (II) of the definition of A_b can be used to give another argument that $A_b = C(\mathbb{T})$. If we regard $A = A(\mathbb{D}) \approx A(\mathbb{T})$ as a subalgebra of $C(\mathbb{D})$ rather than $C(\mathbb{T})$ we still obtain $A_b = A_{wc} = C(\mathbb{D})$ but the arguments are more involved. On the other hand the presence of *singular* analytic measures in $A^{\perp}$ can sometimes be used to show that A_b is a proper subalgebra as is shown in the next section for the polydisc algebra.

2. THE POLYDISC ALGEBRA

Let $A = A(\mathbb{T}^n)$ be the polydisc algebra and take $n = 2$ for simplicity. J. Cima and W. Wogen [5] have recently shown that $A_b = A$ by a slicing argument which uses a well chosen weakly null sequence $\{\psi_n\}$ in A but makes no use of $A^{\perp}$. In contrast to their argument we will show the more limited fact that the Bourgain algebra A_b does not contain the function $g(z, w) = \bar{z}$ by a method which uses a suitable family of singular analytic measures. Both methods employ the same basic weakly convergent subsequence.

We know that for $\psi_n \in A$ and $g \in C(\mathbb{T}^2)$

$$(1) \qquad \inf \| \psi_n g - f \| = \sup \left| \int \psi_n g \, d\mu \right|$$

where the inf is over $f \in A$ and the sup is over $\mu \in A^{\perp}$, $\|\mu\| \leq 1$.

Now $g \in A_b$ means that $\inf \| \psi_n g - f \| \to 0$ as $n \to \infty$ for every sequence $\psi_n \in A$ which converges to zero weakly. We wish to show that $g(z, w) = \bar{z}$ does not belong to A_b. Let $\psi_n \in A$ be the weakly null sequence

(2) $$\psi_n(z, w) = [(w + w_n)/2]^{k_n}$$

where w_n is a fixed sequence in $\mathbb{T}$ tending to 1 and where k_n is a sequence of integers tending to infinity rapidly enough to insure that $\psi_n(z, w)$ converges pointwise to 0.

We will find a constant $\delta > 0$ for which

(3) $$\sup \left| \int \psi_n g \, d\mu \right| \geq \delta > 0, \text{ for all } n.$$

The sup is over $\mu \in A^{\perp}$, $\|\mu\| \leq 1$. Note that in order to determine a lower bound for the sup it is enough to estimate $\left| \int \psi_n g \, d\mu_{\alpha,\lambda} \right|$ for a suitably chosen subfamily of measures $\mu_{\alpha,\lambda}$.

By standard contour integration one can show the following

<u>LEMMA</u>: $\int_{-\infty}^{\infty} e^{ict}(1 - it)^{-2}\,dt = 0$ if $c \geq 0$ and $= 2\pi c e^{c}$ if $c < 0$.

It follows that the measure $\mu_{\alpha,\lambda}$ defined by

(4) $$f \to \int f \, d\mu_{\alpha,\lambda} = (1/\pi) \int_{-\infty}^{\infty} f(e^{it}, e^{i(\alpha t+\lambda)})(1 - it)^{-2}\,dt$$

is (a) singular with respect to Lebesque measure on $\mathbb{T}^2$, (b) non-zero, (c) belongs to $A^{\perp}$ provided α is a positive number, and (d) has norm independent of α. If α/π is irrational then $\mu_{\alpha,\lambda}$ is supported on a coset of the winding line on the torus and is analytic with respect to a larger algebra.

We will show (3) for the given g and ψ_n by estimating

(5) $\quad | \int \psi_n \, g \, d\mu_{\alpha,\lambda} | =$

$$(1/\pi) \left| \int_{-\infty}^{\infty} e^{-it} \, \psi_n \left(e^{it}, e^{i(\alpha t+\lambda)}\right) (1 - it)^{-2} \, dt \right|.$$

It is important to note that for a given n we are free to choose the measure $\mu_{\alpha,\lambda}$; thus $\alpha = \alpha_n$ and $\lambda = \lambda_n$ both depend upon n and we will write μ_n in place of $\mu_{\alpha,\lambda}$. The large power of 2 in the denominator of ψ_n will be compensated for by a correspondingly large binomial expansion of $(1 + 1)^{k_n}$.

Now

$$\psi_n(z, w) = \psi_n\left(e^{it}, e^{i(\alpha t+\lambda)}\right) = \left[\left(e^{i(\alpha t+\lambda)} + w_n\right) / 2\right]^{k_n}$$

is an analytic trignometric polynomial of the form

$$(1/2)^{k_n}\left[a_o \, e^{i\alpha k_n t} + a_1 \, e^{i\alpha(k_n-1)t} + \dots + a_{k_n-1} \, e^{i\alpha t} + a_{k_n} \right]$$

with $k_n + 1$ terms. Here $a_q = e^{i\lambda(k_n-q)} \, w_n{}^q \, C(k_n, q)$ where $C(k_n, q)$ is the binomial coefficient. Multiply this polynomial by $g(z, w) = e^{-it}$ to obtain

$$(1/2)^{k_n} \left[a_o \, e^{i(\alpha k_n-1)t} + a_1 \, e^{i(\alpha(k_n-1)-1)t} + \dots + a_{k_n-1} \, e^{i(\alpha-1)t} + a_{k_n} \, e^{-it} \right]$$

For a fixed n (and hence fixed k_n) we can choose $\alpha = \alpha_n > 0$ so small that

$$-1 \le \alpha(k_n - q) - 1 = c_q \le -(1/2)$$

for $0 \le q \le k_n$. Note that $-1 \le c_{k_n} \le \dots \le c_2 \le c_1 \le c_o \le -(1/2)$.

We use our expansion for $\psi_n\, g$ and the lemma to calculate the integral in (5):

$$E_n = \int_{-\infty}^{\infty} e^{-it}\, \psi_n\,(e^{it},\ e^{i(\alpha t+\lambda)})\ (1 - it)^{-2}\ dt$$

$$= (1/2)^{k_n} \sum_{q=0}^{k_n} (2\pi)\ c_q\ e^{c_q}\,[\,e^{i\lambda(k_n-q)}\ w_n{}^q]\ C(k_n, q).$$

For fixed n we choose $\lambda = \lambda_n$ so that $e^{i\lambda} = w_n$. Hence $e^{i\lambda(k_n-q)}\ w_n{}^q = e^{i\lambda k_n} = w_n{}^{k_n}$ is independent of q and so

$$E_n = (1/2)^{k_n}\ w_n{}^{k_n} \sum_{q=0}^{k_n} (2\pi)\ c_q\ e^{c_q}\ C(k_n, q).$$

To estimate E_n keep in mind that c_q is negative: $-1 \le c_q \le -(1/2)$ and $1/e \le e^{c_q} \le 1/\sqrt{e}$. Consequently

$$E_n / w_n{}^{k_n} \le 2\pi\,[-(1/2)][\,1/\sqrt{e}\,]\ (1/2)^{k_n} \sum_{n=0}^{k_n} C(k_n, q)$$

$$\le -\pi / \sqrt{e}$$

by the binomial theorem. Hence $|\ E_n / w_n{}^{k_n}\ | \ge \pi / \sqrt{e} = \delta > 0.$

Since the polydisc algebra satisfies the Dunford-Pettis property [1] we can apply the proposition to conclude that the polydisc algebra is not tight.

3. REMARKS AND QUESTIONS

We can regard $H^{\infty}(\mathbb{T})$ as a subalgebra of a C(X) space and easily verify that $\bar{z} \in H^{\infty}(\mathbb{T})_b$. Consequently, $H^{\infty}(\mathbb{T}) + C \subseteq H^{\infty}(\mathbb{T})_b$ and it was shown by J. Cima, S. Janson and K. Yale [3] that $H^{\infty}(\mathbb{T})_b = H^{\infty}(\mathbb{T}) + C$. The proof uses Chang-Marshall theory and does *not* provide an argument for the theorem of D. Sarason [12] that $H^{\infty}(\mathbb{T}) + C$ is a closed subalgebra of $L^{\infty}(\mathbb{T})$. The fact that $H^{\infty}(\mathbb{T})_b \neq L^{\infty}(\mathbb{T})$ together with the deep result of J. Bourgain [2] that $H^{\infty}(\mathbb{T})$ has the Dunford-Pettis property provides another argument showing that $H^{\infty}(\mathbb{T})$ is not tight [6].

There is another Bourgain algebra denoted by A_B and contained in A_b whose definition requires a construction in the second dual $C(X)^{**}$ (see [4] for details). J. Bourgain [1] showed that the ball algebra $A(\mathbb{B}^n)$ has the Dunford-Pettis property and his argument can be intrepreted as showing $A(\mathbb{B}^n)_B = A(\mathbb{B}^n)_b = C(\mathbb{B}^n)$. From [6, Theorem 2.1] we have $A(\mathbb{B}^n)_{wc} = C(\mathbb{B}^n)$.

Another class of interesting examples is given by the algebras A(K) of analytic functions on compact abelian groups with ordered duals Γ. It is important to distinguish between archimedian and non-archimedian orderings on Γ. We will tacitly exclude certain types of "trivial" orderings, e.g. Γ = integers. Refer to [6, Sec. 18] for technical details. Very little is known about the Bourgain algebras or the Dunford-Pettis property for A(K). On the other hand from [6] we know that tightness usually fails for A(K). In contrast to the situation on the circle, W. Rudin [11] has shown that $H^{\infty}(K) + C$ is *not* a closed subalgebra of $L^{\infty}(K)$ if the ordering on Γ is total. The Bourgain algebra $H^{\infty}(K)$ has not been determined and there does not appear to be a natural candidate. P. Gorkin, K. Izuchi and R. Mortini [9] have found a number of interesting results for Douglas algebras considered as subalgebras of $C(\mathcal{M}(H^{\infty}))$ where $\mathcal{M}(H^{\infty})$ is the maximal ideal space of H^{∞}. For example, if B is a Douglas algebra ($H^{\infty}(\mathbb{T}) \subseteq B \subseteq L^{\infty}(\mathbb{T})$), then $B_{bb} = B_b$. P. Ghatage, S. Sun and D. Zheng [8] have extended the characterization of $H^{\infty}(\mathbb{T})_b$ from the circle $\mathbb{T}$ to the disc $\mathbb{D}$ by regarding $H^\infty(\mathbb{D})$ as a subalgebra of the algebra generated by $H^\infty(\mathbb{D})$ and $\overline{H^\infty(\mathbb{D})}$.

Very little is known about the properties of A_b for a general uniform algebra A. For example, can one determine $A(X)_b$ as a subalgebra of $C(X)$ from knowledge of $A(\partial X)_b$ as a subalgebra of $C(\partial X)$ where ∂X is the Shilov boundary of A? The condition $A_{wc} = A_b$ seems to prevail; how is it related to the Dunford-Pettis property?

The following table summarizes our discussion.

A	DPP	A_{wc}	A_b	A_B
$A(\mathbb{T})$	Yes	$C(\mathbb{T})$	$C(\mathbb{T})$	$C(\mathbb{T})$
$A(\mathbb{T}^n)$	Yes	$A(\mathbb{T}^n)$	$A(\mathbb{T}^n)$	$A(\mathbb{T}^n)$
$A(\mathbb{B}^n)$	Yes	$C(\mathbb{B}^n)$	$C(\mathbb{B}^n)$	$C(\mathbb{B}^n)$
$A(K)$ *archimedian*	?	$A(K)$	?	?
$A(K)$ *non-archimedian*	?	$\neq C(K)$	$A(K) \neq$	$A(K) \neq$
$H^\infty(\mathbb{T})$	Yes	$H^\infty(\mathbb{T})+C$	$H^\infty(\mathbb{T})+C$	$H^\infty(\mathbb{T})+C$

I wish to thank Joe Cima and Warren Wogen for many helpful converstations about the ideas in this paper.

REFERENCES

[1] J. Bourgain, *The Dunford-Pettis property for the ball algebras, the polydisc-algebra and the Sobolev spaces,* Studia Math. 77 (1984), 245-253.

[2] J. Bourgain, *New Banach space properties of the disc algebra and* H^∞, Acta Math. **152** (1984), 1-48.

[3] J. Cima, S. Janson and K. Yale, *Completely continuous Hankel operators on* H^∞ *and Bourgain algebras,* Proc. Amer. Math. Soc. **105** (1989), 121-125.

[4] J. Cima and R. Timoney, *The Dunford-Pettis property for certain planar uniform algebras,* Michigan Math. J. **34** (1987), 99-104.

[5] J. Cima and W. Wogen, private communication.

[6] B. Cole and T. W. Gamelin, *Tight uniform algebras,* J. Funct. Anal. **46**(1982), 158-220.

[7] J. Diestel, *A survey of results related to the Dunford-Pettis property,* Proc. Conference Integration, Topology and Geometry in Linear Spaces, W. Graves (ed.), Contemp. Math., Vol. 2, Amer. Math. Soc., Providence, R. I., 1980, pp 15-60.

[8] P. Ghatage, S. Sun and D. Zheng, *A remark on Bourgain algebras on the disk,* preprint.

[9] P. Gorkin, K. Izuchi and R. Mortini, *Bourgain algebras of Douglas algebras,* preprint.

[10] A. Pelczynski, *Banach Spaces of Analytic Functions and Absolutely Summing Operators,* CBMS regional conference series, Vol. 30 (1977).

[11] W. Rudin, *Spaces of the type* $H^\infty + C$, Ann. Inst. Fourier Grenoble 25(1975), 99-125.

[12] D. Sarason, *Generalized interpolation in* H^∞, Trans. Amer. Math. Soc. 127(1967), 179-203.

Index